The Archaeology of Human-Environment Interactions

The impacts of climate change on human societies, and the roles those societies themselves play in altering their environments, appear in headlines more and more as concern over modern global climate change intensifies. Increasingly, archaeologists and paleoenvironmental scientists are looking to evidence from the human past to shed light on the processes which link environmental and cultural change. Establishing clear contemporaneity and correlation, and then moving beyond correlation to causation, remains as much a theoretical task as a methodological one.

This book addresses this challenge by exploring new approaches to human-environment dynamics and confronting the key task of constructing arguments that can link the two in concrete and detailed ways. The contributors include researchers working in a wide variety of regions and time periods, including Mesoamerica, Mongolia, East Africa, the Amazon Basin, and the Island Pacific, among others. Using methodological vignettes from their own research, the contributors explore diverse approaches to human-environment dynamics, illustrating the manifold nature of the subject and suggesting a wide variety of strategies for approaching it. This book will be of interest to researchers and scholars in Archaeology, Paleoenvironmental Science, Ecology, and Geology.

Daniel A. Contreras is an archaeologist focused on human-environment interactions in the past, particularly anthropogenic and geomorphic components of dynamic landscapes and environmental change. He pursues these interests in contexts ranging from complex polities in Andean South America and Mesoamerica to early Neolithic societies experimenting with domestication and village life in Southwest Asia. He is currently a LabEx OT-Med postdoctoral researcher in the Institut Méditerranéen de Biodiversité et d'Ecologie (IMBE) and Groupement de recherche en économie quantitative d'Aix-Marseille (GREQAM) at Aix-Marseille Université.

Routledge Studies in Archaeology

For a full list of titles in this series, please visit www.routledge.com

Recent titles:

16 **The Maritime Archaeology of a Modern Conflict**
Comparing the Archaeology of German Submarine Wrecks to the Historical Text
Innes McCartney

17 **The Archaeology of Bronze Age Iberia**
Argaric Societies
Edited by Gonzalo Aranda Jiménez, Sandra Montón-Subías and Margarita Sánchez Romero

18 **Debating Archaeological Empiricism**
The Ambiguity of Material Evidence
Edited by Charlotta Hillerdal and Johannes Siapkas

19 **Archaeology's Visual Culture**
Digging and Desire
Roger Balm

20 **Marking the Land**
Hunter-Gatherer Creation of Meaning in their Environment
Edited by William A Lovis and Robert Whallon

21 **The Archaeology of Human-Environment Interactions**
Strategies for Investigating Anthropogenic Landscapes, Dynamic Environments, and Climate Change in the Human Past
Edited by Daniel A. Contreras

22 **Life of the Trade**
Events and Happenings in Niumi's Atlantic Center
Liza Gijanto

22 **Exploring the Materiality of Food 'Stuffs'**
Transformations, Symbolic Consumption and Embodiment(s)
Edited by Louise Steel and Katharina Zinn

The Archaeology of Human-Environment Interactions

Strategies for Investigating Anthropogenic Landscapes, Dynamic Environments, and Climate Change in the Human Past

Edited by
Daniel A. Contreras

Routledge
Taylor & Francis Group
LONDON AND NEW YORK

First published 2017
by Routledge

2 Park Square, Milton Park, Abingdon, Oxfordshire OX14 4RN
52 Vanderbilt Avenue, New York, NY 10017

Routledge is an imprint of the Taylor & Francis Group, an informa business

First issued in paperback 2019

Library of Congress Cataloging-in-Publication Data
Names: Contreras, Daniel A.
Title: The archaeology of human-environment interactions : strategies for investigating anthropogenic landscapes, dynamic environments, and climate change in the human past / edited by Daniel A. Contreras.
Description: Milton Park, Abingdon, Oxon ; New York, NY : Routledge, 2016. | Series: Routledge studies in archaeology ; 21 | Includes bibliographical references and index.
Identifiers: LCCN 2016002715 | ISBN 9781138901735 (hardback : alkaline paper) | ISBN 9781315697697 (ebook)
Subjects: LCSH: Environmental archaeology—Research. | Social archaeology—Research. | Human ecology—History—To 1500. | Nature—Effect of human beings on—History—To 1500. | Landscape changes—Social aspects—History—To 1500. | Climatic changes—Social aspects—History—To 1500. | Social change—History—To 1500. | Paleoecology—Research. | Paleoclimatology—Research.
Classification: LCC CC81. A74 2016 | DDC 301/.01—dc23
LC record available at http://lccn.loc.gov/2016002715

ISBN: 978-1-138-90173-5 (hbk)
ISBN: 978-0-367-87268-7 (pbk)

Typeset in Sabon
by Apex CoVantage, LLC

Contents

Figures

Tables

Preface

This volume was born out of a symposium entitled "Correlation Is Not Enough: Building Better Arguments in the Archaeology of Human-Environment Interactions" that I organized at the 79th Annual Meeting of the Society for American Archaeology in Austin, Texas, in April 2014. As ever at meetings, the participants did not have enough time to interact as we might have liked, but I realized as the session finished that I had been thoroughly engaged throughout, and that the papers had without exception been thoughtful and stimulating. I was pleased to find that the contributors agreed, and consequently it has been possible to re-unite the group to share this collection of scholarship with others.

Unfortunately not all of the participants were able to contribute chapters to the volume, and I am sorry that neither the thought-provoking work of Emily McClung de Tapia and Chris Roos nor the thoughtful commentary of discussant Carlos Cordova are included here. Conversely, it's been a pleasure to add Alex Baer's work to the collection, and I'm pleased that the Island Pacific's rich history of research into past human-environment interactions is represented.

Thanks to both the commitment of the authors and the editorial support at Routledge, the papers have been subject to peer-review, and represent in my estimation exemplars of what archaeology can contribute to the study of human-environment interactions.

Daniel A. Contreras
Aix-en-Provence, August 2015

Contributors

Alexander Baer is an archaeologist working for Pacific Legacy, Inc. His research examines social complexity and ecology from an interdisciplinary perspective, employing method and theory from anthropological archaeology alongside biology, phylogeography, and geology. He has conducted field projects throughout the Caribbean, the American Southwest, and Polynesia, where his current work in Hawai'i, Easter Island, and Mangareva explore dynamic human-environment interactions and long-term sustainability.

Aleksander Borejsza is an archaeologist at the Universidad Autónoma de San Luis Potosí, Mexico. He has excavated in Mexico, the United Kingdom, Poland, and Spain. He uses archaeology and earth science to study past agriculture and rural life, the Preceramic and Formative periods of Mesoamerica, and late Quaternary fluvial environments.

Anna T. Browne Ribeiro is Assistant Professor of Anthropology at the University of Louisville, Kentucky. For the past decade she has engaged in archaeological and ethnographic research in the Brazilian Amazon, with a particular focus on "Terra Preta de Índio." She is co-director of the interdisciplinary project *Origens, Cultura e Ambiente*, housed at the Museu Parense Emílio Goeld (MPEG) in Brazil, which integrates archaeology with ethnography, ethnobiology, and geology to assemble an environmental history of the Xingu-Amazon confluence region. In 2013 she was awarded funding by the National Geographic Society for her project "History of a Crossroads: An Amazonian City in Deep Time." Her current work interleaves deep historiography of Amazonia and tropical places with data-driven geoarchaeological and anthropological research, with a focus on human-environment interactions. Dr. Browne Ribeiro received her Ph.D. in Anthropology at the University of California, Berkeley, in 2011, was a Social and Biological Sciences Post-Doctoral Fellow at Ohio State University from 2011–2013 and then moved to the MPEG as a post-doctoral fellow bridging the cultural anthropology and archaeology departments. From 2014–2015 she was a Fellow at the John W. Kluge Center in the Library of Congress. She has published in the *Latin American Antiquity, Journal of Archaeological Science* and the *Archaeological Review from Cambridge* and has contributed book chapters to edited volumes dealing with travel writing and placemaking.

Ari Caramanica holds an M.A. from Harvard University, where she is a Ph.D. candidate in the Department of Anthropology. Her research is focused on the sociopolitical impacts of borderland occupation and the reconstruction of agricultural landscapes of prehispanic coastal Peru using remote sensing techniques and paleobotanical analysis. Her archaeological experience includes work in the United States, the Andes, and currently, the Chicama Valley of the north coast of Peru.

Brian F. Codding is Assistant Professor of Anthropology at the University of Utah. He received his B.S. from the Social Sciences Department at California Polytechnic State University, San Luis Obispo, his M.A. and Ph.D. from the Department of Anthropology at Stanford University. His research examines variability in human-environment interactions across ethnographic and archaeological contexts in Western North America and Australia, particularly through the framework of behavioral ecology.

Daniel A. Contreras is an archaeologist focused on human-environment interactions in the past, particularly anthropogenic and geomorphic components of dynamic landscapes and environmental change. He specializes in geoarchaeological and geospatial research that integrates archaeological field research, geomorphic mapping, sediment analyses, and remote-sensing data in a variety of cultural contexts, ranging from complex polities in Andean South America and Mesoamerica to early Neolithic societies experimenting with domestication and village life in Southwest Asia. These interests have led to fieldwork in Peru, Jordan, Mexico, Guatemala, Turkey, Greece, and Honduras. He is currently a post-doctoral researcher for the LabEx OT-Med project Adaptation of Mediterranean Economies of the Past to Hydroclimatic Changes (AMENOPHYS), hosted in the Institut Méditerranéen de Biodiversité et d'Ecologie (IMBE) at Aix-Marseille Université.

Frances Hayashida is Associate Professor of Anthropology at the University of New Mexico. Her research focuses on the long-term ecology, economy, and politics of irrigation agriculture in the late prehispanic Andes. She has worked for many years on the north coast of Peru, and currently co-directs an interdisciplinary study of land use in the high-altitude Atacama with researchers from the United States, Chile, and Spain.

Matthew D. Jones is Associate Professor in Quaternary Science in the School of Geography at the University of Nottingham, United Kingdom. Matt's research focuses on the interaction of people and the environment, particularly the quantity and quality of the water within it, including projects in modern-day Jordan, Turkey, and Iran. Matt's other interests lie in developing methods to better quantify by how much water availability has changed through time, particularly through the use of oxygen isotope palaeohydrology.

Terry L. Jones is Professor of Anthropology and Chair of the Department of Social Sciences at California Polytechnic State University, San Luis Obispo, where he has taught for the last 17 years. He received his Ph.D. in Anthropology from the University of California, Davis, in 1995, and he also holds an M.A. in Cultural Resources Management from Sonoma State University (1982). He has worked as a professional archaeologist for 30 years, mostly on the central California coast where he studies hunter-gatherer ecology and maritime adaptations. He has published over 80 articles and book chapters on California prehistory as well as monographs and edited volumes.

Arthur A. Joyce is Professor of Anthropology at the University of Colorado, Boulder. Since 1986, he has conducted interdisciplinary archaeological research in Oaxaca on issues of political dynamics, religion, landscape, and ecology. He is author of *Mixtecs, Zapotecs, and Chatinos: Ancient Peoples of Southern Mexico* (Wiley-Blackwell, 2010) and *El Pueblo de la Tierra del Cielo: Arqueología de la Mixteca de la Costa* (Centro INAH Oaxaca, 2014, with Jamie Forde) as well as editor of *Polity and Ecology in Formative Period Coastal Oaxaca* (University Press of Colorado 2013). He has held research fellowships from the American Museum of Natural History, Fulbright, Dumbarton Oaks, and the American Council of Learned Societies.

Michele L. Koons is Curator of Archaeology at the Denver Museum of Nature & Science. She studies ancient complex societies and is especially interested in ancient political dynamics, social networks, and how people of the past interacted with their environment.

Danielle Macdonald is Assistant Professor in the Department of Anthropology at the University of Tulsa. She is the co-director (with Lisa Maher) of the Kharaneh IV project, evaluating hunter-gatherer aggregation and Late Pleistocene cultural diversity in Jordan. In addition, she studies the applications of surface metrology to archaeological datasets, developing new quantitative methods for lithic microwear analysis.

Lisa A. Maher is Assistant Professor of Anthropology at the University of California, Berkeley. Her work focuses on the interplay between social practices and environmental change at the transition from hunter-gatherer to farmer in the Near East and North Africa. She has been directing field projects in Jordan since 2001 and involved in projects in the United Kingdom, Syria, Libya, Azerbaijan, Israel, Canada, and the United States. She currently co-directs the Epipalaeolithic Foragers in Azraq Project.

Cheryl Makarewicz is Professor of Zooarchaeology and Stable Isotope Science at the Christian-Albrechts University, Kiel. Her research explores the origins of agriculture and animal domestication processes in the Near East, and the emergence of pastoralism in Inner Asia.

Louise Martin is Reader in Zooarchaeology at the UCL Institute of Archaeology. Her research focuses on exploring human hunting and herding practices in prehistory, alongside mammal domestication studies, primarily through zooarchaeological and ecological approaches. Her main fieldwork and research activities centre on the southern Levant, Anatolia, and southern Arabia, where she is interested in understanding human interactions with herd animals, especially in the Epipalaeolithic and Neolithic periods.

Louise Purdue received her Ph.D. in Environmental Archaeology at the University of Nice, France, in 2011, pursued a post-doctorate at Arizona State University, and is currently a CNRS researcher in Nice, France (CEPAM-UMR 7264). Her research revolves around the geoarchaeological, chronological, and environmental study of water systems in semi-arid, semi-tropical, and temperate environments, in parallel with the reconstruction of fluvial systems and settlement patterns. She has worked in the U.S. Southwest and in Yemen, and her current projects are located in Guatemala, the United Arab Emirates, and southern France in order to study the diversity of human adaptations in environments where water is scarce. While her research lies in the area of past socioenvironmental dynamics, she is also interested in the mathematical and computer modeling of socionatural systems, in order to discuss the robustness and durability of agricultural communities faced with environmental change.

Tobias Richter is Associate Professor at the Department of Cross-Cultural and Regional Studies at the University of Copenhagen, Denmark. His research focuses on the archaeology of the transition from hunting and gathering to agriculture, particularly in southwest Asia. Tobias currently directs fieldwork in Jordan targeting late Epipalaeolithic and early Neolithic sites in the Harra Desert.

Joshua Wright is Visiting Assistant Professor of Anthropology at Oberlin College. He has studied the landscape archaeology of early nomadic pastoralists in Northern Mongolia. Currently he carries out research in Southeastern Mongolia and Northeast China that spans from the Early Bronze Age to the Medieval period.

Introduction

1 Correlation Is Not Enough

Building Better Arguments in the Archaeology of Human-Environment Interactions

Daniel A. Contreras

The impacts of climate change on human societies, and the roles those societies themselves play in altering their environments, appear in headlines more and more as concern over modern global climate change intensifies. Increasingly, archaeologists and paleoenvironmental scientists are looking to evidence from the human past to shed light on the processes which link environmental and cultural change. As they do so, they are emphasizing the complexity of the dynamics underlying both human responses to environmental changes and anthropogenic impacts on environments.

Although archaeology has a long history of interest in human impacts on environments and environmental influences on human activities, construction of detailed arguments about causality in such interactions remains a persistent challenge. Limitations in chronological resolution of archaeological data and paleoenvironmental archives pose a methodological problem, while equifinality remains an interpretative challenge. Establishing clear contemporaneity and correlation, and then moving beyond correlation to causation, remains as much a theoretical task as a methodological one. The contributors to this volume embrace this challenge, and the case studies presented provide a series of snapshots of researchers working, in diverse but related ways, to tease apart the links between human and environmental dynamics around the globe.

The problem that initially brought these authors together was one that for all its apparent simplicity, and in spite of its modern resonance, remains vexing: in what ways did past humans interact with their environments, and how did each affect the other? More particularly, how can we not only generate data about, but also build compelling explanations of, these interactions?

This volume reunites this group, with the addition of a few other like-minded colleagues, to explore these questions through engagement with concrete research problems. As will be evident to the reader, on display throughout is a shared conviction that the questions involved are vital, and a shared recognition that the problems that are endemic to such research need confronting in their own right.

The chapters that follow illustrate this with a variety of case studies, and demonstrate in the accompanying methodological vignettes how some of the research tactics employed can be generalized in the construction of arguments about past human-environment interactions. These case studies exemplify the ways in which diverse methodologies can be mobilized to address human-environment dynamics. Tools used include alluvial geoarchaeology, geospatial modeling, micromorphological study of anthropogenic and alluvial sediments, various paleobotanical and geochemical analyses, and investigation of on-, near-, and off-site paleoenvironmental archives. The use of these methods sheds light, through the case studies presented here, on human-environment interactions around the globe and spanning the time period from the Last Glacial Maximum through the second millennium C.E. (~500 B.P.).

Rather than rehearse each chapter in detail in this introduction, I provide some broad context and draw out some of the themes shared by the studies that follow. The authors use examples from their own research to explore theoretical and methodological approaches to human-environment dynamics and to address the key task of constructing arguments that can link humans and their environments in concrete and detailed ways. These include ways of operationalizing theoretical frameworks, means of demonstrably relating human activity to environmental conditions, exploration of paleoenvironmental proxies that can be concretely linked to human activity, and foci on particular times and places that lend themselves particularly well to exploring such issues. The contributors include researchers working in a wide variety of regions and time periods; each focuses on the particular, while in aggregate they provide a cross-section of strategies for studying long-term human-environment interactions more generally. The regions explored include Mesoamerica (Borejsza and Joyce), Mongolia (Wright), southwest Asia (Contreras and Makarewicz, Jones and colleagues), the U.S. Southwest (Purdue), the Central Andes (Caramanica and Koons), the Amazon Basin (Browne Ribeiro), California (Codding and Jones), and the Island Pacific (Baer) (see Figure 1.1).

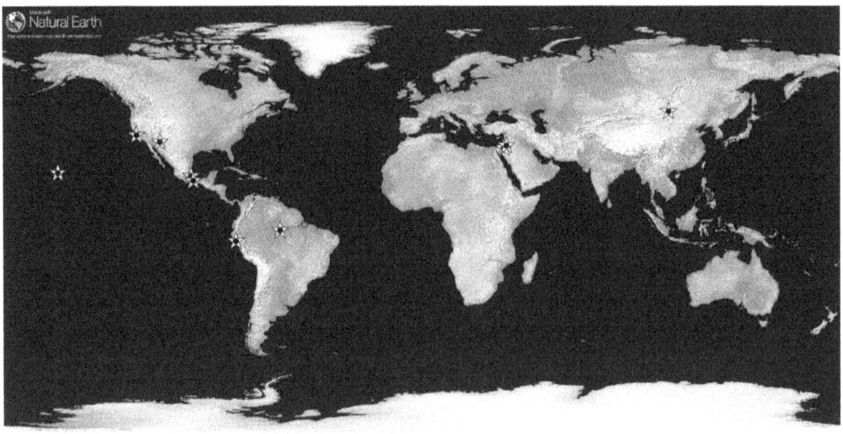

Figure 1.1 Locations of studies included in this volume.

Archaeological Perspectives on Humans and Their Environments

The geographic and temporal breadth of these contributions is testament to how fundamental the theme of human-environment interaction is to anthropological archaeology. This book builds on a distinguished research tradition, and its existence should be taken not as a critique of what has gone before so much as a testament to how difficult—though not, the contributors here would suggest, intractable—the problems of untangling human-environment interactions are. The focus in this collection is on innovation in research tactics—i.e., the ways in which established methods and combinations thereof can be used to approach problems of human-environment interaction.

In theoretical terms, these papers belong to the third of three strands of archaeological literature relating humans to their environments. These comprise (1) investigations of the structuring effects of environments on their human inhabitants, (2) analyses of human impacts on environments, and (3) studies of mutually constitutive human-environment interactions. By focusing particularly on the mechanisms that link human activity and environmental conditions, these papers fall in the latter category, making the directionality of influence a question for research rather than upholding the primacy of either environmental determinism or human agency as points of theoretical principle.

These strands of thought in the archaeology of human-environment interaction both coexist in contemporary archaeology and recapitulate the historical development of archaeological thought. The idea that environmental factors had a structuring effect on the human past was explored nearly from the beginnings of archaeology's development as a field of study; "environmental determinism" and "environmental possibilism" were both based in such a concept (cf. Trigger 1989:Ch.7). This was reinforced in Americanist archaeology in the mid-twentieth century by ecological perspectives (e.g., Butzer 1964; Watson and Watson 1969) and particularly the influence of Julian Steward's cultural ecology (cf. Steward 1972) on such seminal settlement pattern studies as the Virú Valley Survey (Willey 1953) and the Basin of Mexico Survey (Sanders et al. 1979). Although early investigations into human *impacts* on past environments were contemporary with these developments, formative studies like those of George Perkins Marsh (1864) and Walter Lowdermilk (1953) remained relatively isolated from archaeology (see historical overview in Goudie 2013). It was historical geographers like Carl Sauer (e.g., 1941), rather than archaeologists, who developed the ideas about widespread anthropogenic influence that culminated in the 1956 publication of *Man's Role in Changing the Face of the Earth* (Thomas et al. 1956).

These ideas had little apparent impact on archaeology until the final third of the twentieth century, when the discipline really began to embrace study of anthropogenic environmental change as fundamental to understanding the human past (cf. Redman 1999:Ch.2). At approximately the same time, environmental history also developed as a field during this latter part of the twentieth century (cf. Worster 1990; McNeill 2003). Still today, however, archaeology and its sibling fields of historical geography and environmental history remain surprisingly isolated from one another (see McNeill's (2003) review of environmental history and Williams's (1994) treatment of its attenuated relationship to historical geography; it is perhaps telling that archaeology only barely figures in either discussion). Paleoenvironmental and paleoclimatic scientists, meanwhile, collaborated with social scientists from various disciplines, and at least some have begun to advocate a more integrated approach (O'Sullivan 2008; Caseldine and Turney 2010). I do not attempt to comprehensively review even the intellectual history of archaeology here, but provide below a brief conceptual overview.

Increase in the quantity and quality of paleoenvironmental data has produced a resurgence of interest in studies of environmental influence on past societies since the latter decades of the twentieth century, and paleoclimatologists and paleoenvironmental scientists as well as geographers and archaeologists have produced a broad array of studies claiming to recognize such effects (so many that I am not aware of any single summary, but for overviews of particular regions, see Wright 1993 on the Near East; Roberts et al. 2004 for the Mediterranean; Yaeger and Hodell 2008 for the Maya region; Weninger et al. 2009 for the Eastern Mediterranean; Grosjean et al. 2007 for the South-Central Andes; Contreras 2010 for the Central Andes; and Spriggs 2010 for the Island Pacific).

This has been widely criticized as a revival of discredited environmentally determinist perspectives on the human past. However, the label is not used consistently and encompasses a wide range of approaches with a long history (Coombes and Barber 2005); the "revival" certainly comprised a more restrained vision than the original manifestations, which could be totalizing in their explanatory ambitions. Rather than arguing that culture was *always* the product of environment, salient papers in recent decades focused on the proposed and perhaps inevitable impacts of pronounced climatic changes evident in paleoclimatic proxies (e.g., Brenner et al. 2001; deMenocal 2001; Mayewski et al. 2004; Brooks 2006; Kennett et al. 2012; Kaniewski et al. 2015).

Environmentally determinist perspectives are most commonly manifest in studies relating cultural "collapse" to environmental change in various scenarios (too many to list here, but see useful reviews in Tainter 2006; Butzer 2012; and Middleton 2012), but also are inherent to varying degrees in interpretations that relate cultural trajectories to changing affordances provided by environmental shifts. Although the term "environmental determinist" is one of deprecation within anthropological archaeology, studies so labeled range from truly climatic determinist to more nuanced studies that examine climatic or environmental influence without embracing simple determinism. The latter include analyses of past human reactions to changing climatic conditions that focus on contingency, behavioral diversity, and human adaptation (recent examples include Jones et al. 1999; Anderson et al. 2007; Costanza et al. 2007; Sandweiss and Quilter 2009; and Cooper and Sheets 2012), perhaps most programmatically laid out in the research program of human behavioral ecology (cf. Bird and O'Connell 2006).

Critiques of environmental determinism have focused on its privileging of environmental drivers of human behavior at the expense of social ones (e.g., Brumfiel 1992), the lack of room for human agency in its narratives (e.g., Erickson 1999 in archaeology; and, more obliquely, Cronon 1990 in environmental history), and its often simplistic accounts of social, political, and cultural change (e.g., McIntosh et al. 2000; Butzer and Endfield 2012; Middleton 2012). A more subtle criticism that highlights how little explored the mechanisms of "determinism" are points to the absence of political ecology in accounts of environmental effects on human societies (e.g., Fisher and Thurston 1999; Ensor et al. 2003). With some interesting exceptions (e.g., van Buren 2001; Billman and Huckleberry 2008), even archaeologists (much less paleoenvironmental scientists) have largely ignored the fairly obvious point that such effects would be felt differentially in any non-egalitarian society and would have complex effects. Meanwhile, a more fundamental deconstruction questions the Cartesian separation of humans from environments implicit in such analyses (e.g., Ingold 2000; Head 2008). The more extreme of these critiques, however, have often sought to redress the balance by disengaging archaeology from *any* focus on past human environments, implicitly or explicitly arguing that human-environment interactions should be secondary to human-human interactions in any analysis of human behavior. Such a reaction risks throwing the environmental baby out with the determinist bathwater, creating an archaeology in which past humans inhabit spaces which are either featureless or limitlessly malleable.

One reaction that has engaged with rather than backgrounded human-environment interaction centered on the unidirectionality of environmentally determinist explanations. Reformulating the prevailing view—which perhaps owes an uncomfortable amount to notions of the Noble Savage living in harmony with his surroundings (cf. Denevan 1992, *inter alia*)—that pre-Modern humans were simultaneously living lightly on the land and at the mercy of their environments, this renewed emphasis on human impacts on past environments has focused on identifying ever-increasing antiquity and the magnitude of anthropogenic influence (cf. Hayashida 2005). That such a human footprint is global and dates back millennia is apparent from the geographic and temporal span of the literature: exemplary studies include work on Central Europe (e.g., Bork and Lang 2003) and the Mediterranean (e.g., Barker 1995; Bintliff 2002; Butzer 1982; 1996), the Island Pacific (e.g. Kirch 1997; 2005), Mesoamerica (e.g., Denevan 1992; Fisher et al. 2003; Heine 2003; Beach et al. 2015), and Amazonia (e.g., Heckenberger et al. 2003; Erickson 2008), and span the Holocene. They have continued to proliferate since Redman's syntheses (e.g. 1999; Redman et al. 2004), and have been argued to comprise evidence of human impact on global processes (e.g., Ruddiman 2003; Ruddiman et al. 2016). This has dovetailed with the embrace of niche construction perspectives that focus particularly on the potential evolutionary significance of anthropogenic effects on the environments to which humans have adapted (e.g., Smith 2007; Laland and O'Brien 2010), as well as the application of such perspectives to discussions of the antiquity of the Anthropocene (e.g., Braje and Erlandson 2013; Smith and Zeder 2013). Moreover, while environmental determinist perspectives have been salient in popular science literature, notably in the form of Jared Diamond's (1998) *Guns, Germs, and Steel*, discussion of the surprising antiquity and ubiquity of human impact is ably represented by Charles Mann's (2005) *1491*.

Studies of environmental influence and anthropogenic impact have also recently been complemented by perspectives that focus on the ways in which humans and their environments mutually constitute one another, characterizing bidirectional influences rather than unidirectional causality. This development has been explicit and programmatic in historical ecology (cf. Crumley 1994; Balée 2006; Balée and Erickson 2006; Thompson and Waggoner 2013), but is also evident in literature addressing "socionatural" or "socioecological" aspects of the human past (e.g., van der Leeuw and Redman 2002; de Vries and Goudsblom 2002; Hornborg and Crumley 2007; Fisher et al. 2009; Barton et al. 2012; Giosan et al. 2012; Mayle and Iriarte 2014) and "human ecodynamics" (e.g., McGlade 1995; Kirch 2007; Kohler et al. 2007; Varien et al. 2007; Barton et al. 2011). A focus on mutual influence is also evident in attempts to integrate archaeological data and perspectives into discussions of sustainability and resilience (e.g., Minnis 1999; Redman and Kinzig 2003; Redman 2005; Costanza et al. 2007; de Vries 2006; Dean 2010; Morrison 2015; van de Noort 2011; Turner and Sabloff 2012; Kidder and Liu 2014).

The Temptation of Correlation (and What to Do About It)

In spite of this diversity of perspectives, and interest in human-environment interactions that stretches back to the early years of archaeology as a discipline, the articulation of specific linkages between paleoenvironmental and cultural change remains a rarely realized and tantalizing goal. It offers a means of better explicating cultural trajectories, a tool for examining human ecological footprints, and a strategy for untangling intertwined human and environmental histories in the long term. So central is the theme of human-environment interaction to anthropological archaeology that it comprises one of the five pillars of the recently articulated "Grand Challenges for Archaeology" (Kintigh et al. 2014). It also animates current trans-disciplinary debates over the character and timespan of the "Anthropocene," defined roughly as the time period during which human influence on the globe has been significant (e.g., Crutzen and Steffen 2003; Erlandson and Braje 2013; Ruddiman 2013; Smith and Zeder 2013; Zalasiewicz et al. 2015).

The debates over the definition of the Anthropocene are revealing about both interdisciplinary discourse about past human-environment interactions and archaeology's interaction with its sibling disciplines. They turn fundamentally on the establishment of correlation: the debate is global in scale and coarse synchronicity of human activity and environmental effect is considered argument enough to fix the period's origins in time. Studies focused on the antiquity of anthropogenic impact are in this sense analogous to claims of environmentally driven collapse: both take as their points of departure the identification of suggestive correlations at large scales. Where investigations of human impact identify particular environmental shifts that may be broadly coincident with shifts in human population or behavior and argue that the latter brought about the former (e.g., Doughty et al. 2010; Ellis et al. 2013; Erlandson 2014), the more abundant studies of environmental influence maintain that similar correlations are evidence of environmental impact on human societies (e.g., Binford et al. 1997; Brenner et al. 2001; Sandweiss et al. 2009; Weninger et al. 2009; Clare and Weninger 2010; Medina-Elizalde and Rohling 2012; Lachniet et al. 2012, among many). Of course either process might in principle operate; the shortcomings of these arguments are not in their logic, but their evidentiary sufficiency and chronological precision. Nevertheless, their critics notwithstanding, such papers should not necessarily be read as arguments that correlation should be understood to be causation; rather the identification of correlation is at once a statement of hope and an admission of defeat. It is a statement of hope in that reportage of climate-culture correlation is driven by a conviction that it *should* be possible to develop the putative links further, and an admission of defeat in that it remains unclear *how* those links can be developed.

Two primary factors contribute to this difficulty: problems of spatial and temporal scale and resolution, and problems of articulation of mechanism. The former make it difficult to relate archaeological and paleo-climatic/environmental data more than generally, while in the absence of the latter, influence must remain inchoate, causality vague, and effects linked to causes only by commonsensical assertion.

The juxtaposition of archaeology and paleoenvironmental science highlights just how fundamental questions of scale and resolution are. Environmental parameters are generally discussed in the terms of paleoclimate reconstructions and environmental modeling—that is, spatial scales that are regional and temporal scales that are generally at best centennial. Of course, remarkable paleoenvironmental archives like long dendroecological sequences or varved lacustrine deposits can provide temporal resolution that reaches sub-annual, and deposits from small catchments can offer very local signals, but the bulk of the effort in paleoenvironmental science has been to achieve regional and long-term relevance, pushing the focus towards coarser scales of analysis. In contrast, archaeological explanation relies fundamentally on anthropological models of behavior—i.e., understandings of human activity that are grounded in decision-making at local and annual or decadal scales and catchments generally defined by distances reasonable for pedestrian travel. Articulating analyses that focus on distinct scales, with varying resolutions, is vital to characterization of relationships between local and regional data, fundamental to understanding relationships between samples and populations, and often central to relating archaeological and paleoclimatic and paleoenvironmental data. As a result, it has been the focus of both practical and theoretical consideration in archaeology (e.g., Stein 1993; Lock and Molyneaux 2006; Robb and Pauketat 2013). One of the chief challenges of the multi-scalar analyses that necessarily result is that of articulating particular mechanisms of human-environment interaction in order to provide a means of moving between scales.

The contributors to this volume use their research to address these problems of scale, resolution, and mechanism. For instance, they ask how (and if) regional paleoclimatic shifts affected everyday life in particular locales, whether distinct groups responded in similar ways to environmental phenomena, what other imperatives than agricultural ones might drive anthropogenic landscape modifications, and what widespread environmental effects are created by small-scale human activities. As importantly, they ask *how*—how can we identify changes, link scales of analysis, understand which environmental parameters were significant to a landscape's inhabitants, differentiate intentional and incidental human impacts, and address challenges of equifinality. The methodological strategies are diverse, illustrating the manifold nature of the subject and suggesting a wide variety of strategies for approaching it.

With regard to temporal scale and resolution, many of the chapters that follow return to basic questions: how reliable are putative correlations, and how can causal relationships be established if correlation is logically insufficient? Even as they seek more robust means of identifying and characterizing links between environmental and human trajectories, they return to the fundamental archaeological problem of establishing precise chronologies. For instance, the limits of regional periodization schemes that imply rigid and precisely dated divides have been recognized for decades (e.g., Rowe 1962)—but they continue in widespread use due to their convenience. As several of the contributors here emphasize, however, the abundance of such schemes in no way means that they are an adequate basis on which to build arguments *even for correlation*, much less causation. Conversely, paleoenvironmental chronologies—generally based, when regular laminations like varves, growth layers, etc. are not available, on age-depth models anchored with ^{14}C dates—themselves are often chronologically imprecise. Moreover, the interpretations of proxies themselves may be subject to critique, with potentially catastrophic consequences for archaeological interpretations (for cautionary examples of both chronological and interpretive problems see, e.g., Calaway 2005; Meadows 2005; and Maher et al. 2011).

As Jones and colleagues demonstrate in Chapter 5 with their from-the-ground-up approach to the chronology of the Azraq Basin in the Late Pleistocene, it can be necessary to revisit the basic building blocks of chronology, thinking hard about radiocarbon dates and how probabilities distributed over ranges of time relate to one another. This highlights one relatively recent development that none of the chapters in this volume have the data to address: the increasing use of Bayesian chronological modeling in developing both paleoenvironmental and archaeological chronologies (cf. Bayliss 2009; Bronk Ramsey 2009). The resulting improved archaeological chronologies hold significant promise for the elaboration of human-environment interaction, but need to be matched by (or perhaps directly integrated with) paleoenvironmental chronologies of relevant archives.

Building such chronologies for archaeological and paleoenvironmental records is fundamental to any understanding of human-environment interactions. That the resulting correlation-based narratives are challenged as inadequate does not diminish the significance of correlation, but rather highlights its epistemological fragility and its logical insufficiency. These limitations point to the need to use correlation as a prompt rather than an answer, one which demands further investigation of chronology and relationships between humans and the environments they inhabit, as well as construction of robust arguments that can link humans and environments through mechanisms that can be specifically articulated and investigated.

In addition to working with chronology-building, several of the contributing authors focus particularly on more direct means of linking archaeological and paleoenvironmental data. Jones and colleagues in Chapter 5, Contreras and Makarewicz in Chapter 4, and Caramanica and Koons in Chapter 6 all explore the analysis of on-, near-, and off-site paleoenvironmental archives as a means of directly articulating archaeological and paleoenvironmental data. While this by no means obviates the need for chronological control, the potential for identifying markers that can link distinct archives by means other than chronological correlation is tantalizing. Purdue's use, in Chapter 3, of micromorphological markers of distinct climatic regimes, in which she uses source-to-sink sediment analysis to identify not just the sources of sediment but the processes implicated in its transport, illustrates the rich potential for cross-scale links that rest on more than correlation.

Borejsza and Joyce's focus in Chapter 2 on alluvial geoarchaeology shifts the axis of site and off-site, making it temporal as well as spatial. Following Schumm (1991), they describe interpretive problems of convergence and divergence—that is, difficulties of what kind of processes to infer from palimpsests of landscape evidence of human-environment interaction. They suggest—and have elaborated elsewhere (Borejsza et al. 2014)—that alluvial geoarchaeology is a particularly promising source of evidence and argument. Inasmuch as it focuses on archives that juxtapose archaeological and paleoenvironmental data directly, alluvial geoarchaeology might appear to transcend problems of scale, resolution, and linkage. As Borejsza and Joyce emphasize, however, even as alluvial stratigraphic archives may provide both archaeological and paleoenvironmental evidence, the scales of the processes reflected in these archives may vary both spatially and temporally, and must be assessed on a case-by-case basis.

Multi-scalar concerns also explicitly motivate Browne Ribeiro's argument in Chapter 7 that regional questions about anthropogenic dark earth in Amazonia (e.g., is it a marker of a *particular* and relatively synchronous cultural phenomenon?) can only be answered by addressing local questions (e.g., how were particular deposits of *terra preta* formed?). In parallel, on the other side of tropical South America, Caramanica, and Koons argue that regional-scale geographic typologies, not only coarse in scale but also projecting present observed conditions relentlessly into the past, can obscure as much about past human-environment interaction as they reveal. While the Pampa de Mocán on Peru's North Coast is classified as desert and has been archaeologically interpreted as such, their direct investigation suggests that this characterization is inaccurate for at least some past periods, and interpretations derived from it are misleading about the scope and character of past human occupation of the area.

Related issues of scale—the differentiation of environments at micro- or meso-scales within a regional environmental mosaic that might be glossed as uniform by coarse-grained mapping—are apparent in Contreras and Makarewicz' and Jones and colleagues' investigations of the local manifestations of regional environmental patterns in the Levant, and underlie Codding and Jones's assertion in Chapter 8 that complex behavioral patterns may be explained by examining the interaction of environmental diversity and dynamism with hypotheses about human-environmental interaction derived from human behavioral ecology.

These are complemented by more inductive approaches like the ethnomicrogeography that Wright proposes in Chapter 9, which uses an analysis of patterning in the environmental characteristics of Mongolian pastoralist camps as a tool for archaeological interpretation. The soil geochemical analyses that Baer highlights in Chapter 10, similarly, provide a means of assessing which characteristics of the agricultural landscape were important to the early state on Maui, demonstrating that particular elite decisions were apparently responsive simultaneously to political imperatives (state expansion) and environmental assessments (land was chosen for productive potential rather than because of proximity, cosmological significance, etc.). These studies use systematic characterizations of environments and distributions of archaeological sites as a means to attack the problem of mechanism: by examining in aggregate the factors influential in past human decisions about interaction with environments (i.e., where to establish camps or intensify agricultural production), they approach emic systems of environmental classification, identifying some of the ways in which environments were understood by their inhabitants.

Learning How to Study Past Human-Environment Interaction

These cases illustrate several prominent themes that are coming to the fore in the study of human-environment interactions, including most saliently the importance of research that sets out specifically to address questions of scale and resolution and examine linking mechanisms. The research reported here is *designed* to link environmental and cultural change, rather than simply identifying broad correlations and speculating about potential links. As such, it is interdisciplinary even in conception, rather than an attempt to marry distinct archaeological and paleoenvironmental data. However, if there is any indisputable lesson to be taken from the history of research on past human-environment interactions described above, it is that the topic is as theoretically and methodologically challenging as it is important. It would clearly be a mistake to think of even the most exemplary study as prescriptive; both humans and environments are diverse enough that their study cannot be formulaic, and the right questions and methods for any particular context are rarely if ever obvious. The studies included here are illustrations of archaeologists grappling with those problems and wanting their readers to learn both from their successes and the difficulties encountered.

The authors provide a variety of examples of how to frame questions in ways that make them answerable, of ways of thinking about what data may hold answers, and of means of acquiring that data.

Three approaches to addressing problems of human-environment interaction stand out:

1. A focus on interaction: Rather than embracing the convenience of environmental determinism or fetishizing human agency, the authors in this volume set out to detail human-environment relationships that are conceptualized as complex, dynamic, multivalent, and at least potentially mutually influential.
2. A shift in the questions asked: Rather than asking *whether* environments affected their inhabitants, or *whether* inhabitants impacted their environments, the authors ask instead *how* humans and their environments interacted.
3. The mobilization of diverse data at temporal and spatial scales that are theoretically appropriate for explanation of human behavior.

To help make these strategies clear and accessible, the contributors each accompany their chapter with a methodological vignette, outlining one of the principal methods they've employed and why. Of course this is not a methodological handbook and cannot be comprehensive. Various substantial volumes attempt this for environmentally oriented archaeology generally (e.g., Evans and O'Connor 1999; Dincauze 2000; Branch et al. 2005; Reitz and Shackley 2012), and a profusion of specialist literature for particular methods. Amongst the missing here are both methods long standard as means of investigating human-environment interaction and more recently developed methods that provide new tools for addressing the issue. The former include zooarchaeology, archaeo- and pedo-anthracology, dendroecology, and study of a wide variety of organic and inorganic environmental proxies from diverse paleoenvironmental archives (e.g., diatoms, ostracods, organic C, $\delta^{18}O$, mineral inputs), while the latter include stable isotope studies of human and faunal remains, faunal and soil/sediment aDNA, further environmental proxies (particularly geochemical ratios and biomolecules), Bayesian chronological modeling (discussed above), and agent-based modeling. Of course both of those lists could be expanded. More methods are developed or adapted to archaeological and paleoenvironmental purposes regularly, though their utility in many ways remains a function of a basic archaeological challenge: it is fundamentally dependent on the quality of understanding of the contexts from which samples are derived, or—in the case of paleoenvironmental data—of understanding how those contexts relate to human activity.

In any case, their omission here is by no means a slight on their utility. Rather, this collection focuses not on breadth of coverage of methodology *per se*, but on the application of methods to a particular problem of anthropological archaeology: the interactions between humans and their environments at multiple spatial and temporal scales. It highlights research tactics, and the methodological vignettes are designed to delineate the appeal and utility of various methods for those interested in questions of human-environment interaction, and direct interested readers to the relevant literature if they wish to learn more. If the chapters do not neatly demonstrate simple applications of single methods, that reflects the ambition of the questions they address: the mobilization of diverse methodologies in the service of difficult questions is the norm. At the same time, the authors provide capsule introductions to some of the methods that they find contribute significantly to their research into human-environment interactions. Which methods may be productively employed in other research contexts is more a function of the evidence available and the particular questions of interest than it is of any inherent suitability of methods to problem; it is hoped that readers will take inspiration from the research approaches herein, while applying whichever methods are best suited to their own contexts.

The studies included in this volume also point towards a next step that remains beyond their scope: they suggest the need for an iterative process of tacking between local and regional, and perhaps also etic and emic, in exploring past human-environment interactions. Browne Ribeiro explicitly advocates a return to the local *in order to* address the regional, while Contreras and Makarewicz, Jones and colleagues, and Caramanica and Koons, by demonstrating that regional data can produce locally misleading results, argue that the regional may not always be directly applicable to the local. These chapters highlight the way in which problems of scale are problems of sampling; we must confront the twin questions of how representative the local/small-scale may be, on the one hand, and of how much diversity the regional/large-scale encompasses, on the other.

The scale of analysis in much of the work that seeks to relate humans to their environments is regional, especially where evidence of past climate is concerned (in fact many climate studies *strive* for regional relevance). In contrast, the chapters that follow emphasize the local because that is the scale of human consequences of climate change, human perception and experience, and human response. They should not be understood, however, as constituting a call for a (re)turn to the local. One of the cogent critiques of the impact of postmodern thought in archaeology has been that its attack on the validity of generalization produced a retreat to the particular. Here, rather, the local is not so much a refuge as a necessary waystation, a *means* to understanding the more general. In order to understand both local/human and regional/environmental processes, and the feedbacks between them, it is necessary to (a) work *across* scales, incorporating various disciplinary specialties, and (b) work *between* scales, exploring mechanisms through which various human and environmental processes articulate.

As the chapters here suggest, amongst the most salient of those mechanisms are various subsistence practices, through which humans interface directly with the particularity and variability of environmental productivity (Codding and Jones, Contreras and Makarewicz, Jones et al.) and which may involve substantial anthropogenic modifications to environments, both deliberate and strategic and unintended (Baer, Borejsza and Joyce, Browne Ribeiro, Caramanica and Koons, Purdue). At the same time, as Hayashida emphasizes in her concluding comments, it would be reductive to presume that human motivations and activities were confined to the direct and relatively visible interface of subsistence production. Here, too, multi- or trans-scalar perspectives are necessary, as social, political, and economic networks in which local populations are embedded may effect human-environment interactions in tandem with local processes.

Whether much of this literature whose surface I have skimmed here—or indeed the studies included in this volume—escapes nature/culture binarism is perhaps debatable. However, the focus on more complex *inter*actions that Head (2008) advocates can only be pursued by simultaneously investigating both archaeological and paleoenvironmental data (in spite of traditional disciplinary divisions of labor and attention) and seeking to elucidate specific mechanisms of interaction while remaining open-minded about their character. As such, the chapters that follow not only tackle complex problems, but also point the way towards how we may attempt to generate understandings of past human-environment interactions in which sociopolitically embedded humans strive to meet their subsistence and social goals in settings comprising dynamic environmental mosaics. This may have surprising repercussions: environmental historian Ted Steinberg has argued (2002) that considering environment more fully will enhance understanding of agency/structure dynamics, and may in fact be necessary to their analysis. The archaeology of human-environment interactions, in other words, may be vital to understanding past sociopolitical dynamics as well as socioenvironmental ones.

References Cited

Anderson, David G., Kirk A. Maasch and Daniel H. Sandweiss (editors) 2007 *Climate Change and Cultural Dynamics: A Global Perspective on Mid-Holocene Transitions*. Academic Press, London.

Balée, William 2006 The research program of historical ecology. *Annual Review of Anthropology* 35(1):75.

Balée, William and Clark L. Erickson 2006 *Time and Complexity in Historical Ecology: Studies in the Neotropical Lowlands*. Columbia University Press, New York.

Barker, Graeme 1995 *A Mediterranean Valley: Landscape Archaeology and Annales History in the Biferno Valley*. Leicester University Press, London.

Barton, C. Michael, Julien Riel-Salvatore, John M. Anderies and Gabriel Popescu 2011 Modeling human ecodynamics and biocultural interactions in the late Pleistocene of western Eurasia. *Human Ecology* 39(6):705–725.

Barton, C. Michael, Isaac Ullah, Sean M. Bergin, Helena Mitasova and Hessam Sarjoughian 2012 Looking for the future in the past: Long-term change in socioecological systems. *Ecological Modelling* 241:42–53.

Bayliss, Alex 2009 Rolling out revolution: Using radiocarbon dating in archaeology. *Radiocarbon* 51(1):123–147.

Beach, Tim, Sheryl Luzzadder-Beach, Duncan Cook, Nicholas Dunning, Douglas J. Kennett, Samantha Krause, Richard Terry, Debora Trein and Fred Valdez 2015 Ancient Maya impacts on the Earth's surface: An Early Anthropocene analog? *Quaternary Science Reviews* 124:1–30.

Billman, Brian R. and Gary Huckleberry 2008 Deciphering the politics of prehistoric El Niño events on the north coast of Peru. In *El Niño, Catastrophism, and Culture Change in Ancient America*, edited by Daniel H. Sandweiss and Jeffrey Quilter, pp. 111–138. Dumbarton Oaks Research Library and Collection, Cambridge.

Binford, M., A. L. Kolata, M. Brenner, J. W. Janusek, M. T. Seddon, M. B. Abbott and Jason H. Curtis 1997 Climate variation and the rise and fall of an Andean civilization. *Quaternary Research* 47:235–248.

Bintliff, John 2002 Time, process and catastrophism in the study of mediterranean alluvial history: A review. *World Archaeology* 33(3):417–435.

Bird, Douglas W. and James F. O'Connell 2006 Behavioral ecology and archaeology. *Journal of Archaeological Research* 14(2):143–188.

Borejsza, Aleksander, Charles Frederick, Luis Morett Alatorre and Arthur Joyce 2014 Alluvial stratigraphy and the search for preceramic open-air sites in highland Mesoamerica. *Latin American Antiquity* 25(3):278–299.

Bork, Hans-Rudolf and Andreas Lang 2003 Quantification of past soil erosion and land use/land cover changes in Germany. In *Long Term Hillslope and Fluvial System Modelling*, edited by Andreas Lang, K. Hennrich and R. Dickau, pp. 231–239. Springer, Heidelberg.

Braje, Todd J. and Jon M. Erlandson 2013 Looking forward, looking back: Humans, anthropogenic change, and the Anthropocene. *Anthropocene* 4:116–121.

Branch, Nick, Matthew Canti, Peter Clark and Chris Turney 2005 *Environmental Archaeology: Theoretical and Practical Approaches*. Routledge, New York.

Brenner, Mark, David A. Hodell, Jason H. Curtis, Michael F. Rosenmeier, Michael W. Binford and Mark B. Abbott 2001 Abrupt climate change and pre-Columbian cultural collapse. In *Interhemispheric Climate Linkages*, edited by Vera Markgraf, pp. 87–104. Academic Press, New York.

Bronk Ramsey, Christopher 2009 Bayesian analysis of radiocarbon dates. *Radiocarbon* 51(1):337–360.
Brooks, Nick 2006 Cultural responses to aridity in the Middle Holocene and increased social complexity. *Quaternary International* 151(1):29–49.
Brumfiel, Elizabeth 1992 Distinguished lecture in archaeology: Breaking and entering the ecosystem—gender, class, and faction steal the show. *American Anthropologist* 94(3):551–568.
Butzer, Karl W. 1964 *Environment and Archeology: An Introduction to Pleistocene Geography*. Aldine Publishing Company, Chicago.
——— 1982 *Archaeology as Human Ecology: Method and Theory for a Contextual Approach*. Cambridge University Press, New York.
——— 1996 Ecology in the long view: Settlement histories, agrosystemic strategies, and ecological performance. *Journal of Field Archaeology* 23(2):141–150.
——— 2012 Collapse, environment, and society. *Proceedings of the National Academy of Sciences* 109(10):3632–3639.
Butzer, Karl W. and G. H. Endfield 2012 Critical perspectives on historical collapse. *Proceedings of the National Academy of Sciences* 109(10):3628–3631.
Calaway, Michael J. 2005 Ice-cores, sediments and civilisation collapse: A cautionary tale from Lake Titicaca. *Antiquity* 79(306):778–790.
Caseldine, C. J. and C. Turney 2010 The bigger picture: Towards integrating palaeoclimate and environmental data with a history of societal change. *Journal of Quaternary Science* 25(1):88–93.
Clare, Lee and Bernhard Weninger 2010 Social and biophysical vulnerability of prehistoric societies to rapid climate change. *Documenta Praehistorica* 37:269–284.
Contreras, Daniel A. 2010 Landscape and environment: Insights from the prehispanic central Andes. *Journal of Archaeological Research* 18(3):241–288.
Coombes, Paul and Keith Barber 2005 Environmental determinism in Holocene research: causality or coincidence? *Area* 37(3):303–311.
Cooper, Jago and Payson D. Sheets (editors) 2012 *Surviving Sudden Environmental Change: Answers from Archaeology: Understanding Hazards, Mitigating Impacts, Avoiding Disasters*. University Press of Colorado, Boulder, CO.
Costanza, Robert, Lisa J. Graumlich and Will Steffen (editors) 2007 *Sustainability or Collapse? An Integrated History and Future of People on Earth*. The MIT Press, Cambridge.
Cronon, William 1990 Modes of prophecy and production: Placing nature in history. *The Journal of American History* 76(4):1122–1131.
Crumley, Carole L. 1994 *Historical Ecology: Cultural Knowledge and Changing Landscapes*. School of American Research Press, Santa Fe.
Crutzen, Paul J. and Will Steffen 2003 How long have we been in the Anthropocene era? *Climatic Change* 61(3):251–257.
Dean, Rebecca M. (editor) 2010 *The Archaeology of Anthropogenic Environments*. 37. Center for Archaeological Investigations, Carbondale, IL.
deMenocal, Peter B. 2001 Cultural responses to climate change during the Late Holocene. *Science* 292(5517):667–673.
Denevan, William M. 1992 The pristine myth: The landscape of the Americas in 1492. *Annals of the Association of American Geographers* 82(3):369–385.
Diamond, Jared M. 1998 *Guns, Germs, and Steel: A Short History of Everybody for the Last 13,000 Years*. W.W. Norton & Company, New York.
Dincauze, Dena F. 2000 *Environmental Archaeology: Principles and Practice*. Cambridge University Press, Cambridge.

Doughty, C. E, A. Wolf and C. B. Field 2010 Biophysical feedbacks between the Pleistocene megafauna extinction and climate: The first human-induced global warming? *Geophysical Research Letters* 37(15):L15703.

Ellis, Erle C., Dorian Q. Fuller, Jed O. Kaplan and Wayne G. Lutters 2013 Dating the Anthropocene: Towards an empirical global history of human transformation of the terrestrial biosphere. *Elementa: Science of the Anthropocene* 1:1–6.

Ensor, Bradley E., Marisa O. Ensor and Gregory W. De Vries 2003 Hohokam political ecology and vulnerability: Comments on Waters and Ravesloot. *American Antiquity* 68(1):169–181.

Erickson, Clark L. 1999 Neo-environmental determinism and agrarian "collapse" in Andean prehistory. *Antiquity* 73:634–642.

——— 2008 Amazonia: The historical ecology of a domesticated landscape. In *The Handbook of South American Archaeology*, edited by Silverman Helaine and William H. Isbell, pp. 157–183. Springer, New York.

Erlandson, Jon M. 2014 Shell middens and other anthropogenic soils as global stratigraphic signatures of the Anthropocene. *Anthropocene* 4:24–32.

Erlandson, Jon M. and Todd J. Braje 2013 Archeology and the Anthropocene. *Anthropocene* 4:1–7.

Evans, John G. and Terence Patrick O'Connor 1999 *Environmental Archaeology: Principles and Methods*. Sutton Pub Limited, Stroud.

Fisher, Christopher T., J. Brett Hill and Gary M. Feinman (editors). 2009 *The Archaeology of Environmental Change: Socionatural Legacies of Degradation and Resilience*. University of Arizona Press, Tucson.

Fisher, Christopher T., Helen P. Pollard, Isabel Israde-Alcántara, Victor H. Garduño-Monroy and Subir K. Banerjee 2003 A reexamination of human-induced environmental change within the Lake Pátzcuaro Basin, Michoacán, Mexico. *Proceedings of the National Academy of Sciences* 100(8):4957–4962.

Fisher, Christopher T. and Tina L. Thurston 1999 Special Section: Dynamic landscapes and socio-political process: The topography of anthropogenic environments in global perspective. *Antiquity* 73(281):630–631.

Giosan, Liviu, Dorian Q. Fuller, Kathleen Nicoll, Rowan K. Flad and Peter D. Clift 2012 *Climates, Landscapes, and Civilizations*. American Geophysical Union, Washington, DC.

Goudie, Andrew S. 2013 *The Human Impact on the Natural Environment: Past, Present, and Future*, 7th ed. Wiley-Blackwell, Oxford.

Grosjean, Martin, Calogero M. Santoro, Lonnie G. Thompson, Lautaro Núñez and Vivien G. Standen 2007 Mid-Holocene climate and culture change in the South Central Andes. In *Climate Change and Cultural Dynamics: A Global Perspective on Mid-Holocene Transitions*, edited by David G. Anderson, Kirk A. Maasch and Daniel H. Sandweiss, pp. 51–115. Elsevier, Amsterdam.

Hayashida, Frances 2005 Archaeology, ecological history, and conservation. *Annual Review of Anthropology* 34:43–65.

Head, L. 2008 Is the concept of human impacts past its use-by date? *The Holocene* 18(3):373–377.

Heckenberger, M. J., A. Kuikuro, U. T. Kuikuro, J. C. Russell, M. Schmidt, C. Fausto and B. Franchetto 2003 Amazonia 1492: Pristine forest or cultural parkland? *Science* 301(5640):1710.

Heine, K. 2003 Paleopedological evidence of human-induced environmental change in the Puebla–Tlaxcala area (Mexico) during the last 3,500 years. *Revista Mexicana de Ciencias Geológicas* 20(3):235–244.

Hornborg, Alf and Carole L. Crumley 2007 *The World System and the Earth System: Global Socioenvironmental Change and Sustainability since the Neolithic.* Left Coast Press, Walnut Creek, CA.
Ingold, Tim 2000 *The Perception of the Environment: Essays on Livelihood, Dwelling, and Skill.* Routledge Press, New York.
Jones, Terry L., Gary M. Brown, L. Mark Raab, Janet L. McVickar, W. Geoffrey Spaulding, Douglas J. Kennett, Andrew York and Phillip L. Walker 1999 Environmental imperatives reconsidered: Demographic crises in Western North America during the medieval climatic anomaly. *Current Anthropology* 40(2): 137–170.
Kaniewski, David, Joël Guiot and Elise Van Campo 2015 Drought and societal collapse 3200 years ago in the Eastern Mediterranean: A review. *Wiley Interdisciplinary Reviews: Climate Change* 6(4):369–382.
Kennett, Douglas J., Sebastian F. M. Breitenbach, Valorie V. Aquino, Yemane Asmerom, Jaime Awe, James U. L. Baldini, Patrick Bartlein, Brendan J. Culleton, Claire Ebert, Christopher Jazwa, Martha J. Macri, Norbert Marwan, Victor Polyak, Keith M. Prufer, Harriet E. Ridley, Harald Sodemann, Bruce Winterhalder and Gerald H. Haug 2012 Development and disintegration of Maya political systems in response to climate change. *Science* 338(6108):788–791.
Kidder, Tristram R. and Haiwang Liu 2014 Bridging theoretical gaps in geoarchaeology: Archaeology, geoarchaeology, and history in the Yellow River valley, China. *Archaeological and Anthropological Sciences*:1–18.
Kintigh, Keith W., J. H. Altschul, M. C. Beaudry, Robert D. Drennan, Ann P. Kinzig, Timothy A. Kohler, W. Frederick Limp, H. D. G. Maschner, William K. Michener, Timothy R. Pauketat, P. Peregrine, J. A. Sabloff, T. J. Wilkinson, H. T. Wright and M. A. Zeder 2014 Grand challenges for archaeology. *Proceedings of the National Academy of Sciences* 111(3):879–880.
Kirch, Patrick V. 1997 Microcosmic histories: Island perspectives on "global" change. *American Anthropologist* 99(1):30–42.
——— 2005 Archaeology and global change: The Holocene record. *Annual Review of Environmental Resources* 30:409–440.
——— 2007 Hawaii as a model system for human ecodynamics. *American Anthropologist* 109(1):8–26.
Kohler, Timothy A., C. David Johnson, Mark Varien, Scott Ortman, Robert Reynolds, Ziad Kobti, Jason Cowan, Kenneth Kolm, Schaun Smith and Lorene Yap 2007 Settlement ecodynamics in the prehispanic central Mesa Verde region. In *The Model-Based Archaeology of Socionatural Systems*, edited by Timothy A. Kohler and S. van der Leeuw, pp. 61–104. School for Advanced Research Press, Santa Fe, NM.
Lachniet, Matthew S., Juan Pablo Bernal, Yemane Asmerom, Victor Polyak and Dolores Piperno 2012 A 2400 yr Mesoamerican rainfall reconstruction links climate and cultural change. *Geology* 40(3):259–262.
Laland, Kevin N. and Michael J. O'Brien 2010 Niche construction theory and archaeology. *Journal of Archaeological Method and Theory* 17(4):303–322.
van der Leeuw, S. and C. Redman 2002 Placing archaeology at the center of socionatural studies. *American Antiquity* 67(4):597–605.
Lock, Gary and Brian Leigh Molyneaux 2006 *Confronting Scale in Archaeology: Issues of Theory and Practice.* Springer, New York.
Lowdermilk, Walter C. 1953 *Conquest of the Land Through 7,000 Years.* Soil Conservation Service Bulletin 99. U.S. Department of Agriculture, Washington, DC.

Maher, Lisa A., Edward Bruce Banning and Michael Chazan 2011 Oasis or Mirage? Assessing the role of abrupt climate change in the prehistory of the Southern Levant. *Cambridge Archaeological Journal* 21(1):1–30.
Mann, Charles C. 2005 *1491: New Revelations of the Americas before Columbus*. Vintage Books, New York.
Marsh, George Perkins 1864 *Man and Nature, or Physical Geography as Modified by Human Action*. C. Scribner, New York.
Mayewski, Paul A., E. E. Rohling, J. Curt Stager, W. Karlén, K. A. Maasch, L. David Meeker, E. A. Meyerson, F. Gasse, S. van Kreveld and K. Holmgren 2004 Holocene climate variability. *Quaternary Research* 62(3):243–255.
Mayle, Francis E. and Jose Iriarte 2014 Integrated palaeoecology and archaeology—A powerful approach for understanding pre-Columbian Amazonia. *Journal of Archaeological Science* 51:54–64.
McGlade, James 1995 Archaeology and the ecodynamics of human-modified landscapes. *Antiquity* 68:113–132.
McIntosh, Roderick J., Joseph A. Tainter and Susan Keech McIntosh 2000 *The Way the Wind Blows: Climate, History, and Human Action*. Columbia University Press, New York.
McNeill, J. R. 2003 Observations on the nature and culture of environmental history. *History and Theory* 42:5–43.
Meadows, John 2005 The Younger Dryas episode and the radiocarbon chronologies of the Lake Huleh and Ghab Valley pollen diagrams, Israel and Syria. *The Holocene* 15(4):631–636.
Medina-Elizalde, M., and E. J. Rohling 2012 Collapse of Classic Maya civilization related to modest reduction in precipitation. *Science* 335(6071):956–959.
Middleton, Guy D. 2012 Nothing lasts forever: Environmental discourses on the collapse of past societies. *Journal of Archaeological Research* 20(3):257–307.
Minnis, Paul E. 1999 Sustainability: The long view from archaeology. *New Mexico Journal of Science* 39:23–41.
Morrison, Kathleen D. 2015 Provincializing the Anthropocene. *Seminar* 673:75–80.
O'Sullivan, Patrick 2008 The "collapse" of civilizations: What palaeoenvironmental reconstruction cannot tell us, but anthropology can. *The Holocene* 18(1):45–55.
Redman, Charles L. 1999 *Human Impact on Ancient Environments*. University of Arizona Press, Tucson.
Redman, Charles L., Steven R. James, Paul R. Fish and J. Daniel Rogers (editors) 2004 *The Archaeology of Global Change: The Impact of Humans on Their Environment*. Smithsonian Institution Press, Washington, DC.
Redman, Charles L. and A. P. Kinzig 2003 Resilience of past landscapes: Resilience theory, society, and the Longue Durée. *Conservation Ecology* 7(1):14.
Reitz, Elizabeth Jean and Myra L. Shackley 2012 *Environmental Archaeology*. Springer, New York.
Robb, John and Timothy R. Pauketat 2013 From moments to millennia: Theorizing scale and change in human history. In *Big Histories, Human Lives: Tackling Problems of Scale in Archaeology*, edited by John Robb and Timothy R. Pauketat, pp. 3–33. School for Advanced Research Press, Santa Fe, NM.
Roberts, Neil, Tony Stevenson, Basil Davis, Rachid Cheddadi, Simon Brewster, and Arlene Rosen 2004 Holocene climate, environment and cultural change in the circum-Mediterranean region. In *Past Climate Variability through Europe and Africa*, edited by Richard W. Battarbee, Françoise Gasse, and Catherine E. Stickley, pp. 343–362. Springer, Dordrecht.

Rowe, John Howland 1962 Stages and periods in archaeological interpretation. *Southwestern Journal of Anthropology* 18(1):40–54.

Ruddiman, William F. 2003 The anthropogenic greenhouse era began thousands of years ago. *Climatic Change* 61(3):261–293.

——— 2013 The Anthropocene. *Annual Review of Earth and Planetary Sciences* 41(1):45–68.

Ruddiman, W. F., D. Q. Fuller, J. E. Kutzbach, P. C. Tzedakis, J. O. Kaplan, E. C. Ellis, S. J. Vavrus, C. N. Roberts, R. Fyfe, F. He, C. Lemmen and J. Woodbridge 2016 Late Holocene climate: Natural or anthropogenic? *Reviews of Geophysics* 54(1):93–118.

Sanders, W. T., J. R. Parsons and R. S. Santley 1979 *The Basin of Mexico: Ecological Processes in the Evolution of a Civilization*. Academic Press, New York.

Sandweiss, Daniel H. and Jeffrey Quilter 2009 *El Niño, Catastrophism, and Culture Change in Ancient America*. Dumbarton Oaks Research Library and Collection, Washington, DC.

Sandweiss, Daniel H., Ruth Shady Solís, Michael E. Moseley, David K. Keefer and Charles R. Ortloff 2009 Environmental change and economic development in coastal Peru between 5,800 and 3,600 years ago. *Proceedings of the National Academy of Sciences* 106(5):1359–1363.

Sauer, Carl O. 1941 Foreword to historical geography. *Annals of the Association of American Geographers* 31(1):1–24.

Schumm, Stanley A. 1991 *To Interpret the Earth: Ten Ways to Be Wrong*. Cambridge University Press, Cambridge.

Smith, Bruce D. 2007 Niche construction and the behavioral context of plant and animal domestication. *Evolutionary Anthropology: Issues, News, and Reviews* 16(5):188–199.

Smith, Bruce D. and Melinda A. Zeder 2013 The onset of the Anthropocene. *Anthropocene* 4:8–13.

Spriggs, Matthew 2010 Geomorphic and archaeological consequences of human arrival and agricultural expansion on Pacific Islands: A reconsideration after 30 years of debate. *Terra Australis* 32(Altered Ecologies: Fire, climate and human influence on terrestrial landscapes):239–252.

Stein, Julie K. 1993 Scale in archaeology, geosciences, and geoarchaeology. *Geological Society of America Special Papers* 283:1–10.

Steinberg, Ted 2002 Down to earth: Nature, agency, and power in history. *American Historical Review* 107(3):798–820.

Steward, Julian Haynes 1972 *Theory of Culture Change: The Methodology of Multilinear Evolution*. University of Illinois Press, Urbana.

Tainter, Joseph A. 2006 Archaeology of overshoot and collapse. *Annual Review of Anthropology* 35(1):59–74.

Tucson. 2005 Resilience theory in archaeology. *American Anthropologist* 107(1):70–77.

Thomas, William L., Carl O. Sauer, Marston Bates and Lewis Mumford (editors) 1956 *Man's Role in Changing the Face of the Earth*. The University of Chicago Press, Chicago.

Thompson, Victor D. and James C. Waggoner (editors) 2013 *The Archaeology and Historical Ecology of Small Scale Economies*. University Press of Florida, Gainesville.

Trigger, Bruce G. 1989 *A History of Archaeological Thought*. Cambridge University Press, Cambridge.

Turner, B. L, II, and J. A. Sabloff 2012 Classic Period collapse of the Central Maya Lowlands: Insights about human-environment relationships for sustainability. *Proceedings of the National Academy of Sciences* 109(35):13908–13914.
Van Buren, Mary 2001 The archaeology of El Niño events and other "natural" disasters. *Journal of Archaeological Method and Theory* 8(2):129–149.
Van de Noort, R. 2011 Conceptualising climate change archaeology. *Antiquity* 85:1039–1048.
Varien, M. D, S. G. Ortman, T. A. Kohler, D. M. Glowacki and C. D. Johnson 2007 Historical ecology in the Mesa Verde region: Results from the village ecodynamics project. *American Antiquity* 72(2):273–299.
de Vries, Bert 2006 In search of sustainability: What can we learn from the past? In *The World System and the Earth System: Global Socioenvironmental Change and Sustainability Since the Neolithic*, edited by Alf Hornborg and Carole L Crumley, pp. 243–257. Left Coast Press, Walnut Creek, CA.
de Vries, Bert and Johan Goudsblom (editors) 2002 *Mappae Mundi: Humans and their Habitats in a Long-Term Socio-Ecological Perspective: Myths, Maps and Models*. Amsterdam University Press, Amsterdam.
Watson, Richard A. and Patty Jo Watson 1969 *Man and Nature: An Anthropological Essay in Human Ecology*. Harcourt, Brace & World New York.
Weninger, Bernhard, Lee Clare, Eelco Rohling, Ofer Bar-Yosef, Utz Böhner, Mihael Budja, Manfred Bundschuh, Angelica Feurdean, Hans Georg Gebe, Olaf Jöris, Jörg Linstädter, Paul Mayewski, Tobias Mühlenbruch, Agathe Reingruber, Gary O. Rollefson, Daniel Schyle, Laurens Thissen, Henrieta Todorova, and Christoph Zielhofer 2009 The impact of rapid climate change on prehistoric societies during the Holocene in the Eastern Mediterranean. *Documenta Praehistorica* 36:7–59.
Willey, Gordon R. 1953 *Prehistoric Settlement Patterns in the Virú Valley, Peru*. Bureau of American Ethnology Bulletin 155. U.S. Government Printing Office, Washington, DC.
Williams, M. 1994 The relations of environmental history and historical geography. *Journal of Historical Geography* 20(1):3–21.
Worster, Donald 1990 Transformations of the earth: Toward an agroecological perspective in history. *The Journal of American History* 76(4):1087–1106.
Wright, H. E. 1993 Environmental determinism in Near Eastern prehistory. *Current Anthropology* 34(4):458–469.
Yaeger, Jason and David A. Hodell 2008 The Collapse of Maya civilization: Assessing the interaction of culture, climate, and environment. In *El Niño, Catastrophism, and Culture Change in Ancient America*, edited by Daniel H. Sandweiss and Jeffrey Quilter, pp. 197–251. Dumbarton Oaks, Washington, DC.
Zalasiewicz, Jan, Colin N. Waters, Mark Williams, Anthony D. Barnosky, Alejandro Cearreta, Paul Crutzen, Erle Ellis, Michael A. Ellis, Ian J. Fairchild, Jacques Grinevald, Peter K. Haff, Irka Hajdas, Reinhold Leinfelder, John McNeill, Eric O. Odada, Clément Poirier, Daniel Richter, Will Steffen, Colin Summerhayes, James P. M. Syvitski, Davor Vidas, Michael Wagreich, Scott L. Wing, Alexander P. Wolfe, An Zhisheng and Naomi Oreskes 2015 When did the Anthropocene begin? A mid-twentieth century boundary level is stratigraphically optimal. *Quaternary International* 383:196–203.

Case Studies

2 Convergence and Divergence as Problems of Explanation in Land Use Histories

Two Mexican Examples

Aleksander Borejsza and Arthur A. Joyce

Abstract

Similar results that arise from different processes and causes (convergence or equifinality) and different results that arise from similar processes and causes (divergence) are two common problems of explanation in any historical science. Two cases rooted in geoarchaeological fieldwork in highland Mexico exemplify such problems. A widespread entrenchment of the stream network of the Nochixtlan Valley occurred at ca. A.D. 1000. Potential proximate causes include changes in runoff, sediment delivery from slopes, adjustments of stream gradient, or channel form. Each hints in turn at a plethora of converging ultimate causes in realms such as climate, demography, agriculture, or warfare, but choosing between them is impossible without widening research to include issues usually considered to be beyond the scope of environmental archaeology. Tlaxcala experienced cycles of demographic explosion and collapse between 500 B.C. and A.D. 100, and again between A.D. 1200 and 1600. Each cycle was accompanied by agricultural intensification and disintensification, but the form they took were very different. In the first cycle, population growth resulted in severe land degradation, in the second in widespread land improvement. The divergence seems to be due to differences in cultural context and accumulated historical experience. In order to satisfactorily resolve problems of explanatory convergence and divergence, environmental archaeologists must constantly shift attention between process and history, and actively contribute to endeavors such as chronology building, on-site excavation, and the long-term curation of collections.

Introduction

This chapter sets out to demonstrate, by example, how easy it is to make up multiple and conflicting stories that purport to explain the alluvial record in terms of human land use. Our ulterior aim is to show how similar epistemological misgivings can be across the "Cartesian divide of nature and culture," in this case the divide that separates disciplines as disparate as fluvial geomorphology and cultural anthropology. We use fluvial geomorphology to understand how different physical processes shape the alluvial stratigraphic record and cultural anthropology to understand how and why farmers decide between alternative forms of land use. We cross the divide between them, almost imperceptibly, when we assess the physical effects of agricultural activities such as tillage, grazing, or the building of terraces.

In our experience, many fellow archaeologists suffer from an inferiority complex that leads them to believe that the natural sciences rest on an infinitely more solid theoretical foundation, and are therefore more resolute when relating cause and effect. Stanley Schumm's *To Interpret the Earth: Ten Ways to Be Wrong* (1991) is adept at dispelling this belief and inspired the title of our contribution. Schumm explores some of the common pitfalls of relating cause and effect in the earth sciences, especially when the effect is a fragmentary stratigraphic record and the cause something that existed in the past, but need not exist any longer in the present, a conundrum familiar to most archaeologists. Two problems Schumm specifically writes about are convergence and divergence. We illustrate each with an example drawn from our research in highland Mexico. The convergence example is negative, in that we cannot yet confidently choose between alternative explanations. The divergence example seems more positive, in that we are able to discount some explanations advanced by our predecessors. We end by pondering what makes the difference between the two examples, and why we have progressed more in one study area than the other.

Convergence in the Nochixtlan Valley: The Causes of Stream Incision at A.D. 1000

Convergence or equifinality refers to a situation where different causes produce similar effects (Schumm 1991:58). The effect we are concerned with is one that Schumm himself studied throughout his career, that of stream incision (e.g., Schumm 1957; 1999). Incised streams are common throughout the highlands of formerly Spanish America (Figure 2.1), and there is a corresponding wealth of vernacular terms—*arroyos*, *barrancas*, *barrancos*, *cañadas*, *quebradas*—that identify them in place names. They flow confined between almost vertical walls, and adjust to changing external inputs by filling up with sediment and raising the elevation of their channel (i.e., aggrading their narrow floodplain) or by incising (degrading, downcutting, trenching). If they remain stable, i.e., retain the same elevation, their floodplains develop soils. Because the channel experiences minor lateral shifts with each incision, the walls or cutbanks often expose the stratigraphic record of previous filling cycles (Borejsza et al. 2014a:fig. 1).

The cutting is usually much more rapid than the filling and can have dramatic effects on human land use. It drops water tables, makes irrigation impracticable, undermines bridges and dams, creates obstacles to movement, and may increase the magnitude and frequency of devastating floods. The causes of stream incision are thus of not only academic, but also practical concern. For example, a spate of stream incisions that began in the

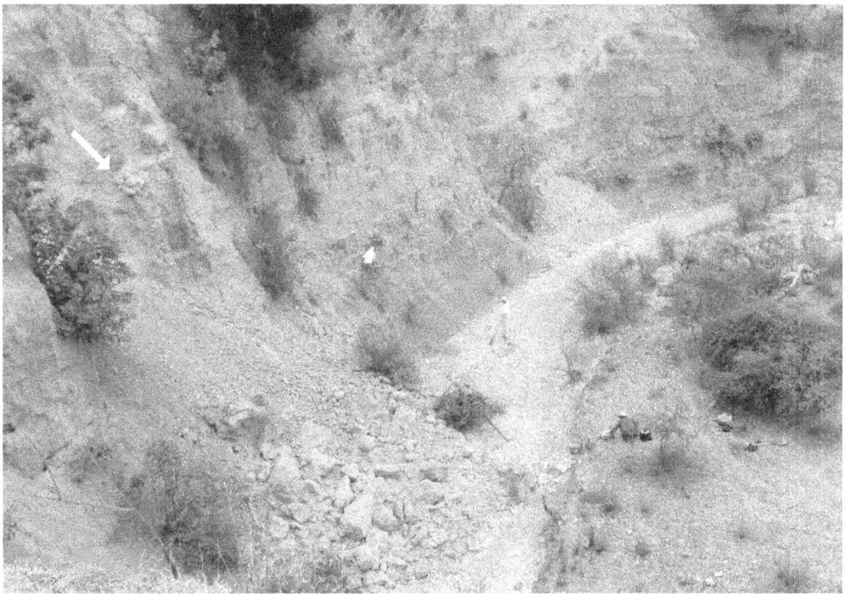

(a)

(b)

Figure 2.1 Incised streams in highland Mexico: (a) Yuzanu in the Nochixtlan Valley. The arrow points to the retaining wall of an agricultural terrace which crossed the channel at 150 B.C. (b) Barranca Tenexac in north-eastern Tlaxcala. This reach incised at the onset of the Holocene, and then again at ca. A.D. 100. Note the proximity of slope and channel.

southwestern United States roughly at the time of Anglo-American takeover and adversely affected the settlers' agricultural enterprises gave birth to the notion of an "arroyo problem" and a vast literature debating its exact timing and causes (Bryan 1925; Cooke and Reeves 1976; Elliott et al. 1999; Waters and Haynes 2001).

In the stratigraphic record, one recognizes an incision by an erosional unconformity that laterally separates alluvial fills of differing age. By definition, the incision leaves behind a fluvial terrace, i.e., a surface that is no longer flooded. The terraces, however, need not be paired, nor form obvious steps. Instead, younger fills are often "inset" or "nested" within older ones. The age of an incision can be bracketed by dating the age of deposition of fills on either side of the unconformity, ideally something from just underneath the surface of the terrace, and something from near the base of the ensuing fill. This is the procedure we tried to follow in our research in the Nochixtlan Valley in the Mixteca Alta region of the state of Oaxaca (Mueller et al. 2012).

An incision spread throughout the stream network that drains this 500 km^2 valley about a thousand years ago. Radiocarbon dates on the organic matter of buried palaeosols, artifact inclusions, and the position of a few small archaeological sites with respect to modern and former stream channels suggest that the incision started at the mouth of the valley close to A.D. 1000 and reached the smallest headwater tributaries some two centuries later. We have documented other incisions throughout the late Quaternary, but this one seems to have been particularly pervasive and rapid, especially considering the relatively large size and complexity of the fluvial system in question.

The three or four centuries centered on A.D. 1000 loom large in the imagination of scholars concerned with both environmental change and the culture history of Mesoamerica (e.g., Diehl and Berlo 1989; Sodi 1990; Demarest et al. 2004; Manzanilla 2005). They seem packed with dramatic events that revolve around ecological disaster and political turmoil. The most severe droughts of the Holocene are placed within this timeframe, both in central Mexico and the Yucatan (Metcalfe et al. 2000; Stahle et al. 2011). They have been held responsible for the "Great" Maya collapse of the Terminal Classic (see Kennett and Beach 2014), as well as for the contraction of the agricultural frontier on the northern fringe of Mesoamerica, and the resulting southward migration of people viewed by the old urban civilizations of central Mexico as barbarians (Braniff 1989; Beekman and Christensen 2003:148). In both the Mixteca Alta and the neighboring Valley of Oaxaca, this is a time of major shifts in settlement patterns and sociopolitical organization (Spores 1972; Kowalewski et al. 1989; 2009; Winter 1989; 1994; Blomster 2008), synthesized in terms like "balkanization" (Flannery and Marcus 1983) or "collapse and reemergence" (Joyce 2010). Do they bear any relation—as either cause or consequence—to our stream incision?

Georarchaeological research suggests that the dramatic cultural transformations centered on A.D. 1000 may indeed bear some relation to stream incision in Nochixtlan. Figures 2.2 and 2.3 map out some of the *plausible*

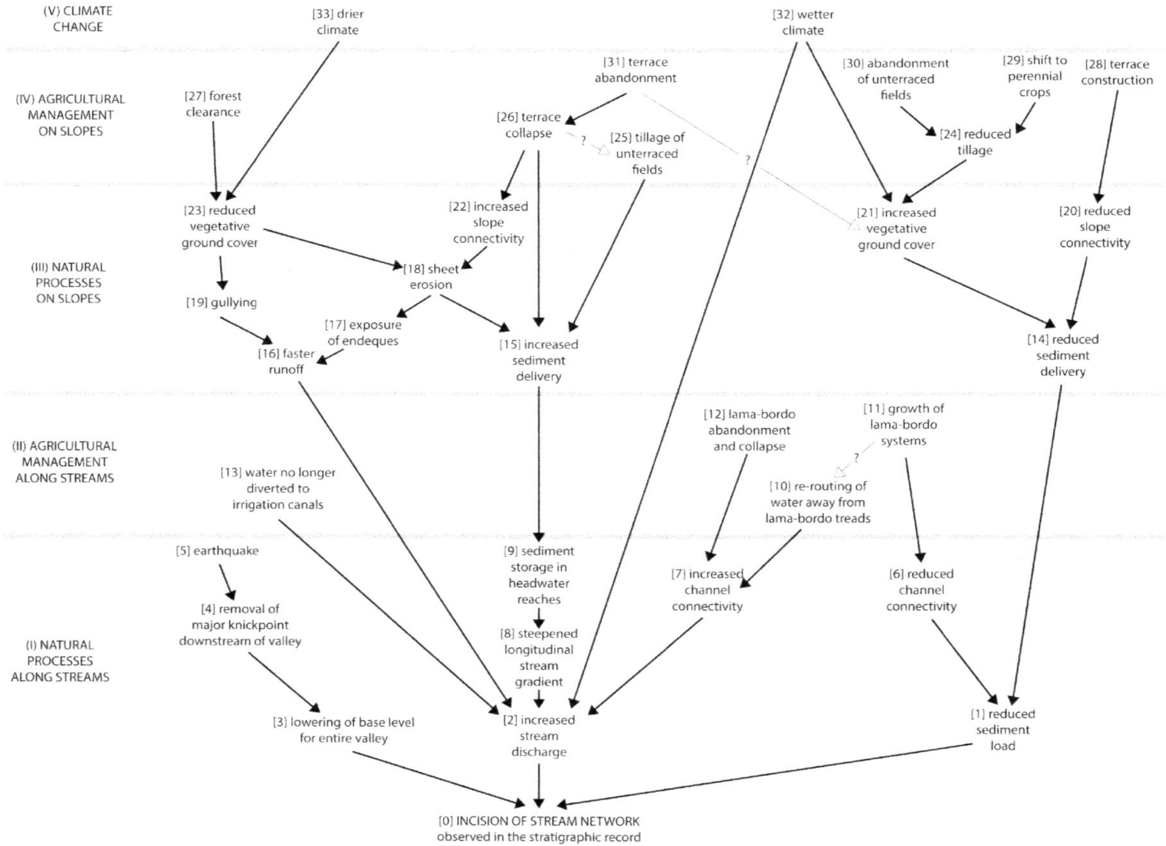

Figure 2.2 Convergent causes of stream incision in the Nochixtlan Valley.

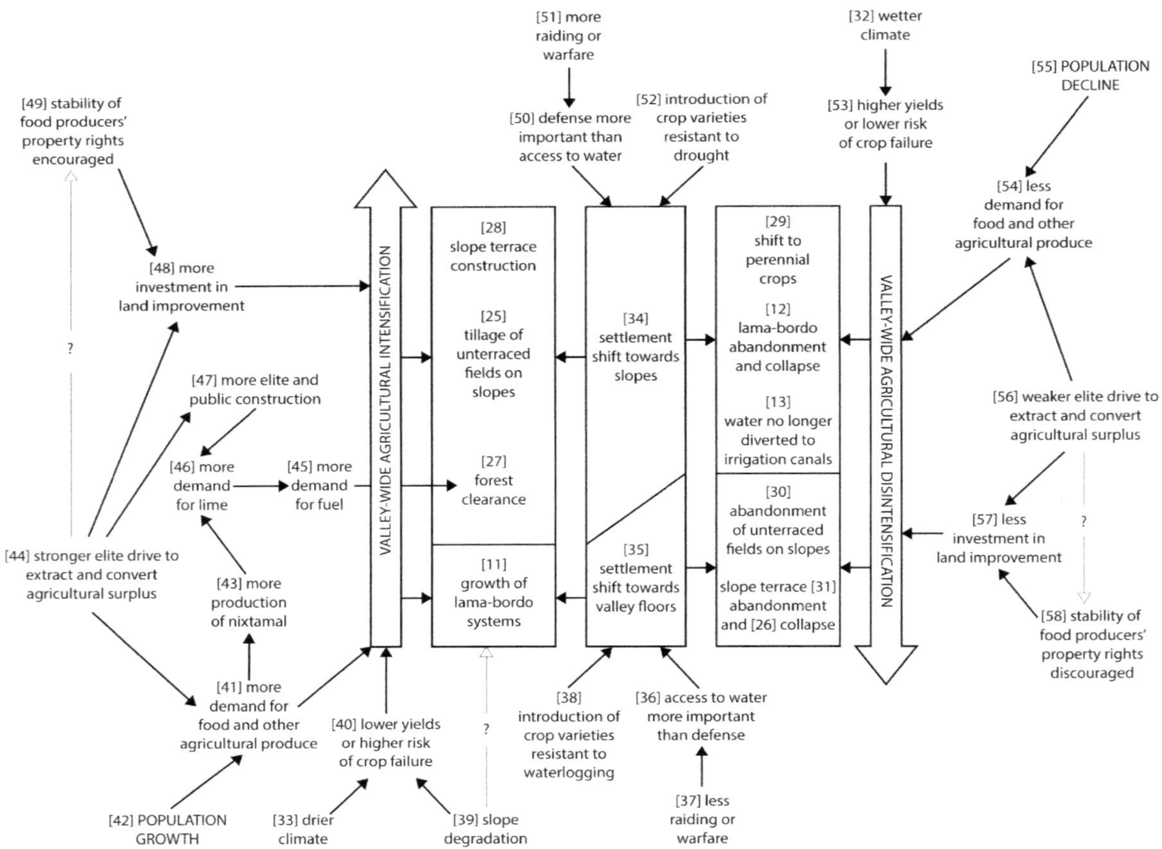

Figure 2.3 Convergent causes of different land use choices in the Nochixtlan Valley.

causal relationships triggering incision. Because flowing water is the physical force that cuts a new channel, the proximate causes of incision (entries [1,2,3] in Figure 2.2) are hydrologic. They are essentially a function of the balance between water discharge and sediment load. Reduced sediment loads [1] or increased average or peak discharges [2] could have tilted the balance in the direction of stream incision (Bull 1991:fig. 1.4; Knighton 1998:formulae 6.9–6.18; Schumm 1999:table 2.2D). Yet another possibility is the lowering of the base level at some point farther downstream. Formed by the confluence of the two major branches of the Nochixtlan Valley, the Río Verde descends to the Pacific Ocean through narrow canyons cut into bedrock. The breaching of a major obstruction somewhere along the way [4] at A.D. 1000, perhaps triggered by an earthquake [5], could have suddenly dropped the base level.

If we focus on the more mundane factor of increased discharge [2], it could be caused in turn by more rainfall [32]; faster runoff from slopes [16]; increased connectivity (Fryirs 2013) of the stream network [7]; a steepened longitudinal stream gradient [8] (Schumm 1977:77–137; Bull 1979; 1997; Patton and Schumm 1981); or the abandonment of irrigation networks, with the water that was formerly diverted to fields now re-integrated in the overall stream discharge [13]. If we follow instead the lead of reduced sediment load [1], it may take us to reduced sediment delivery from slopes [14], often linked to their reduced geomorphic connectivity [20], perhaps brought about by the construction of agricultural terraces on slopes [28].

A different kind of agricultural terrace, the *lama-bordos* (Spores 1969; Pérez Rodríguez 2006; Mueller et al. 2012; Leigh et al. 2013; Pérez Rodríguez and Anderson 2013) may also have played a critical part. This ancient agricultural technology, known to date in the Mixteca as far back as 1500 B.C., is a form of cross-channel terracing (for agricultural terrace typology and related terminology see Whitmore and Turner 2001:133–64; Frederick and Krahtopoulou 2000). Farmers block a barranca with an obstacle of brush or stone rubble, which slows the movement of water and traps nutrient-rich fine sediment and plant litter. As new courses are added to the terrace riser, the tread grows in both thickness and surface area. The scale of these works in the Mixteca is truly monumental, with risers several meters tall, and flights of treads spanning several kilometers. Thousands of breached stone risers of different ages jut out from barranca walls (Figure 2.1(a)). We would expect the growth of lama-bordo systems [11] to result in reduced channel connectivity [6], and reduced sediment load downstream [1]. Their abandonment and collapse [12] would, conversely, result in increased connectivity [7], as the barranca re-established its natural continuous channel. This reasoning, however, is complicated by a technological variant we observed in some lama-bordos. Instead of blocking the barranca completely, farmers may shape the risers and the cultivation surfaces in such a way that excess water is shunted to one side of the field [10]. If repeated over a whole flight of terraces, this procedure promotes the appearance of a narrow but high-energy channel that hugs one side of the valley. The growth of *such* lama-bordos may instead lead to high channel connectivity and discharge [7→2].

It is clear from this discussion and from the tree-like shape of Figure 2.2 that, as we move from proximate to more distant causes, we can imagine a dizzying array of equifinal processes that converge on the same effect of stream incision. What is even more disconcerting is that there are pairs of exact opposites in the diagram: reduced and increased channel or slope connectivity [6/7, 20/22]; construction and abandonment of hill slope or cross-channel agricultural terraces [28/31, 11/12]; a wetter and a drier climate [32/33]. What we find interesting and somewhat reassuring is that there is no neat separation of natural factors near the base of the tree or (agri)cultural factors in the higher branches, nor vice versa. It is thus neither a simplistic scheme of people controlling nature, nor a case of environmental determinism. The progression from what we have marked as levels I to IV can be thought instead, in geomorphic terms, as moving from sediment sink (or outlet) to source. We shall see further on what it means in terms of archaeological practice.

So far, we have not even left the realm of the technological and rather mechanistic aspects of agriculture to explore those ultimate causes that may lie in the realm of human decision-making. If alluvial geoarchaeology is to shed some light on *why* people changed the way they used the land, we will need to graft, on each of the terminal branches of the tree in level II or IV, some segments cut from Figure 2.3. Several of the level II and IV entries of Figure 2.2 are therefore repeated in the two side rectangles of Figure 2.3. The left-hand rectangle [11,25,27,28] groups together those changes in land use that one would expect to have taken place, in isolation or in tandem, when the valley as a whole underwent a cycle of agricultural intensification. Conversely, those in the right-hand rectangle [12,13,26,29,30,31] can be thought of as effects of disintensification. Most of the new entries in this diagram [numbers higher than 33] are concerned with the possible causes of either, and therefore converge on one of the two thick arrows labeled "intensification" and "disintensification."

A smaller set of entries, however, converges on the central rectangle, which lists two opposed trends in settlement location, preferring either the slopes or the valley floors. Archaeological settlement surveys in the Mixteca Alta (Spores 1972; Kowalewski et al. 2009) reveal shifting trends towards one or the other. Each could lead to changes in land use dictated by the distance and safety of walking to this or that field, rather than by valley-wide changes in agricultural intensity. For example, fields (terraced or not) might expand at the expense of forest on relatively marginal land on slopes [25,27,28] because of valley-wide population growth [42] and rising demand for agricultural produce [41]. Just as well, however, they may reflect an intensification of raiding or warfare [51] motivating people to seek out defensible hilltop locations for settlement [34], while relinquishing more fertile valley floor land [50]. This is why the two opposing trends in settlement pattern [34/35] do not map one-on-one with intensification or disintensification.

We have two main sources of inspiration for the entries in Figure 2.3. One is the vast body of theory on agricultural intensification and landesque capital (Boserup 1965; Brookfield 1972; 1984; Netting 1993; Morrison 1994; Allen and Ballard 2001; Håkanson and Widgren 2014). Much of it has been assimilated and hotly debated by Mesoamerican archaeologists (e.g., Sanders et al. 1979; Sanders and Nichols 1988; Killion 1992; Johnston 2003; Marcus and Stanish 2006; Thurston and Fisher 2007). Netting's writings were explicitly used as a frame of reference by Pérez Rodríguez (2003; 2006) in her analysis of Postclassic houses and terraces at Nicayuju, a few kilometers west of the Nochixtlan Valley. The other source is works that report on archaeological fieldwork in the Mixteca Alta (for an introduction see Joyce 2010; Pérez Rodríguez 2013), particularly the regional and intra-settlement surveys whose authors pay sustained attention to problems of land use (Spores 1969; 1972; Balkansky 1998; Balkansky et al. 2004; Kowalewski et al. 2009; Pérez Rodríguez et al. 2011; Pérez Rodríguez and Anderson 2013). A few entries may be our original ideas, at least in the regional context.

Pairs of exact opposites crop up in Figure 2.3, as well, e.g., population growth and decline [42/55]. This time there are so many opposites that the upper right and lower left parts of Figure 2.3 are almost mirror images of each other. The most frustrating aspect of this is that we find each entry at least somewhat plausible for some point in time between A.D. 800 and A.D. 1200. For some entries, this is due to our almost complete ignorance of certain topics, for example, the changes in elite-commoner relationships and land tenure arrangements in this time interval [44,49,56,58] or the date of appearance of crop varieties such as the *maíz de cajete* that does well in seasonally waterlogged lama-bordo soils [38]. For most others, it is due to the lack of a well-dated ceramic chronology and differences of opinion regarding this particular interval in Oaxacan archaeology (Winter 1989; 1994; Marcus 1990; Blomster 2008; Joyce 2010:248–52; Feinman and Nicholas 2011; Faulseit 2013).

Some will perceive our Figures 2.2 and 2.3 as an overgrown "heuristic" tree, others as a reminiscence of flow diagrams fashionable in the heyday of archaeologists' romance with systems theory in the 1970s. Neither amounts to an explanation. Before we offer a cure to reduce convergence, let us express caution against a common placebo, that of *quantitative* modeling. A suggestion we sometimes hear from colleagues when alluvial geoarchaeology fails to provide the definitive answer is to estimate crucial parameters of the fluvial system at some point in the past; for the problem at hand, it would probably be its erosive power (Bull 1979). This is to be done on the basis of certain assumptions about the size and shape of the stream network, channel form, the amount of sediment and runoff contributed from the catchment with varying coverage of forest, arable, and eroded land, etc. We feel that such exercises only obfuscate our ignorance. Most fluvial systems we study are open-ended, and we can only hope to catch, in the stratigraphic record, glimpses of the state they assumed along this or that reach.

What many archaeologists probably do not know is that fully quantitative modeling is elusive even at the short timescales amenable to direct observation, i.e., those of "process" rather than "historical" geomorphology. Geomorphologists and engineers tasked with such modeling routinely use coarse approximations, assumptions of negligibility, competing equations (e.g., Knighton 1998:101–5) and "fudge" factors, and even visual comparisons with picture books of representative reaches (e.g., Barnes 1967).

We think that the only logical way forward is to gradually prune our tree, by proving that some parts of Figures 2.2 and 2.3 are false in the study area or in the timeframe under consideration. This can be achieved in two different ways, depending on whether we lop mid-way at the branches, i.e., the arrows of Figures 2.2 and 2.3, or at the nodes, i.e., the textual entries. The first course of action amounts to testing the strength of different causal links. Is it really justified, for example, to draw an arrow connecting terrace abandonment and terrace collapse [31→26]? Numerous case studies suggest that it is indeed, and that once farmers cease to maintain terrace risers, slopes revert to their natural gradient (e.g., Llerena et al. 2004; Tarolli et al. 2014; see also section on Tlaxcala below). Large volumes of sediment are mobilized in the process [15], as gullies cascade from one breached riser to another and terrace fills are flushed out towards footslopes and stream channels. Van Andel and Runnels (1987:146–9), however, discuss certain scenarios—sudden and total abandonment, not accompanied by a conversion of arable land to pasture—under which terraces may be stabilized and preserved by rapidly invading secondary vegetation. Some Mediterranean (Bevan et al. 2013) and Andean (Denevan et al. 1987) examples suggest that this may indeed be the case with particularly well-built terraces on certain geological substrates. If so, we would be more justified in drawing an arrow connecting terrace abandonment with increased ground cover [31→21]. Such arguments are essentially about process, and the answers will come from long-term familiarity with the study area, as well as a lot of comparative reading in process geomorphology (for Figure 2.2) and in anthropological theory (for Figure 2.3). Few of us are willingly immerse ourselves in both literatures.

The second course of action is more about history, and about eliminating some of the entries, preferably in a manner independent of the alluvial record itself. Chronological refinement is fundamental in this task. There are three related aspects of it, to which we refer as moment, duration, and historical sequence. The first requires asking when a certain factor did *not* operate. If we can show, for example, that an expansion of agricultural terracing is an unlikely proposition at A.D. 1000, we can eliminate the growth of terrace systems along channels [11] and on slopes [28], as well as anything that we grafted onto them in Figure 2.3. As we have just pointed out, however, terraces are a very fragile landform, so that the absence of terraces dated to ca. A.D. 1000, especially when our fieldwork is limited to a few hillslope locations, cannot be taken as incontrovertible proof of the absence of terrace construction at that moment. Moreover, the associations of agricultural terraces with surface sherd scatters, and even with artifacts included in terrace fills, can be very misleading. Reliable dating is only possible through excavation, an examination of stratigraphic relationships, and radiometric dating (e.g., Zaro and Umire 2005; Krahtopoulou and Frederick 2008; Smith et al. 2013).

Duration is what geomorphologists refer to as response time (Bull 1991:12). One of the perils of drawing flow diagrams is creating an illusion of clockwork-like synchroneity, emphasizing process over history. In fact, some arrows represent an almost instantaneous response, while others represent events that unfolded over several centuries. The branch leading from forest clearance to gullying [27→23→19] can be extremely rapid. Werner (1981; 1988:photos 12–16) presents a sequence of photographs taken on the slopes of the Pico de Orizaba where the clearance of a forest patch to plant potatoes in 1973 led to deep gullying within a year, and the appearance of a 40m-deep barranca by 1978. On the other hand, the branch leading from terrace abandonment to the steepening of the stream gradient [31→26→15→9→8] is something that could take several centuries in a 500 km^2 catchment. In the former case we would want to demonstrate chronological correlation, i.e., forest clearance at ca. A.D. 1000. In the latter, we would want to demonstrate substantial precedence, i.e., terrace abandonment at A.D. 800 or earlier.

Finally, we must consider the historical sequence in which different factors may operate, and the different states that the fluvial system may have assumed before the incision. The alluvial record itself offers some clues here. The deposits at the top of the fluvial terrace created by the A.D. 1000 incision are often thick packets of rapidly aggraded sand. They strongly suggest that the condition "reduced sediment load" [1] did not apply before the incision, and could lead us to ignore the right-hand branches ending at [28], [29], [30], and probably [11]. But, it is also possible that all the factors related to slope erosion, grouped in the right and center of Figure 2.2 and joining at [2], acted first to initiate the incision, but that once the sediment supply was exhausted, the reduced sediment load helped to propagate and perpetuate it.

As is becoming apparent, we are unlikely to progress much if we limit our study to alluvium, and to the physical processes that govern its deposition (level I). In order to prune, we will have to climb the tree. In less metaphorical terms, this means shifting the location of fieldwork upstream, upslope, and from off-site to on-site locations. In fact, our research strategies in the Nochixtlan Valley are evolving in that direction. Between 1989 and 2009 we paid a number of relatively brief visits to the valley, walking most of the higher-order reaches and focusing on alluvial sequences representative of its major branches (Joyce and Mueller 1997; Mueller et al. 2012). In 2012, during a 10-week field season, we concentrated our efforts in one of them, walked several previously unexplored headwater reaches, and dug into several lama-bordos, thus reaching level II. The next logical step would be to study colluvial deposits and excavate some of the terraces on slopes (levels III and IV).

Because the latter were likely used for both cultivation and habitation, we would be entering the realm of regular on-site archaeology, with all the logistical complications that this implies. Though necessary, climbing the tree comes at a cost. Our observations become more finely resolved, but their scope is progressively reduced. By our reckoning, in about five days it is possible to clean, log in all the necessary detail, and sample a cutbank exposure of 10m-thick alluvium spanning 10,000 years at the mouth of a 20 km^2 catchment. In the same five days, we can uncover a lama-bordo spotted in a cutbank in a manner sufficient to fully understand its three-dimensional geometry, record and sample its multiple cultivation surfaces, and find datable inclusions. But, the lama-bordo will likely span a few centuries at most, and will provide only a localized vignette of agricultural land use. In order to ascertain how representative it is, we will have to study several others. Moving to higher levels of Figure 2.2, the density of artifacts rises, as do the chances of finding skeletal materials, imposing a slower pace of exploration, longer hours of post-excavation analysis, and the duty to plan for their curation.

For the time being, we are unable to weigh in on the issue of climate change (level V) at A.D. 1000, precisely because the alluvial record may have been impacted by non-climatic yet equifinal causes. Filling that void with other proxy records of climate change is difficult. Climate reconstructions in Mexico have so far relied largely on lake sediments and glacial landforms (for overview see Borejsza and Frederick 2010:supplementary file 1; Caballero Miranda and Ortega 2011:vol. 1). The former are extremely scarce (e.g., Goman et al. 2011), the latter inexistent in the Mixteca Alta. There is hope in recent studies of tree rings (Villanueva et al. 2011; Stahle et al. 2012) and speleothems (Bernal et al. 2011) but those from Oaxaca and neighboring states cover timespans either before A.D. 710 or after A.D. 1280. A tree ring series from Queretaro (Stahle et al. 2011) records a prolonged drought from A.D. 897 to 922, but the climate of that locale is weakly correlated with that of Oaxaca. Given the rugged topography of the Mixteca, it would be preferable to have local data, for example, from the few remaining bald cypresses that we have seen in valleys to the west of Nochixtlan.

Where we get completely disoriented is in Figure 2.3. It would help immensely if we had good independent data on whether population was growing or declining at A.D. 1000. The Nochixtlan Valley was surveyed, with state-of-the-art techniques, in the late 1960s (Spores 1972), but the ceramic periodization that guided this work had a Las Flores phase lasting from A.D. 500 or 600 to A.D. 1000 or 1100, followed by a similarly protracted Natividad phase, spanning the remaining four of five centuries before Spanish conquest. Kowalewski et al. (2009) push Las Flores back to A.D. 200–900/950, and subdivide it into Transición, Early, and Late, but make clear that the Late subphase (starting at A.D. 550) is difficult to distinguish, to some extent because of conservatism in ceramic styles, and to some extent because there seem to be few Late Las Flores occupations in the portion of the Mixteca Alta that they surveyed. This included the western third of the Nochixtlan Valley, for which they infer a population decline between A.D. 550 and 900/950, though less pronounced than farther west. The Natividad phase remains undivided. There are many sites in the central and eastern Nochixtlan Valley, however, that have both Las Flores and Natividad sherds, and with the current level of chronological resolution and spatial coverage, we can argue that valley-wide population grew through A.D. 1000; that it remained roughly the same while people migrated from west to east; or, pushing data to their extreme, that widespread abandonment at A.D. 550 was followed by re-occupation as late as A.D. 1300. Until this is sorted out, agricultural intensification remains as plausible as disintensification in the time window of interest to solving our problem of explanatory convergence.

Since Spores's survey there have been a number of excavations, in the Mixteca Alta and neighboring regions, of stratified deposits that correspond in age to Late Las Flores or Early Natividad (Joyce et al. 2001; Pérez Rodríguez 2006; Lind 2008; Markens et al. 2008; Winter 2008; Faulseit 2013; Robles 2014), as well as explicit attempts to refine the ceramic chronology of that time interval, at least in the Valley of Oaxaca (Martínez et al. 2000; Markens 2008). What is badly needed in the Mixteca Alta are excavations designed to improve the regional ceramic sequence and anchor its chronology with multiple radiocarbon dates from stratified contexts rich in refuse that has suffered no redeposition since discard. However, it will not be possible to re-calibrate the valley-wide survey results for Nochixtlan without further fieldwork, even if there were somebody willing to face the drudgery of a few years in the pot shed. The reason is as prosaic as it is predictable: the original survey collections were discarded in one of the periodic curation crises afflicting the competent authorities in one of Mexico's most intensively excavated states (Ronald Spores, personal communication to Borejsza 2013).

Divergence in Tlaxcala: The Effects of Population Growth and Collapse at 500 B.C.–A.D. 100 and A.D. 1200–1600

Divergence refers to a situation where similar causes produce different effects (Schumm 1991:62). It can refer to a situation where similar processes, operating in the same place, but at different points in time, elicit an alternative response. On closer inspection, it is usually possible to solve the problem of divergence by showing that the causes and processes in question were not quite the same, or that they were acting on a (geomorphic or cultural) system that had assumed, by that time, a different state. Scientists like to refer to the latter solution as "historical contingency" (Gould 1991:283–306). This time we will start with a demographic cause, consider how it affected farmers' choices with respect to land use, and how their decisions, in turn, affected the transfers of sediment on slopes and along streams. We will dwell a little less on the regional specifics, and a little more on their theoretical implications, because we have published the results of our Tlaxcalan fieldwork more fully (Borejsza 2006; 2014; Lesure et al. 2006; 2013; Borejsza et al. 2008; 2011; 2016; Borejsza and Frederick 2010; Lesure 2014).

The long-distance transport of staple foods was largely impracticable in the Mexican highlands before Spanish conquest (Drennan 1984a-b; Cowgill 1993) and, with the exception of meat-on-the-hoof, remained limited until the building of the railways in the nineteenth century. Non-comestible crops (e.g., cotton) were few in prehispanic times, and occupied a limited surface area in the highlands. Under these circumstances, agricultural land use was largely geared to the satisfaction of regional subsistence needs, and the volume of food produced closely tied to the number of inhabitants of a particular region, a variable archaeologists feel quite confident estimating from surface surveys (but see O'Brien and Lewarch 1992). This correlation rests, above all, on the truism that more people equal more mouths to be fed. But, because in the time period we are concerned with the bulk of the population made a living working the land, and because labor-saving innovations were few, the correlation also rests on the relatively immovable ratio of labor input to *regional* output in agriculture. We would thus expect rising population numbers to result in agricultural intensification: an expanding surface area under cultivation, or more intensive forms of agriculture.[1] Conversely, falling population numbers would result in disintensification: a shrinking surface area under cultivation or more extensive forms of agriculture.

As mentioned in the section on convergence, scholars of the most varied denominations, and Mesoamerican archaeologists in particular, have debated *ad nauseam* the relationships of demographic variables, agricultural intensification, and social complexity. The correlation we have just spelled out—of population growth with agricultural intensification—is perhaps one proposition that most would agree on, this rising limb of the curve receiving more attention than the falling one. In contrast, the question of how *land degradation*—and its opposite, *land improvement*—relate to the above was, until recently, severely under-theorized in the social sciences (Blaikie and Brookfield 1987:xvii–xx,1–2). The geomorphic technicalities of measuring land degradation were no doubt one of the main deterrents to the development of theory. Research on the closely related and now trendy topics of agricultural sustainability, resilience, and landesque capital is filling that void (Altieri 1995; Redman 1999; Brookfield 2001; Gunderson and Holling 2002; Håkanson and Widgren 2014), but little of it has trickled down to Mesoamerican archaeology (the writings of Fisher, Pérez Rodríguez, and associates discussed below are two commendable exceptions).

The earliest and still the most common explanations of land degradation sought its causes in population growth and the intensification of agriculture and other extractive economic activities (logging, mining, etc.). Blaikie and Brookfield (1987) refer to them as PPR ("pressure of population on resources" or "pressure of production on resources") explanations. Tlaxcala has its exponent of PPR in Klaus Heine (1976; 1978; 2003), who correlated population highs, inferred from archaeological settlement surveys, with episodes of accelerated soil erosion, inferred from thick alluvial fills and, to a lesser extent, from observations made on slopes, often in the vicinity of severely eroded archaeological sites. Conversely, he saw population lows as times of respite, during which the ecosystem regenerated, and soil erosion subsided. His conclusions were in tune with those of Cook (1949; 1963), Butzer and Butzer (1993; 1995), and some interpretations of sediment delivery to Lake Patzcuaro (O'Hara et al. 1993).

Blaikie and Brookfield (1987), however, take a less mechanistic and more humanistic ("agency"-focused in recent archaeological jargon) view of the matter, trying to discern what social conditions make good or bad stewards of land. As they point out, it is possible to stand PPR on its head, by arguing that it is precisely in situations where land is plentiful that farmers and landlords become careless and wasteful in its management. Conversely, crowding, land scarcity, and other pressures can promote ingenious land improvement and conservation, a view epitomized in the titles of two influential books: Netting's (1993) *Smallholders, Householders: Farm Families and the Ecology of Intensive, Sustainable Agriculture* and Wilken's (1987) *Good Farmers: Traditional Agriculture and Resource Management in Mexico and Central America*. It is the collapse of the fragile agroecosystems dependent on the continuity of high inputs of human labor that smallholders create which is to be feared as an environmental disaster. In search of a matchingly mnemonic acronym, we will refer to this strand of thought as SSS ("sustainable smallholder stewards") explanations. Nobody has systematically tried to enlist the archaeological data from Tlaxcala to argue in its favor, even though the *Good Farmers* were informed by Wilken's own experiences in the region, and followed by other works documenting the ingenuity of Tlaxcalan smallholders in adapting to a changing natural and political environment (e.g., Altieri and Trujillo 1987; González 1993; 2008). Melville's (1994) account of the "plague of sheep" grazing on the ruins of intensive Indian farms in the Mezquital is in tune with SSS, contrasting as it does the harm inflicted by introduced livestock with implicitly sustainable prehispanic land use.

More recent work in two other areas of highland Mexico explicitly embraces SSS. At Patzcuaro, an examination of sediments and agricultural features in three transects from slope to lakeshore (Fisher et al. 2003; Fisher 2005; 2007) leads its authors to contest the conclusions of the work of O'Hara et al. (1993) cited above. In terms of our tree metaphor, this amounts to dissonance between coarse-resolution work at the sediment sink near the base of the trunk and higher-resolution vignettes higher up the branches. Both this body of work and the publications of Pérez Rodríguez and associates in the Mixteca Alta (Pérez Rodríguez 2006; Pérez Rodríguez et al. 2011; Pérez Rodríguez and Anderson 2013) emphasize the success of dense prehispanic populations in building sophisticated systems of terraces for agriculture and habitation, and in maintaining them for centuries. They cast periods of population decline in a negative role, resulting in terrace collapse, loss of productive soil, and the release of large volumes of sediment. In the 500 years since the arrival of Europeans, the negative mix is also said to include the encroachment of strangers on native landholdings, and the introduction of extensive and harmful forms of land use geared to export markets or short-term profits, rather than long-term land improvement. The same authors affirm that the drama of population growth and decline, correlated with the build-up and destruction of landesque capital, was played out more than once.

The Tlaxcalan data provide an opportunity to test PPR and SSS as divergent—in fact, diametrically opposed—explanations of the effects of population growth and decline on the productive capacity of farmland (Figure 2.4). In the 3,000-year history of farming in Tlaxcala, there are two particularly striking cycles of demographic boom-and-bust, convincingly attested by settlement pattern studies at different scales. The demographic crash of the second cycle is also vividly described in early historical sources.

Sedentary farming communities begin to appear in Tlaxcala at ca. 1000 B.C., but it is after 500 B.C. that their proliferation becomes particularly striking. There are hundreds of village sites of that period, spaced at intervals of less than 4 km from one another, and arranged in a three-tiered hierarchy (Plunket and Uruñuela 2012:18). At the top of this hierarchy are centers with monumental public architecture and several thousand inhabitants, but even at the second-tier sites there are certain signs of affluence expressed in architectural elaboration and the conspicuous consumption of food and of exotic ritual paraphernalia (Carballo 2009; 2012; Carballo et al. 2014; Borejsza et al. 2014b). Virtually all these sites are abandoned for good between ca. 100 B.C. and A.D. 150. Local movement from uplands to basin floors may account for some abandonments, but can hardly mask substantial regional depopulation, attributed to voluntary or "assisted" migration towards the new urban metropoleis of Teotihuacan, Cholula, or Cantona, burgeoning just beyond Tlaxcala's borders.

The second cycle starts at ca. A.D. 1200 with the arrival of Nahuatl-speaking immigrants (Smith 1984) and the founding of new villages. Between that moment and Spanish conquest in A.D. 1519, population densities soared above 100 inhabitants per km^2, making Tlaxcala one of the most densely settled areas in the Americas and underwriting the military might of the province, which successfully resisted incorporation in the Aztec empire, and fielded more than 20,000 soldiers for Cortés's siege of Tenochtitlan. In contrast to the more nucleated villages of a millennium-and-a-half earlier, much of the population lived in small dispersed farmsteads. As a result, the sherd scatter of that period is so continuous that we are at pains to draw the limits of "sites" and count them. Successive outbreaks of smallpox and other introduced diseases reduced that population by 80–90 percent within a century of A.D. 1519. Survivors flocked to larger settlements, often at the

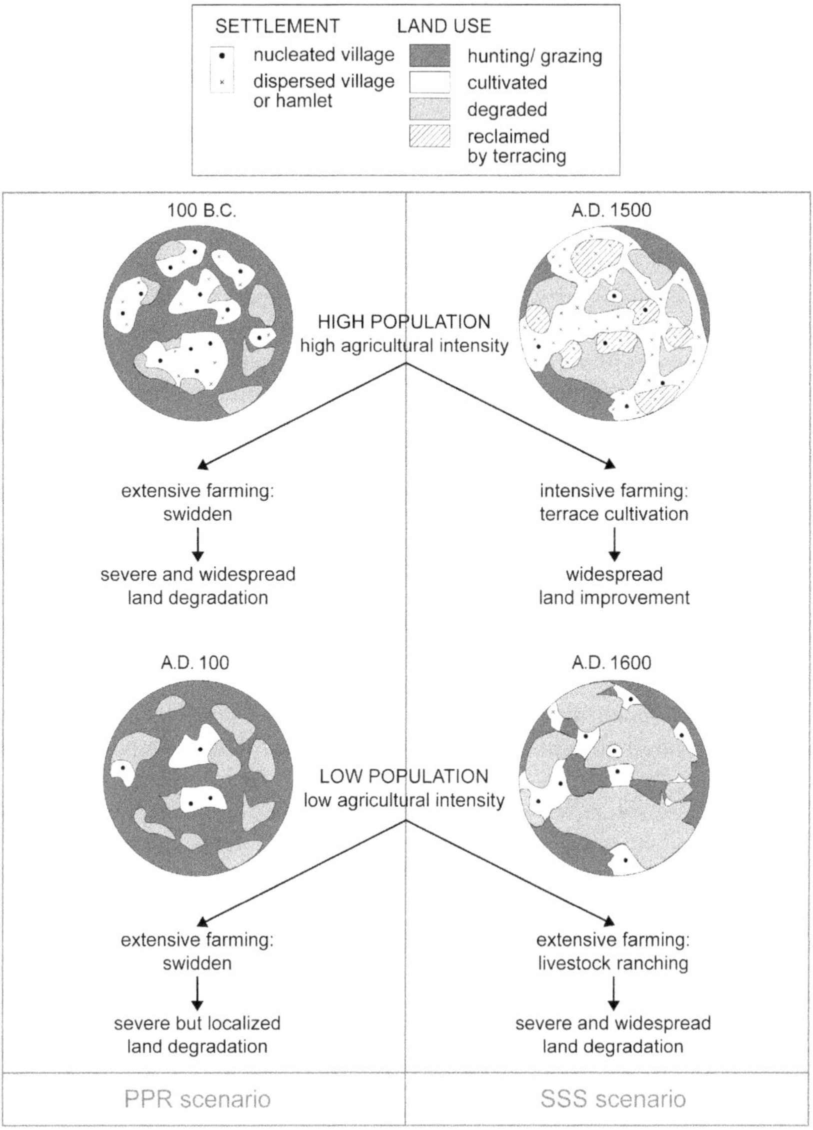

Figure 2.4 Divergent effects of population growth and decline in Tlaxcala. The circle represents the same notional portion of the landscape at four different moments in time. The dominant form of agricultural slope management is indicated at each moment.

foot of previously inhabited hills, or migrated to towns and cities, inside and outside the province.

The two demographic cycles were obviously not the same, but they display remarkable symmetry in terms of settlement and population history.

Did agricultural and geomorphic changes follow suit? There is empirical evidence that both demographic cycles were indeed accompanied, as theory would predict, by a rise and fall in regional agricultural intensity. Extensive and intensive agricultural practices were different in each cycle, however—both in their relative importance, and in the very nature of the farming techniques involved. Swidden on unterraced slopes, with fallow periods of a few years' length ("bush fallow" in Boserupian terms)—a rather extensive form of agriculture—dominated *throughout* the first cycle. We infer this from a very peculiar type of alluvial deposit found in some headwater reaches, which contains hundreds of laminae of charred plant matter, including pine—several species of which act as pioneer plants in Tlaxcala—shrubs, and non-woody species. At the same time, excavating several hundred meters of stratigraphic trenches at five Tlaxcalan villages with occupations ranging from 900 B.C. to A.D. 150 has failed to produce a shred of evidence of contemporaneous agricultural terracing (Borejsza 2006; Borejsza et al. 2008; 2011; 2016). However, the charcoal-rich alluvia, which start at ca. 300 B.C., continued to aggrade in different drainages for several centuries after A.D. 100. We are not in a position to quantify how much of this alluvium aggraded before and after the depopulation: we have not surveyed all streams in Tlaxcala, and the alluvial record is in any case too fragmentary to undertake any quantitative estimates. We surmise that, post-depopulation, swidden continued in the same form, but in different locations, and on a reduced surface area.

In contrast, bench terraces faced with field stone and blocks of tepetate (a silica-indurated subsoil) are ubiquitous during the last stretch of prehispanic history, between ca. A.D. 1200 and 1520. Most dispersed farmsteads stood on such terraces, separated by fields that could be carefully tended and enriched in human waste and other domestic refuse. Terracing permeates the entire settlement hierarchy, culminating in the fortified capital city of Tlaxcallan (Fargher et al. 2010; 2011). By analogy with other contemporaneous cities in the Mexican highlands (Smith 2008; Pérez Rodríguez et al. 2011; Smith et al. 2013), we suspect that even at Tlaxcallan, many terraces supported highly productive infields or urban gardens. Both fieldwork and archival research reveal that much of this intensive agricultural landscape was abandoned in the century following Conquest because of the acute shortage of labor induced by epidemics (see Borejsza 2014). With relocation of settlement away from slopes, the surviving peasants often found themselves living too far to walk to the fields in question, and were no longer able to claim their usufruct rights. The native landlords reacted by investing in sheep ranching, or by divesting themselves of their ultimate property rights in a thriving black market with mostly Spanish buyers (Assadourian 1991). The grazing of livestock required few field hands, and was therefore, by definition, an extremely extensive form of land use. The plow and European small grains opened up for cultivation some of the basin floors inimical to prehispanic agriculture (because of heavy soils and exposure to frost), but also promoted forms of agriculture more extensive than what was prevalent before Conquest.

The two cycles diverge even more in their impact on the quantity and quality of available farmland, and in the sediment transfers that each of them induced. The charcoal-rich alluvium of the first cycle is testimony not only to the practice of swidden, but also to the severe land degradation that it provoked. The thickness of the alluvial units in question (more than 10 m in catchments of a few square kilometers), their pace of aggradation (several meters per century), and the presence of redeposited peds of top- and subsoil, indicate that entire hillsides had been stripped of soil and transformed into barren badlands. At the site of La Laguna, erosional unconformities and colluvial deposits provide independent evidence of substantial soil loss, probably before 400 B.C. (Borejsza et al. 2008; Borejsza and Carballo 2014; Borejsza and Rodríguez 2014). If the cultivated area had contracted by A.D. 100, as we think it did, the scope and pace of degradation must have subsided.

In contrast, we have not come across any alluvium clearly contemporaneous with the second cycle of population growth. It seems that once in place, terraces reduced the transfer of sediment from slopes to the fluvial system to negligible proportions. With the withdrawal of maintenance after Conquest, the vast majority of terrace risers collapsed, releasing the enormous amount of sediment that had been stored in terrace fills. Many barren hillsides strewn with rubble from risers and house foundations, and with sherds dated to A.D. 1200–1600, bear witness to this episode of land degradation. But, intriguingly, there is not much alluvium that can convincingly be correlated with the time of population decline, either. It seems that, instead of being delivered to the fluvial system, terrace fills were locally redeposited on slopes, forming a colluvial blanket of varying thickness. On the other hand, the sediment that did make it to the fluvial system was apparently transported by streams with excess transport capacity to sinks beyond our study areas, and beyond Tlaxcala itself.

The most severe land degradation thus seems to coincide with population growth and agricultural intensification during the first cycle, but with population collapse and disintensification during the second one. The first follows a PPR, the second a SSS scenario. In the most general sense, we think that the divergence is due to the fact that, while some cultural and geomorphic transformations are indeed cyclical, others have cumulative and often irreversible effects, and still others represent unique events. At A.D. 1 or 100 there was still a sizable pool of relatively intact farmland. Depopulation relieved some of the regional pressure on resources and allowed the remaining farmers to perpetuate deleterious swiddening practices. By A.D. 1200 the pool of available farmland had shrunk to such an extent that it was necessary to innovate. The development of agricultural terracing systems was not only a response to renewed population growth, but also to the cumulative effects of more than two millennia of land degradation. Many terrace fills can indeed be seen to rest on top of tepetate surfaces, which had not "regenerated" naturally since they had first been exposed. The bench terraces, however—geomorphically unstable and predicated on the existence of a numerous and disciplined workforce—could not survive the next population collapse. As this coincided with the appearance of new crops, animals, technologies, and market opportunities, instead of reverting to some version of swidden, regional land use suffered its most profound metamorphosis since the transition from foraging to farming.

To understand the even more divergent responses of geomorphic systems to land degradation, it is necessary to step back in time several millennia before agriculture. Barranca-type streams are subject to some intrinsic thresholds (Schumm and Parker 1973; Schumm 1977) and their history may predispose them to incision or in-filling at any particular moment in time. In other words, stream baselevel and other physical constraints make indefinite incision or in-filling impossible. The headwaters of Tlaxcalan stream networks had suffered a deep incision at the onset of the Holocene. It substantially reduced the longitudinal gradient of streams, and made them predisposed towards in-filling during the Early and Middle Holocene (Figure 2.5). By the time farmers arrived, many streams were close to the threshold of incision

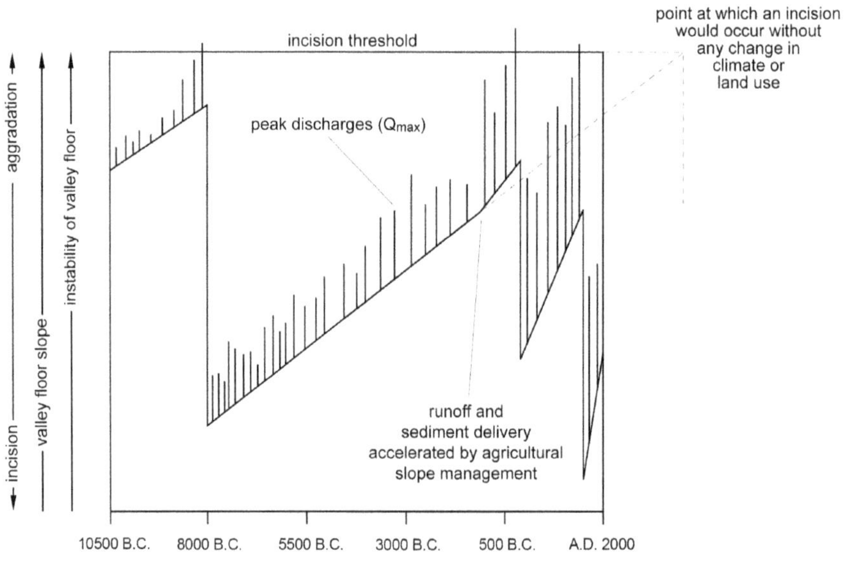

Figure 2.5 Stream aggradation and the threshold of incision in Tlaxcala. Modified from Schumm (1977:fig. 4–15) to fit the regional circumstances. Aggradation tends to oversteepen the longitudinal gradient (slope) of the valley floor, decreasing its stability, i.e., the probability that it will incise. The stream can reach the threshold of incision through progressive aggradation, or an exceptionally large flood. Most Tlaxcalan streams incised at the transition from the Pleistocene to the Holocene, and then entered a long aggradational phase. Anthropogenic slope erosion in the Late Holocene increased both the pace of aggradation, and the magnitude of floods, thereby increasing the frequency of cut-and-fill cycles. The number and timing of Late Holocene cut-and-fill cycles varies from one drainage to the next.

again. The deposition of several of the swidden-induced alluvia was in fact preceded by deep incision, perhaps in response to runoff increased by initial land clearance, by the earliest phases of swidden itself, or by an episode of particularly intense rainfall and peak discharges that took a stream over the threshold of incision. This "made room" for the massive charcoal-rich fills of ca. 500 B.C.–A.D. 1000.

Agriculture, though of changing intensity, irreversibly increased runoff and sediment delivery from slopes, and accelerated cut-and-fill cycles to a degree unprecedented during the Late Quaternary. Moreover, the peculiarities of land use history in neighboring 10–50 km^2 drainages made the behavior of their streams diverge from one another. Though generalization becomes difficult under these conditions, it is our impression that at A.D. 1600, many stream gradients had been oversteepened by deposition in headwater reaches, making them capable of transporting the sediment derived from terrace collapse beyond our study areas. On the other hand, the mentioned colluvial blanket also suggests that segments of terrace risers still in place reduced overall geomorphic connectivity on slopes, and slowed or impeded sediment delivery to streams. If the water reaching the streams was relatively free of sediment, it would constitute yet another physical factor tilting the balance towards incision, not in-filling.

Even though there may be some measure of cyclicity in cultural and geomorphic processes, history did not repeat itself in Tlaxcala. The cultural and geomorphic systems are each very complex on its own terms. Their internal connectivity, as well as the degree to which they are connected ("coupled") with one another, changes through time. The response of one to the other is often delayed by decades or centuries. They move on different timescales and are rarely in-phase. It is thus naive to expect a signature stratigraphic deposit of cultural florescence or collapse.

The Case for a Delayed-Return Environmental Archaeology

We feel that we are closer to a satisfactory explanation in Tlaxcala than in the Nochixtlan Valley. The comparison is not altogether fair, because divergence seems epistemologically a problem easier to solve than convergence. However, it is no accident that we sound more confident in describing land use histories in Tlaxcala, and perhaps annoyingly non-committal about Nochixtlan. The Tlaxcalan dataset is of course perfectible, but sufficient to enter the fray of theoretical debates. In Nochixtlan the main use of recognizing the multiplicity of possible explanations is, at this stage, bringing into focus the priorities of further research. What makes the difference is that we have reached, in Tlaxcala, a critical mass of research data, generated by ourselves and our predecessors. It allows us to assume—such as in the case of early agricultural terraces or late alluvial fills—that the absence of evidence is evidence of absence. Our research in Tlaxcala has also benefited immensely from insertion in a larger project (Lesure 2014) that undertook excavations at several sites and completely revised the chronology of the first millennium B.C., on the basis of ceramic analysis of large assemblages of secondary refuse, and an intensive program of radiocarbon dating. Excavation at settlement sites has also allowed us to integrate on- and off-site research more efficiently, i.e., to move more easily between the different levels of the tree of Figure 2.2. This was also possible because, from the outset, and by chance rather than design, we concentrated our efforts near the headwaters of the fluvial system, where the coupling of stream and slope processes is much stronger.

We perceive both theoretical and methodological morals to our Mexican examples. On a theoretical level, we must pay due attention to both history and process. Trigger (1989) sees the entire history of archaeological thought as a pendulum between process and history, and in this sense our explanatory travails in reconstructing human-environment interactions may be no more than a special case of the perennial balancing act performed by all self-critical archaeologists. Kirch (1994:321–3) comes to the same conclusion in a case study much closer in subject matter to our examples, in which he examines the land use history of the island of Futuna in the light of the process of agricultural intensification. Significantly, to demonstrate that this tension is not exclusive to archaeology, Kirch quotes from the writings of Stephen Jay Gould, a palaeontologist who argued animatedly in his writings that explanation in the natural sciences, though never unmoored from the principles of physics or evolutionary theory, had to heed the contingencies of history. Gould (1991:284) jeers at the "rhetoric of inferiority," which ranks scientific disciplines by the timelessness of their laws.

Because the processes we are concerned with are simultaneously cultural and natural, paying due attention to process requires immersion in very disparate sets of theoretical literature, in geomorphology, agronomy, and cultural anthropology (e.g., Joyce and Goman 2012). The relevant literature may not be sufficiently tailored to the natural environment or the agricultural tradition in our study area. Because of the incomplete coverage or insufficient resolution of basic maps and inventories of rocks, soils, plants, or vegetation communities, this is especially true in developing countries, in regions of young relief, volcanic substrates, or high biodiversity: highland Mexico meets all these criteria. Riverine areas are sometimes given short shrift by scholars, especially when they are regarded as ribbons of atypical or "disturbed" conditions in the wider landscape. The only remedy is to devote time to actualistic ecological and agro-ecological research in the study area, in conjunction with the archaeological survey or excavation.

Paying attention to history means looking back, beyond our "favorite" period. Analyzing deposits predating agriculture or even the arrival of humans, as well as digging "off-site," "incidental site," and "non-site" locations[2] can provide baselines against which to judge environmental change induced by human activities. These are recommendations that are well known and sometimes heeded. More than anything else, however, paying attention to history requires investing time and resources in chronological refinement, which comes not only from the application of radiometric dating techniques and the rigorous analysis of the resulting data, but also the development of artifactual typologies and periodizations. Contrary to common belief, once the labor involved is factored in, the latter are probably more costly than the former. But, once in place, good artifactual typologies pay a continuing dividend.

Our methodological moral is thus probably less familiar and more jarring to the ears of environmental archaeologists: we need more on-site research, more excavation, and more involvement with the archaeological "mainstream." There is an increasing trend to dissociate the study of human-environment interactions from bread-and-butter archaeological endeavors such as the digging of settlement sites or sorting of sherds. It may have the allure of originality, of sidestepping the often byzantine dealings needed to obtain an excavation permit, or of laying the burden of artifact curation on somebody else. A parallel trend in the natural sciences is to favor high-tech applications of remote sensing or genetics over the "antiquated" tasks of completing taxonomic inventories or collecting reference specimens. Because for large regions of the planet, the tasks of describing nature or building culture-historical chronologies are by no means complete, this approach risks leaving many of us in explanatory limbo.

Acknowledgments

Our research was authorized by the Instituto Nacional de Antropología e Historia, and funded by a number of institutions, among which we wish to single out the National Science Foundation (most recently grant BCS-0096012), the Wenner-Gren Foundation for Anthropological Research (grant 7063), the Subsecretaría de Educación Superior (PROMEP grant 103.5/10/4412), and the University of Colorado (Innovative Seed Grant program). We thank Raymond Mueller, Charles Frederick, and Richard Lesure for many years of shared research endeavors on which this chapter builds. The photographs used in Figure 2.1 are by Johnrobert Koukopoulos (a) and Elma Ramírez Hernández (b).

Methodological Vignette: Alluvial Geoarchaeology

The appeal of alluvial geoarchaeology for reconstructing human-environment interactions is twofold. Almost anywhere on Earth the floodplains, terraces, and fans formed by flowing water are among the places where people most frequently foraged, farmed, procured raw materials, and built their dwellings. The remains of their activities can be exquisitely preserved by rapid burial during a flood, or through waterlogging by rising water tables. Access to streamside areas, however, may be hindered by dense vegetation, heavily entrenched channels, standing water, or landowners jealous of what often is the most coveted land for farming or development. Simple fieldwalking is usually futile and survey must instead rely on backhoe trenching, or opportunistic exposure in stream cutbanks, quarries, or wells.

Alluvial sediment itself is also an ubiquitous proxy record of environmental change. The climate, vegetation, and land use in a catchment influence the pace and style of sedimentation, as well as the nature of palaeosols formed when sedimentation ceases. Purposeful interference with hydrological processes by means of dams, canals, or agricultural terraces may also leave a stratigraphic imprint. But, the alluvial record is highly fragmented: discontinuous in time and distributed in complex three-dimensional patterns over different landforms.

Alluvial geoarchaeology requires an understanding of geomorphic and soil-forming processes at different scales (Ferring 1992; 2001; Waters 1992:115–84; Frederick 2001). Streams behave very differently depending on whether they are small or large, ephemeral or perennial, flowing over bedrock or unconsolidated sediment, situated in arid or humid zones. Local ecological knowledge gained through long-term fieldwork is therefore an unsurpassed asset.

Once the methodological challenges are met, alluvial environments offer rewards that few other settings can equal. Conditions of preservation may be truly Pompeian, i.e., similar to those in settlements rapidly abandoned because of catastrophic events. Waterlogging will preserve wood and other perishable materials. Rapid burial will preserve activity areas and other spatial associations of the systemic context. Deep burial will ensure minimal post-depositional disturbance. In some systemic contexts—from mobile bands who set up their basecamps near water to irrigation-dependent states—streamside areas are the only places where we can hope to find types of sites crucial to the understanding of a whole way of life. In heavily eroded landscapes, valley bottoms may hold the only sites worth excavating. The ubiquity of alluvium sometimes makes it the only environmental proxy available, while its proximity to settled and farmed areas means that it frequently registers human impacts.

Notes

1 Archaeologists and geographers commonly muddle the notions of *agricultural intensity*, *intensive agriculture*, and *agricultural intensification*. We adhere to the following definitions. Agricultural intensity is the amount of human labor

expended per surface area of land. In order to be meaningful it has to be measured over time intervals long enough to include the longest fallows. Intensive agriculture refers to forms of agriculture that require high inputs of labor per surface area of land. Its opposite is *extensive* agriculture. Individually planted maize in a plot hand-irrigated from a well represents a form of agriculture more intensive than, say, barley broadcast in a dry-farmed plot. Agricultural intensification is the process by which land use becomes more intensive. It does so because more labor is expended on previously cultivated land, or because previously uncultivated land is brought under cultivation. It is possible to reserve the use of "intensification" to the former, using another term, for example, agricultural *expansion*, for the latter. We use the broader definition of intensification in this chapter. The opposite of intensification is *disintensification*. Note that yield—the volume of production per unit of land—is not part of any of the definitions, though within the parameters of pre-industrial agriculture it is correlated with agricultural intensity more often than not. For similar definitions see Boserup (1965) and Brookfield (1972; 1984). For radical alternatives, focusing on outputs rather than inputs, see Turner and Doolittle (1978) or Morrison (1994:115).

2 We take archaeological *sites* to be places with particularly high concentrations of artifacts, features, and ecofacts (Ashmore and Sharer 2000:55–9). Places beyond the always somewhat arbitrary boundaries of sites are *off-site*. We take *incidental sites* to be places where artifacts or ecofacts have been swept together by natural processes, such as a cluster of size- or density-sorted bones in a palaeochannel. This seems to be the meaning attributed to the term by Butzer (2008:406). We take *non-sites* to be places purposefully chosen by the archaeologist to minimize the chances of finding any significant trace of past human activity (e.g., Dillehay 1997; Homburg and Sandor 2011).

References Cited

Allen, Bryant J. and Chris Ballard (editors) 2001 Beyond intensification? Reconsidering agricultural transformations. Special section of *Asia Pacific Viewpoint* 42:157–333.

Altieri, Miguel A. 1995 *Agroecology: The Science of Sustainable Agriculture*. 2nd ed. Westview, Boulder.

Altieri, Miguel A. and Javier Trujillo 1987 The agroecology of corn production in Tlaxcala, Mexico. *Human Ecology* 15:189–220.

Ashmore, Wendy and Robert J. Sharer 2000 *Discovering our Past: A Brief Introduction to Archaeology*. 3rd ed. McGraw-Hill, Boston.

Assadourian, Carlos Sempat 1991 Estructuras indígenas en transición. In *Tlaxcala, una historia compartida*, vol. 9: *siglo XVI*, edited by A. Martínez Baracs and C. S. Assadourian, pp. 69–148. Gobierno del Estado de Tlaxcala, Tlaxcala.

Balkansky, Andrew K. 1998 Urbanism and early state formation in the Huamelulpan Valley of southern Mexico. *Latin American Antiquity* 9:37–67.

Balkansky, Andrew K., Verónica Pérez Rodríguez and Stephen A. Kowalewski 2004 Monte Negro and the urban revolution in Oaxaca, Mexico. *Latin American Antiquity* 15:33–60.

Barnes, Harry H. 1967 *Roughness Characteristics of Natural Channels*. United States Geological Survey, Washington, DC.

Beekman, Christopher S. and Alexander F. Christensen 2003 Controlling for doubt and uncertainty through multiple lines of evidence: A new look at the Mesoamerican Nahua migrations. *Journal of Archaeological Method and Theory* 10:111–164.

Bernal, Juan Pablo, Matthew Lachniet, Malcolm McCulloch, Graham Mortimer, Pedro Morales and Edith Cienfuegos 2011 A speleothem record of Holocene climate variability from southwestern Mexico. *Quaternary Research* 75:104–113.

Bevan, Andrew, James Conolly, S. Colledge, Charles Frederick, C. Palmer, R. Siddall and A. Stellatou 2013 The long-term ecology of agricultural terraces and enclosed fields from Antikythera, Greece. *Human Ecology* 41:255–272.

Blaikie, Piers M. and Harold Brookfield (editors) 1987 *Land Degradation and Society*. Methuen, New York.

Blomster, Jeffrey P. (editor) 2008 *After Monte Albán: Transformation and Negotiation in Oaxaca, Mexico*. University Press of Colorado, Boulder.

Borejsza, Aleksander 2006 *Agricultural Slope Management and Soil Erosion in Tlaxcala, Mexico*. Ph.D. Dissertation, University of California.

Borejsza, Aleksander and David M. Carballo 2014 La Laguna: Overview of site. In *Formative Lifeways in Central Tlaxcala*, vol. I: *Excavations, Contexts, and Chronology*, edited by R. G. Lesure, pp. 83–88. Cotsen Institute of Archaeology Press, Los Angeles.

Borejsza, Aleksander and Charles D. Frederick 2010 Fluvial response to Holocene climate change in low-order streams of central Mexico. *Journal of Quaternary Science* 25:762–781.

Borejsza, Aleksander, Charles D. Frederick, Luis Morett Alatorre and Arthur A. Joyce 2014a Alluvial stratigraphy and the search for Preceramic open-air sites in highland Mesoamerica. *Latin American Antiquity* 25:278–299.

Borejsza, Aleksander, Charles D. Frederick and Richard G. Lesure 2011 Swidden agriculture in the *tierra fría*? Evidence from sedimentary records in Tlaxcala. *Ancient Mesoamerica* 22:91–106.

Borejsza, Aleksander, Richard G. Lesure, Thomas A. Wake, Víctor Emmanuel Salazar Chávez, Gilberto Pérez Roldán and María Teresa Gómez Lomelí 2014b Animal bones as remains of food and indicators of site environment. To appear in *Formative Lifeways in Central Tlaxcala*, vol. II: *Environment and Subsistence*, edited by A. Borejsza and R. G. Lesure. Cotsen Institute of Archaeology Press, Los Angeles. Manuscript.

Borejsza, Aleksander, Emily McClung de Tapia, Lorenzo Vázquez Selem, Cristina Adriano Morán, Renato Castro Govea and Isabel Rodríguez López 2016 Changing rural landscapes of the last three millennia at Santiago Tlalpan, Tlaxcala, Mexico. *Geoarchaeology*. doi: 10.1002/gea.21543

Borejsza, Aleksander and Isabel Rodríguez López 2014 La Laguna: Formative contexts away from the site center. In *Formative Lifeways in Central Tlaxcala*, vol. I: *Excavations, Contexts, and Chronology*, edited by R. G. Lesure, pp. 113–167. Cotsen Institute of Archaeology Press, Los Angeles.

Borejsza, Aleksander, Isabel Rodríguez López, Charles D. Frederick and Mark D. Bateman 2008 Agricultural slope management and soil erosion at La Laguna, Tlaxcala, Mexico. *Journal of Archaeological Science* 35:1854–1866.

Boserup, Ester 1965 *The Conditions of Agricultural Growth: The Economics of Agrarian Change under Population Pressure*. Aldine, Chicago.

Braniff, Beatriz 1989 Oscilación de la frontera norte mesoamericana: un nuevo ensayo. *Antropología* 1:99–114.

Brookfield, Harold C. 1972 Intensification and disintensification in Pacific agriculture: A theoretical approach. *Pacific Viewpoint* 13:30–48.

——— 1984 Intensification revisited. *Pacific Viewpoint* 25:15–44.

——— 2001 *Exploring Agrodiversity*. Columbia University Press, New York.
Bryan, Kirk 1925 Date of channel trenching (arroyo cutting) in the Arid Southwest. *Science* 62:338–344.
Bull, William B. 1979 Threshold of critical power in streams. *Geological Society of America Bulletin* 90:453–464.
——— 1991 *Geomorphic Responses to Climatic Change*. Oxford University Press, Oxford.
——— 1997 Discontinuous ephemeral streams. *Geomorphology* 19:227–276.
Butzer, Karl W. 2008 Challenges for a cross-disciplinary geoarchaeology: The intersection between environmental history and geomorphology. *Geomorphology* 101:402–411.
Butzer, Karl W. and Elisabeth K. Butzer 1993 The sixteenth-century environment of the central Mexican Bajío: Archival reconstruction from Colonial land grants and question of Spanish ecological impact. In *Culture, Form, and Place: Essays in Cultural and Historical Geography*, edited by K. Mathewson, pp. 89–124. Department of Geography and Anthropology, Louisiana State University, Baton Rouge.
——— 1995 Transfer of the Mediterranean livestock economy to New Spain: Adaptation and ecological consequences. In *Global Land Use Change: A Perspective from the Columbian Encounter*, edited by B. L. Turner, pp. 151–193. Consejo Superior de Investigaciones Científicas, Madrid.
Carballo, David M. 2009 Household and status in formative central Mexico: Domestic structures, assemblages, and practices at La Laguna, Tlaxcala. *Latin American Antiquity* 20:473–501.
——— 2012 Public ritual and urbanization in Central Mexico: Plaza and temple offerings from La Laguna, Tlaxcala. *Cambridge Archaeological Journal* 22:329–352.
Carballo, David M., Luis Barba, Agustín Ortiz, Jorge Blancas, Nicole Cingolani, Jorge H. Toledo Barrera, David Walton, Isabel Rodríguez López and Lourdes Couoh 2014 Suprahousehold consumption and community ritual at La Laguna, Mexico. *Antiquity* 88:141–159.
Caballero Miranda, Margarita and Beatriz Ortega Guerrero (editors) 2011 *Escenarios de cambio ambiental: registro del Cuaternario en América Latina*. 2 vols. Universidad Nacional Autónoma de México, México.
Cook, Sherburne F. 1949 *Soil Erosion and Population in Central Mexico*. Ibero-Americana 34. University of California Press, Berkeley.
——— 1963 Erosion morphology and occupational history in western Mexico. *Anthropological Records* 17:281–334.
Cooke, Ronald U. and Richard V. Reeves 1976 *Arroyos and Environmental Change in the American South-West*. Clarendon Press, Oxford.
Cowgill, George L. 1993 Comments on Andrew Sluyter. Long-distance staple transport in western Mesoamerica: Insights through quantitative modeling. *Ancient Mesoamerica* 4:201–203.
Demarest, Arthur A., Prudence M. Rice and Don S. Rice (editors) 2004 *The Terminal Classic in the Maya Lowlands*. University Press of Colorado, Boulder.
Denevan, William M., Kent Mathewson and Gregory Knapp (editors) 1987 *Pre-Hispanic Agricultural Fields in the Andean Region*. 2 vols. British Archaeological Reports, Oxford.
Diehl, Richard A. and Janet C. Berlo (editors) 1989 *Mesoamerica after the Decline of Teotihuacan, A.D. 700–900*. Dumbarton Oaks, Washington.

Dillehay, Tom D. 1997 Research design and methods. In *Monte Verde: A Late Pleistocene Settlement in Chile*, vol. 2: *The Archaeological Context and Interpretation*, edited by T. D. Dillehay, pp. 53–72. Smithsonian Institution Press, Washington.

Drennan, Robert D. 1984a Long-distance movement of goods in the Mesoamerican Formative and Classic. *American Antiquity* 49:27–43.

——— 1984b Long-distance transport costs in pre-Hispanic Mesoamerica. *American Anthropologist* 86:105–112.

Elliott, John G., Allen C. Gellis and Scott B. Aby 1999 Evolution of arroyos: Incised channels of the southwestern United States. In *Incised River Channels: Processes, Forms, Engineering and Management*, edited by S. E. Darby and A. Simon, pp. 153–185. Wiley, Chichester.

Fargher, Lane F., Richard E. Blanton and Verenice Y. Heredia Espinoza 2010 Egalitarian ideology and political power in prehispanic central Mexico: The case of Tlascallan. *Latin American Antiquity* 21:227–251.

Fargher, Lane F., Richard E. Blanton, Verenice Y. Heredia Espinoza, John Millhauser, Nezahualcoyotl Xiuhtecutli and Lisa Overholtzer 2011 Tlaxcallan: The archaeology of an ancient republic in the New World. *Antiquity* 85:172–186.

Faulseit, Ronald K. 2013 *Cerro Danush: Excavations at a Hilltop Community in the Eastern Valley of Oaxaca, Mexico*. Museum of Anthropology, University of Michigan, Ann Arbor.

Feinman, Gary and Linda Nicholas 2011 Monte Albán: una perspectiva desde los límites del Valle de Oaxaca. In *Monte Albán en la encrucijada regional y disciplinaria: Memoria de la Quinta Mesa Redonda de Monte Albán*, edited by N. M. Robles García and A. I. Rivera Guzmán, pp. 241–284. Instituto Nacional de Antropología e Historia, México.

Ferring, C. Reid 1992 Alluvial pedology and geoarchaeological research. In *Soils in Archaeology*, edited by V. T. Holliday, pp. 1–40. Smithsonian Institution Press, Washington.

——— 2001 Geoarchaeology in alluvial landscapes. In *Earth Sciences and Archaeology*, edited by P. Goldberg, V. T. Holliday and C. R. Ferring, pp. 77–106. Kluwer, New York.

Fisher, Christopher T. 2005 Demographic and landscape change in the Lake Pátzcuaro basin, Mexico: Abandoning the garden. *American Anthropologist* 107:87–95.

———2007 Agricultural intensification in the Lake Pátzcuaro Basin: Landesque capital as statecraft. In *Seeking a Richer Harvest: The Archaeology of Subsistence Intensification, Innovation and Change*, edited by T. L. Thurston and C. T. Fisher, pp. 91–106. Springer, New York.

Fisher, Christopher T., Helen P. Pollard, Isabel Israde Alcántara, Victor H. Garduño Monroy and Subir K. Banerjee 2003 A reexamination of human-induced environmental change within the Lake Patzcuaro Basin, Michoacan, Mexico. *Proceedings of the National Academy of Sciences* 100:4957–4962.

Flannery, Kent V. and Joyce Marcus (editors) 1983 *The Cloud People: Divergent Evolution of the Zapotec and Mixtec Civilizations*. Academic Press, New York.

Frederick, Charles D. 2001 Evaluating causality of landscape change: Examples from alluviation. In *Earth Sciences and Archaeology*, edited by P. Goldberg, V. T. Holliday and C. R. Ferring, pp. 55–76. Kluwer, New York.

Frederick, Charles D. and Athanasia Krahtopoulou 2000 Deconstructing agricultural terraces: Examining the influence of construction method on stratigraphy, dating and archaeological visibility. In *Landscape and Land Use in Postglacial*

Greece, edited by P. Halstead and C. D. Frederick, pp. 79–94. Sheffield Academic Press, Sheffield.

Fryirs, K. A. 2013 (Dis)connectivity in catchment sediment cascades: A fresh look at the sediment delivery problem. *Earth Surface Processes and Landforms* 38:30–46.

Goman, Michelle, W. Guerra, Arthur Joyce and D. Dale 2011 The paleoclimate potential and enigma of Laguna Minucua, Oaxaca, Mexico. Paper presented at the 25th Pacific Climate Workshop, Asilomar.

González Jácome, Alba 1993 Management of land, water and vegetation in traditional agro-ecosystems in central Mexico. *Landscape and Urban Planning* 27:141–150.

——— 2008 *Humedales en el suroeste de Tlaxcala: agua y agricultura en el siglo XX*. Universidad Iberoamericana, México.

Gould, Stephen Jay 1991 *Wonderful Life: The Burgess Shale and the Nature of History*. Penguin, London.

Gunderson, L. H. and C. S. Holling (editors) 2002 *Panarchy: Understanding Transformations in Systems of Humans and Nature*. Island Press, Washington.

Håkanson, N. Thomas and Mats Widgren (editors) 2014 *Landesque Capital: The Historical Ecology of Enduring Landscape Modifications*. Left Coast Press, Walnut Creek.

Heine, Klaus 1976 Schneegrenzdepressionen, Klimaentwicklung, Bodenerosion und Mensch im zentralmexikanischen Hochland im jüngeren Pleistozän und Holozän. *Zeitschrift für Geomorphologie* Supplementband 24:160–176.

——— 1978 Mensch und geomorphodynamische Prozesse in Raum und Zeit im randtropischen Hochbecken von Puebla/Tlaxcala, Mexiko. *Verhandlungen des Deutchen Geographentages* 41:390–406.

——— 2003 Paleopedological evidence of human-induced environmental change in the Puebla-Tlaxcala area (Mexico) during the last 3,500 years. *Revista mexicana de ciencias geológicas* 20:235–244.

Homburg, Jeffrey A. and Jonathan A. Sandor 2011 Anthropogenic effects on soil quality of ancient agricultural systems of the American Southwest. *Catena* 85:144–154.

Johnston, Kevin J. 2003 The intensification of pre-industrial cereal agriculture in the tropics: Boserup, cultivation lengthening, and the Classic Maya. *Journal of Anthropological Archaeology* 22:126–161.

Joyce, Arthur A. 2010 *Mixtecs, Zapotecs, and Chatinos: Ancient Peoples of Southern Mexico*. Wiley, Chichester.

Joyce, Arthur A., Laura Arnaud Bustamante and Marc N. Levine 2001 Commoner power: A case study from the classic period collapse on the Oaxaca coast. *Journal of Archaeological Method and Theory* 8:343–385.

Joyce, Arthur A. and Michelle Goman 2012 Bridging the theoretical divide in Holocene landscape studies: Social and ecological approaches to ancient Oaxacan landscapes. *Quaternary Science Reviews* 55:1–22.

Joyce, Arthur A. and Raymond G. Mueller 1997 Prehispanic human ecology of the Río Verde drainage basin, Mexico. *World Archaeology* 29:75–94.

Kennett, Douglas J. and Tim Beach 2014 Archaeological and environmental lessons for the Anthropocene from the Classic Maya Collapse. *Anthropocene* 4:88–100.

Killion, Thomas W. (editor) 1992 *Gardens of Prehistory: The Archaeology of Settlement Agriculture in Greater Mesoamerica*. University of Alabama Press, Tuscaloosa.

Kirch, Patrick V. 1994 *The Wet and the Dry: Irrigation and Agricultural Intensification in Polynesia*. University of Chicago Press, Chicago.

Knighton, David 1998 *Fluvial Forms and Processes: A New Perspective*. Arnold, London.

Kowalewski, Stephen A., Andrew K. Balkansky, Laura R. Stiver Walsh, Thomas J. Pluckhahn, John F. Chamblee, Verónica Pérez Rodríguez, Verenice Y. Heredia Espinoza and Charlotte A. Smith 2009 *Origins of the Ñuu: Archaeology in the Mixteca Alta, Mexico*. University Press of Colorado, Boulder.

Kowalewski, Stephen A., Gary M. Feinman, Laura Finsten, Richard E. Blanton and Linda M. Nicholas 1989 *Monte Alban's Hinterland*, Part II: *Prehispanic Settlement Patterns in Tlacolula, Etla, and Ocotlan, The Valley of Oaxaca, Mexico*. Museum of Anthropology, University of Michigan, Ann Arbor.

Krahtopoulou, Athanasia and Charles D. Frederick 2008 The stratigraphic implications of long term terrace agriculture in dynamic landscapes: Polycyclic terracing from Kythera Island, Greece. *Geoarchaeology* 23:550–585.

Leigh, David S., Stephen A. Kowalewski and Genevieve Holdridge 2013 3400 years of agricultural engineering in Mesoamerica: Lama-bordos of the Mixteca Alta, Oaxaca, Mexico. *Journal of Archaeological Science* 40:4107–4111.

Lesure, Richard G. (editor) 2014 *Formative Lifeways in Central Tlaxcala*, vol I: *Excavations, Contexts, and Chronology*. Cotsen Institute of Archaeology Press, Los Angeles.

Lesure, Richard G., Aleksander Borejsza, Jennifer Carballo, Charles Frederick, Virginia Popper and Thomas A. Wake 2006 Chronology, subsistence, and the earliest Formative of central Tlaxcala, Mexico. *Latin American Antiquity* 17:474–492.

Lesure, Richard G., Thomas A. Wake, Aleksander Borejsza, Jennifer L. Carballo, David M. Carballo, Isabel Rodríguez López and Mauro de Ángeles Guzmán 2013 Swidden agriculture, village longevity, and social relations in Formative central Tlaxcala: Towards an understanding of macroregional structure. *Journal of Anthropological Archaeology* 32:224–241.

Lind, Michael D. 2008 The classic to Postclassic at Lambityeco. In *After Monte Albán: Transformation and Negotiation in Oaxaca, Mexico*, edited by J. P. Blomster, pp. 171–192. University Press of Colorado, Boulder.

Llerena, Carlos A., Moshe Inbar and María A. Benavides (editors) 2004 *Conservación y abandono de andenes*. Universidad Nacional Agraria La Molina, Lima.

Los Angeles 2014 Village and field abandonment in post-Conquest Tlaxcala: A geoarchaeological perspective. *Anthropocene* 3:9–23.

Manzanilla, Linda (editor) 2005 *Reacomodos demográficos del Clásico al Posclásico en el centro de México*. Instituto de Investigaciones Antropológicas, México.

Marcus, Joyce (editor) 1990 *Debating Oaxaca Archaeology*. Museum of Anthropology, University of Michigan, Ann Arbor.

Marcus, Joyce and Charles Stanish (editors) 2006 *Agricultural Strategies*. Cotsen Institute of Archaeology, Los Angeles.

Markens, Robert 2008 Advances in defining the Classic-Postclassic portion of the Valley of Oaxaca ceramic seriation: Occurrence and phyletic seriation. In *After Monte Albán: Transformation and Negotiation in Oaxaca, Mexico*, edited by J. P. Blomster, pp. 49–94. University Press of Colorado, Boulder.

Markens, Robert, Marcus Winter and Cira Martínez López 2008 Ethnohistory, oral history, and archaeology at Macuilxóchitl. In *After Monte Albán: Transformation and Negotiation in Oaxaca, Mexico*, edited by J. P. Blomster, pp. 193–215. University Press of Colorado, Boulder.

Martínez López, Cira, Robert Markens, Marcus Winter and Michael Lind 2000 *Cerámica de la fase Xoo (Monte Albán IIIB-IV) del Valle de Oaxaca*. Centro Regional Oaxaca, Instituto Nacional de Antropología e Historia, Oaxaca.

Melville, Elinor G. K. 1994 *A Plague of Sheep: Environmental Consequences of the Conquest of Mexico*. Cambridge University Press, Cambridge.

Metcalfe, Sarah E., Sarah L. O'Hara, Margarita Caballero Miranda and Sarah J. Davies 2000 Records of Late Pleistocene-Holocene climatic change in Mexico—a review. *Quaternary Science Reviews* 19:699–721.

Morrison, Kathleen D. 1994 The intensification of production: Archaeological approaches. *Journal of Archaeological Method and Theory* 1:111–159.

Mueller, Raymond G., Arthur A. Joyce and Aleksander Borejsza 2012 Alluvial archives of the Nochixtlan valley, Oaxaca, Mexico: Age and significance for reconstructions of environmental change. *Palaeogeography, Palaeoclimatology, Palaeoecology* 321–322:121–136.

Netting, Robert McC. 1993 *Smallholders, Householders: Farm Families and the Ecology of Intensive, Sustainable Agriculture*. Stanford University Press, Stanford.

O'Brien, Michael J. and Dennis E. Lewarch 1992 Regional analysis of the Zapotec empire, Valley of Oaxaca, Mexico. *World Archaeology* 23:264–282.

O'Hara, Sarah L., F. Alayne Street-Perrott and Timothy P. Burt 1993 Accelerated soil erosion around a Mexican highland lake caused by prehispanic agriculture. *Nature* 362:48–51.

Patton, Peter and Stanley Schumm 1981 Ephemeral stream process: Implications for studies of Quaternary valley fills. *Quaternary Research* 15:24–43.

Pérez Rodríguez, Verónica 2003 *Household Intensification and Agrarian States: Excavation of Houses and Terraced Fields in a Mixtec Cacicazgo*. Ph.D. Dissertation, University of Georgia, Athens.

——— 2006 States and households: The social organization of terrace agriculture in Postclassic Mixteca Alta, Oaxaca, Mexico. *Latin American Antiquity* 17:3–22.

——— 2013 Recent advances in Mixtec Archaeology. *Journal of Archaeological Research* 21:75–121.

Pérez Rodríguez, Verónica and Kirk C. Anderson 2013 Terracing in the Mixteca Alta, Mexico: Cycles of resilience of an ancient land-use strategy. *Human Ecology* 41:335–349.

Pérez Rodríguez, Verónica, Kirk C. Anderson and Margaret K. Neff 2011 The Cerro Jazmín archaeological project: Investigating prehispanic urbanism and its environmental impact in the Mixteca Alta, Oaxaca, Mexico. *Journal of Field Archaeology* 36:83–99.

Plunket, Patricia and Gabriela Uruñuela 2012 Where East meets West: The Formative in Mexico's Central Highlands. *Journal of Archaeological Research* 20:1–51.

Redman, Charles L. 1999 *Human Impact on Ancient Environments*. University of Arizona Press, Tucson.

Robles García, Nelly M. 2014 Atzompa y la expansión urbana de Atzompa. *Arqueología mexicana* 21(126):28–35.

Sanders, William T. and Deborah L. Nichols 1988 Ecological theory and cultural evolution in the Valley of Oaxaca. *Current Anthropology* 29:33–80.

Sanders, William T., Jeffrey R. Parsons and Robert S. Santley 1979 *The Basin of Mexico: Ecological Processes in the Evolution of a Civilization*. Academic Press, New York.

Schumm, Stanley A. 1957 Arroyos and the semiarid cycle of erosion. *American Journal of Science* 255:161–174.

——— 1977 *The Fluvial System*. Wiley, New York.
——— 1991 *To Interpret the Earth: Ten Ways to Be Wrong*. Cambridge University Press, Cambridge.
——— 1999 Causes and controls of stream incision. In *Incised River Channels: Processes, Forms, Engineering and Management*, edited by S. E. Darby and A. Simon, pp. 19–33. Wiley, Chichester.
Schumm, Stanley A. and R. S. Parker 1973 Implications of complex response of drainage systems for Quaternary alluvial stratigraphy. *Nature Physical Science* 243:99–100.
Smith, Michael E. 1984 The Aztlan migrations of the Nahuatl chronicles: Myth or history? *Ethnohistory* 31:153–186.
———2008 *Aztec City-State Capitals*. University Press of Florida, Gainesville.
Smith, Michael E., Aleksander Borejsza, Angela Huster, Charles D. Frederick, Isabel Rodríguez López and Cynthia Heath-Smith 2013 Aztec period houses and terraces at Calixtlahuaca: The changing morphology of a Mesoamerican hilltop urban center. *Journal of Field Archaeology* 38:227–243.
Sodi Miranda, Federica (editor) 1990 *Mesoamérica y Norte de México: siglo IX–XII*. 2 vols. Museo Nacional de Antropología e Historia, México.
Spores, Ronald 1969 Settlement, farming technology, and environment in the Nochixtlan Valley. *Science* 166:557–569.
———1972 *An Archaeological Settlement Survey of the Nochixtlan Valley, Oaxaca*. Vanderbilt University, Nashville.
Stahle, David W., Dorian J. Burnette, José Villanueva, Julián Cerano, Falko K. Fye, R. Daniel Griffin, Malcolm K. Cleaveland, Daniel K. Stahle, Jesse R. Edmondson and Kathryn P. Wolff 2012 Tree-ring analysis of ancient baldcypress trees and subfossil wood. *Quaternary Science Reviews* 34:1–15.
Stahle, David W., José Villanueva Díaz, Dorian J. Burnette, Julián Cerano Paredes, R. R. Heim Jr., Falko K. Fye, R. Acuña Soto, Matthew D. Therrell, Malcolm K. Cleaveland and Daniel K. Stahle 2011 Major Mesoamerican droughts of the past millennium. *Geophysical Research Letters* 38:L05703.
Tarolli, Paolo, Federico Preti and Nunzio Romano 2014 Terraced landscapes: From an old best practice to a potential hazard for soil degradation due to land abandonment. *Anthropocene* 6:10–25.
Thurston, Tina L. and Christopher T. Fisher (editors) 2007 *Seeking a Richer Harvest: The Archaeology of Subsistence Intensification, Innovation and Change*. Springer, New York.
Trigger, Bruce G. 1989 *A History of Archaeological Thought*. Cambridge University Press, Cambridge.
Turner II, B. L. and William E. Doolittle 1978 The concept and measure of agricultural intensity. *Professional Geographer* 30:297–301.
van Andel, Tjeerd H. and Curtis Runnels 1987 *Beyond the Acropolis: A Rural Greek Past*. Stanford University Press, Stanford.
Villanueva Díaz, José, Julián Cerano Paredes, Dave W. Stahle, Brian H. Luckman, Matthew D. Therrell, Malcom K. Cleaveland and Peter Z. Fulé 2011 La dendrocronología y reconstrucciones paleoclimáticas en el norte-centro de México. In *Escenarios de cambio climático: registro del Cuaternario en América Latina*, edited by M. Caballero Miranda and B. Ortega Guerrero, vol. I, pp. 47–72. Universidad Nacional Autónoma de México, México.
Waters, Michael R. 1992 *Principles of Geoarchaeology: A North American Perspective*. University of Arizona Press, Tucson.

Waters, Michael R. and C. Vance Haynes 2001 Late Quaternary arroyo formation and climate change in the American Southwest. *Geology* 29:399–402.

Werner, Gerd 1981 Verbreitung, Nutzung und Zerstörung von Böden unter randtropischen Bedingungen im zentralmexikanischen Hochland. *Ibero-Amerikanisches Archiv* 7:3–32.

——— 1988 *Die Böden des Staates Tlaxcala im zentralen Hochland von Mexiko: Untersuchungen über ihre Entwicklung, Verbreitung, Erosion und Nutzung unter dem Einfluß 3000-jährigen Ackerbaus.* Steiner, Wiesbaden.

Whitmore, Thomas M. and B. L. Turner II 2001 *Cultivated Landscapes of Middle America on the Eve of Conquest.* Oxford University Press, Oxford.

Wilken, Gene C. 1987 *Good Farmers: Traditional Agriculture and Resource Management in Mexico and Central America.* University of California Press, Berkeley.

Winter, Marcus 1989 From Classic to Post-Classic in pre-Hispanic Oaxaca, Mexico. In *Mesoamerica after the Decline of Teotihuacan, A.D. 700–900*, edited by R. A. Diehl and J. C. Berlo, pp. 123–130. Dumbarton Oaks, Washington.

——— 1994 The Mixteca prior to the Late Postclassic. In *Mixteca-Puebla: Discoveries and Research in Mesoamerican Art and Archaeology*, edited by H. B. Nicholson and E. Quiñones Keber, pp. 201–221. Labyrinthos, Culver City.

——— 2008 Classic to Postclassic in four Oaxaca regions. In *After Monte Albán: Transformation and Negotiation in Oaxaca, Mexico*, edited by J. P. Blomster, pp. 393–426. University Press of Colorado, Boulder.

Zaro, Gregory and Adán Umire Álvarez 2005 Late Chiribaya agriculture and risk management along the arid Andean coast of southern Perú, A.D. 1200–1400. *Geoarchaeology* 20:717–737.

3 From the River to the Fields

The Contribution of Micromorphology to the Study of Hydro-Agrosystems in Semi-Arid Environments (Phoenix, Arizona)

Louise Purdue

Abstract

This chapter presents a field and laboratory approach for better understanding human-environment interactions in past agricultural societies in semi-arid environments. The field approach aims at studying the formation and evolution of anthropic landscapes based on a geomorphic approach to fluvial dynamics, combined with the geoarchaeological study of irrigation systems, considered as anthropic and environmental structures, and connected agricultural fields. To frame these dynamics chronologically and provide a more precise description of environmental and social dynamics (e.g., dynamics of deposition, flooding events, vegetation cover, fire regime, slash and burn, manuring), systematic sampling for chronology and soil analysis (micromorphology) is conducted in these three connected environments. The selection, description, and quantification of markers in soil thin sections allows for the statistical classification of past dynamics, interpreted based on historical or current references. The multiplication of local studies is key to documenting large-scale socioenvironmental dynamics, which can then be juxtaposed with climatic, archeological, and demographic data. This approach is illustrated through the socioenvironmental study of the hydrosystem of the Salt River in the semi-arid Phoenix Basin (Arizona) and the Hohokam irrigated agrosystem. This case study demonstrates how the field and micromorphological study of two irrigation canals and intertwined fluvial deposits at the headgates of one of the largest canal systems, Canal System 1, in use from the seventh to the fifteenth century A.D., provides vital information about past water management and land use as well as fluvial regime and climate change.

Introduction: State of the Art and Objectives

Water management has contributed to the construction and organization of human communities since the emergence of the first agricultural societies. Societies have built water systems for food production, energy, industry, transport, and/or commerce purposes, and they have dealt with water resources facing the combined effects of social (water need) and environmental constraints (water availability and capacity). As a result, the evolution of water availability and allocation strategies through time and its impact on cultural shifts is a socioenvironmental question necessarily raised by researchers. Approaches to the study of water systems and irrigation communities have traditionally revolved around the fields of theoretical anthropology (e.g., Wittfogel 1957; Mitchell 1973; Netting 1974; Service 1975), ethnography (Hunt and Hunt 1973; Lees 1973; Downing and McGuire 1974; Hunt and Hunt 1976), social history (e.g., Glick 1996; Nabhan 1986), and spatial archaeology (e.g., Gentelle 1980; Bruins et al. 1986; Francfort and Lecomte 2002; Mouton 2004; Ortloff 2009). In parallel, researchers have developed high-resolution tools to study water networks, such as geomatics and geophysics (eg., Jones et al. 2000; Powlesland et al. 2006; Keay et al. 2009), as well as aerial and satellite imagery (e.g., Sanders 1982; Fowler 2002; Pournelle 2003; Wilkinson 2003; Clarke et al. 2005), and have adapted new approaches from such fields as environmental history (e.g., Crook et al. 2008), geography and geomorphology (e.g., Marcolongo and Morandi Bonacossi 1997; Harrower 2006), and hydrology (e.g., Orengo and Alaix 2013).

These various approaches and methods have allowed for the study of water systems on the one hand (water management and related social structures) and water networks on the other (spatial structure and links with the landscape and river systems). To understand past agricultural communities, it is necessary to reconstruct the environment they lived in, adapted to, and modified through time. The agrarian landscape they built, which is defined by agricultural practices and production, has evolved over the long term under the influence of climatic fluctuations and human activity. Some recent large interdisciplinary projects have focused on the interaction between climate, water management, and human adaptation (e.g., *Long Term Vulnerability and Transformation Project* coordinated by M. Nelson (cf. Nelson et al. 2010). However, the study of ancient water systems themselves, as signatures of both the agrosystem and the hydrosystem, is still scarce. These systems are rarely considered as objects which can be studied at various spatial and temporal scales and connected to agricultural fields and practices, as well as local and micro-regional geomorphic changes (e.g., Wilkinson 1998; Berger 2000; Beach et al. 2009; Purdue 2011; Bernigaud 2012).

To reconstruct the dynamic interplay (phases of stability and instability) between the agrosystem and the hydrosystem, diachronically and synchronically, it is necessary to reconstruct fluvial dynamics, irrigation canal function and management, and agricultural practices. This chapter presents an overview of how to study these three connected dynamics and their interactions based on a systematic field geomorphic, geoarchaeological, agronomic, and archaeological approach refined by laboratory analysis. Chronological control is necessary to understand the temporalities of fluvial dynamics and water and soil management, but I focus here more specifically on the geoarchaeological and micromorphological approach. Micromorphology is the description of soils at a microscopic scale, based on the interpretative concepts of petrography and pedology (Kubiëna 1938; Brewer and Sleeman 1960; Bal 1973; Bullock et al. 1985; Fédoroff and Courty 1994; Stoops 2003) and is considered as one the major tools for understanding landscape evolution when correlated with other environmental analyses (Gebhardt 2000). Micromorphology has been used widely in semi-arid fluvial environments, where research has mainly focussed on paleoenvironmental and climatic issues (e.g., Courty 1990; Fédoroff and Courty 1989; 1999; Courty and Fédoroff 1999; Singhvi and Derbyshire 1999). In agrarian contexts, micromorphology has aimed at understanding the properties of cultivated soils (e.g., Courty and Fédoroff 1985; Mathieu and Ruellan 1987; Courty 1990; Verba et al. 1995; Hallaire et al. 1998; Lebedeva-Verba and Gerasimova 2009); these studies often serve as references for recognition of past agricultural practices (e.g., Courty 1990; Verba et al. 2002; Presley et al. 2004) and estimate the impact of long-term agriculture on current soil properties (e.g., Wilson et al. 2002; Goodman-Elgar 2007; 2008; Sandor et al. 2007). The micromorphological study of hydraulic structures, in contrast, is nearly nonexistent in arid environments (Purdue et al. 2010; Purdue 2011) and very rare in temperate environments (e.g., Gebhardt 1988; 1993; Courty 1990; Leroyer and Krier 1991; Berger 2000). The combination of signatures of human practices (canal cleaning, slash and burn, irrigation practice, manuring) and hydroclimatic dynamics (flow, sediment origin, erosion processes, sedimentary load, flood regime, water supply, vegetation development, fire regime) has rarely been used.

This approach and method will be highlighted through the example of Hohokam irrigation systems in the Phoenix Basin in central Arizona. Possibly continuing and elaborating upon a tradition previously established by early agricultural groups, the prehistoric Hohokam people were dependent upon and more or less adapted to arid environmental conditions, especially fluvial dynamics. To survive, they built major irrigation systems as early as A.D. 100 (Henderson 1989), which persisted until A.D. 1450 (Haury 1976; Doyel 1979; Gumerman 1991). Their chronology is articulated around the Formative Period (A.D. 1–600), the Pioneer Period (A.D. 600–750), the Preclassic Period (A.D. 750–1150), and the Classic Period (A.D. 1150–1450). Both social and environmental hypotheses have been put forward to explain these cultural shifts: warfare (LeBlanc 1999) and overpopulation (Ensor et al. 2003) on the one hand, and droughts and floods (Nials et al. 1986; Gregory 1991; Huckleberry 1999; Waters and Ravesloot 2001) or channel metamorphosis and widening/downcutting events (Waters and Ravesloot 2001; Graybill et al. 2006) on the other. The cause of the "collapse," however, is still in debate. A case study (Riverview at Dobson, AZ U:9:135 (ASM)) in Mesa, Arizona, provides an example of how environmental surroundings, fluvial dynamics, and human adaptation can be understood at a local scale to help answer these socioenvironmental questions.

Approach and Methods

The Field Approach

To understand the interactions between the hydrosystem, water systems, and the agrosystem in the field, various spatial and temporal scales should be considered (i.e., geomorphic unit per system, per canal, and per stratigraphic unit; and phases of stability or cultural shift) (Chouquer and Favory 1991; Chouquer 2000). The field approach requires three steps: (1) First, it is necessary to reconstruct the current and past environmental framework based on a topographical, hydrological, and geomorphological study of alluvial archives, which allows for the understanding of lateral and vertical river dynamics through time (aggradation, incision, fluvial migration); (2) Second, hydraulic networks can be studied from a geoarchaeological perspective. Fossil traces are identified and mapped using aerial photographs or historic maps, for instance. Transversal or longitudinal crosscuts of hydraulic features allow the reconstruction of episodes of canal construction and abandonment and the link between these temporalities and the dynamics of the hydrosystem. Indeed, canals are often filled with clastic or geochemical sediments which record paleoecological dynamics (Bertrand 1975) as well as local and regional hydrosedimentary dynamics ranging from short-term events (storms, floods, vegetation development) to long-lasting environmental shifts (fluvial metamorphosis, incision, soil erosion) (Berger 2000). The major limits in the geoarchaeological study of canals are, first, their regular maintenance that removes sedimentary information and, second, the hierarchy of water systems, with headgates acting like sedimentary filters. This approach therefore requires diachronic and synchronous systematic studies; (3) Last, it is possible to identify and study cultivated deposits (more easily discernible towards the terminal end of the network) which are fed with water and sediments from adjacent hydraulic structures. The color, texture, structure, and inclusions in cultivated deposits are a reflexion of human practices as well as the dynamics, management, and efficiency of the irrigation system.

Traditionally, numerous samples for paleoenvironmental studies and chronology are collected in these three environments of study. Sherds recovered in irrigation canals, if well preserved, can provide a more or less precise period of construction or use. However, it is usually necessary to derive radiocarbon dates from organic material. Datable materials from the base of hydraulic structures can provide a date of construction, while datable material from upper fills corresponds to abandonment. Chronological assessments related to each episode of visible maintenance and natural erosion are also crucial. Alternative methods, such as OSL dating, can be adopted (Berger 1986). The period of use of the connected agricultural fields can be estimated based on chronostratigraphy, and numerous sherds are traditionally encountered in this environment. Fluvial deposits should also be dated, by radiocarbon, or OSL when organic material is scarce. Bulk sediment samples in these three environments will allow pursuit of paleoenvironmental studies, such as paleobotanical studies, ostracodes, and diatoms (Palacios-Fest 1994; Palacios-Fest et al. 2001), and soil studies, such as grain size and geochemical analyses. Micromorphology is much less commonly employed in such contexts, but can be used to distinguish events not visible through more common methods (fire signals, origin of the sediments, etc.) (Gebhardt 1988; Leroyer and Krier 1991; Berger 2000).

The Micromorphological Approach

Sampling Strategy, Processing, and Study Protocol

When simultaneously studying fluvial deposits, sediments that fill hydraulic structures, and associated agricultural fields, there are various lines of inquiry which allow one to understand past socioenvironmental dynamics such as sedimentary, pedological, ecological, and anthropic processes. To properly understand these processes, micromorphological sampling is usually continuous in every deposit studied or discontinuous within homogeneous units (Courty et al. 1990). Blocks of $12 \times 6 \times 6$ cm are sampled and wrapped in plaster bands and manila paper to protect them. The samples are then processed according to P. Guilloré's (1983–1987) method. Thin sections are observed with the naked eye, under a binocular, and last under an optical microscope with plane polarized light, cross-polarized light, and incident light for the description of microfabric and microfacies (Goldberg and McPhail 2006).

Topics of Enquiry in Fluvial Deposits, Canal Fills, and Agricultural Fields

Sedimentological, pedological/pedoclimatical, ecological and anthropic processes have their own signatures—markers that enable differentiation and possible interpretation (cf. Table 3.1). When reconstructing sedimentary processes, we aim at understanding the environment of deposition (e.g., alluvial, eolian), the fluvial dynamic (e.g., flooding event, water stagnation, and rhythmic sedimentation), the origin of the sediments, and their hydroclimatic significance. Pedological and pedoclimatical studies include rhythm in sedimentation, episodes of water stagnation, evaporation, and presence of salts. The type of data selected are the microstructure (size, form, and arrangement of aggregates and grains), soil crusts (Evans and Buol 1968; Chen et al. 1980; Bresson and Valentin 1994), and three pedological features: biological features (Courty 1994), textural features, and crystalline features such as oxidation and reduction processes or the accumulation and depletion of carbonated/gypsum/salt crystals (e.g., Sehgal and Stoops 1972; Monger et al. 1991; Courty 1994). Investigations of ecological dynamics include the presence of vegetation and its type, origin, and impact on water level and flow, which provides information on the local to regional vegetation cover, processes of erosion with the deposition of decayed plant matter, and also on the in situ plant growth and its kind (e.g., characeae, cyperaceae). Charcoals are also very relevant as they provide information about regional or local fire regimes (including slash and burn cultivation), its intensity, and the type of vegetation burned. Shells and ostracodes provide signatures of typical water temperature and salinity, while the presence of excremental pellets is indicative of a well-aerated soil, specific humidity, and biological activity (Courty 1990). Lastly, human activities identified at the microscopic scale correspond to signatures of irrigation, slash and burn, tillage, and structures to protect hydraulic features. To identify these specific activities, micromorphologists traditionally study soil structure, porosity, organic matter, illuviation processes (coatings, e.g., Pal et al. 2003) and surface crusts (e.g., Bresson and Valentin 1994), and crystalline features (iron, carbonates, gypsum) (e.g., Alonso et al. 2004).

Data Quantification, Processing, and Interpretation

Qualitative descriptions are often used when thin sections are sampled to answer a specific question. However, in the context of systematic sampling, some quantification is necessary. The chart of semi-quantification proposed by Fitzpatrick in 1980 or the semi-quantification proposed by Harden (1982) and Dorronsoro (1994), which assigns a value of 1 to 100 (multiple of 10) to each interval of proportion or size, can be used on well-identified markers to identify large trends. Markers can also be described based on their presence or absence. Due to the subjectivity of these approaches, it may be preferable to quantitatively describe markers, e.g., by point counting (Eswaran 1968; Murphy and Kemp 1987; Amonette 1994) or by image analysis (e.g., Purdue 2011 for grain size studies).

Once all these markers are described, they can be classified by multivariate statistical analyses to compare and group synchronous sequences of sedimentation and/or paleoenvironmental dynamics (Berger 2000; Purdue 2011). While more and more studies provide keys for interpretation (eg. Stoops et al. 2010), these are not easily transposable to other environmental and archaeological contexts (Courty and Nornberg 1985; MacPhail 1986; 1990; 1992; MacPhail et al. 1987; 1990; Gebhardt 1993; Berger 1996). Also, interpretation often puts aside the principle of equifinality (Valentine and Dalrymple 1976; Pawluk 1978; Kemp et al. 1993). In order to mitigate both issues, it is necessary to compare archaeological data with samples from modern or historical contexts for whose dynamics we have written archives. Models of function and paleoenvironmental dynamics at the scale of the irrigation network and then the valley can be formulated. Ultimately, one can then put forward models of landscape evolution.

Case Study: Hohokam Irrigation Along the Lower Salt River, Arizona

Environmental Background: Climate, Geography, and Geomorphology

The Salt River watershed in Central Arizona covers an area of approximately 35,000 km^2. The Salt River, 320 km long, originates in the White Mountains east of Arizona, in the Fort Apache Indian Reservation, at an average altitude of 3,500 m. The Salt River leaves the mountains near its confluence with the Verde River, its major tributary, and flows through the urban area of Phoenix until its confluence with the Gila River, tributary of the Colorado River, 37 km downstream. This area, referred to as the lower Salt River Valley in the Phoenix Basin, belongs to the Basin and Range Province (Gilbert 1875; Davis 1902), characterized by narrow mountain chains and low altitude fertile plains (300–350 m), gently sloping to the east (Fenneman 1931). The lower Salt River, dry today as a consequence of dam building, presents a channel-into-channel configuration (Gregory and Park 1974; Richards 1982), with flow contained in a braided channel during high discharges and a meandering channel set in a large floodplain (400 m) during dry years. This cut-and-fill floodplain is characterized by overbank aggradation, channel accretion (Nanson and Croke 1992), and gully erosion. Precipitation in this area is bimodal: average annual precipitation reaches 190 mm (average from 1933 to 2015; Western Regional Climate Center 2015) with monsoonal summer precipitation, short and intense due to convective storms, and winter precipitation, long and less intense as a result of frontal storms, between November and March (Sellers 1965). Average monthly precipitation for these two seasons reaches 20 mm.

Intense urbanization and erosion of superficial formations precludes precise geomorphic mapping of the lower Salt River Valley. The most substantial maps were provided by Péwé in 1978 and Pearthree and Huckleberry in 1994, and were completed by Cultural Resource Management projects upstream and downstream of the valley (Onken et al. 2004; Phillips et al. 2004) as well as recent publications (Huckleberry et al. 2013; Purdue 2013). Five terraces are generally described. The Pleistocene Sawik (T1, 15 to 72 m above modern river-bed level) and Mesa terraces (T2, 9–29 m) are morphologically very similar due to their carbonate content. The Blue Point Terrace (3–24 m above modern river-bed level) is composed of coarse gravels and rocks with calcium carbonate coatings. The Holocene Lehi Terrace (T3), composed of fluvial deposits, lies 1.5 to 6 m above the modern stream-bed level. Another lower terrace has been identified in parts of the lower valley (Huckleberry et al. 2013; Purdue 2013). Deposition only occurred on these two most recent alluvial formations during the last 2,000 years. A synthesis of the above-mentioned projects and publications has allowed identification of several phases: (1) Landscape stability and soil development occurred from the first century B.C. to the third century A.D.; (2) Fluvial and eolian accretion is recorded from that date until the ninth century A.D.; (3) A second phase of pedogenesis has been noticed between A.D. 800–950 with the development of cienegas in the direct vicinity of the stream; (4) Increasing flooding events are recorded between A.D. 900–1150 with fluvial morphological evolution towards a braided system; (5) Signatures of a downcutting event (eleventh to twelfth century A.D.) are evident along the Gila River and its tributaries associated with soil development on the upper part of the Lehi Terrace; (6) The chronological interval between the fifteenth to the nineteenth century A.D. witnessed channel fill and floodplain accretion.

Archaeological Background: Hohokam Chronology and Irrigation Systems

Because they lived along the Salt and Gila Rivers in the semi-desert Phoenix Basin in Arizona, the Hohokam (A.D. 100–1450), like both their descendants the Akimel O'Odham (after 1694) and white pioneers (nineteenth century), were dependent on fluvial dynamics. To survive, the Hohokam built major irrigation systems as early as A.D. 100 (Henderson 1989), and continued until A.D. 1450 (Haury 1976; Doyel 1979; Gumerman 1991). It was mainly during the Preclassic Period (A.D. 750–ca. 1150) that major canals were built and expanded (Doyel 1991), in parallel with an increase in settlement size, functional diversity, and social differentiation, the construction of ball courts, and the emergence of regional trade and networks (Gregory 1991). By the end of the Preclassic Period, 16 gravity-fed irrigation systems, some measuring nearly 30 km long, organized as a large "spider web," were in use (Figure 3.1(a)). The transition to the Classic Period (A.D. 1050–1150) seems to represent a break in Hohokam social structure, shown by a major settlement reorganization (village aggregation, construction of compound walls, etc.), evolution in architecture, evolving ceramic production, decreasing trade and connectivity, and diversification in agricultural techniques and architecture (cf. Gumerman 1991). This new equilibrium and short-term reorganization was followed by the society's collapse around the fifteenth century. The Salt River Valley subsequently remained unoccupied for a few centuries.

From a technical standpoint, Hohokam canals were organized in tertiary systems: main canals diverted water from the river and supplied secondary structures, the latter bringing water to tertiary canals, from which it was then evacuated into fields. In general, these tertiary canals were straight and regularly spaced, parallel or perpendicular to secondary canals. Headgates probably controlled the entrance of water into the canals. In the lower Salt River Valley, salvage archaeology conducted over the last 20 years has enabled the creation of a very precise map of the irrigation systems (Howard and Huckleberry 1991) (Figure 3.1(a)). Numerous paleoenvironmental studies have also been conducted in irrigation structures themselves (e.g., Howard and Huckleberry 1991; Doyel et al. 1995; Hackbarth et al. 1995), but these have never been based on a micromorphological analysis.

The Riverview at Dobson Project

The Riverview at Dobson Project (AZ U:9:135(ASM)), which is nearly 32 acres in extent, is located south of the Red Mountain Freeway 202 and east of the Price Road, less than 1 km from the river (Figure 3.1(a)). Fieldwork was carried out in 2005 in collaboration with Arizona Museum of Natural History and Arizona State University. This project, as well as previous projects in this area (Masse 1987; Ackerly and Henderson 1989; Mitchell and Motsinger 1998), unearthed more than 20 canals belonging to one of the biggest irrigation systems along the lower Salt River: Canal System 1. Their concentration in the area is favored by the presence of a narrow and low-elevation strip of land between the alluvial plain (currently the T3 Lehi Terrace) and the T2 Mesa Pleistocene Terrace. The canals excavated were oriented west-east and supplied water to the south and in areas of what is now Mesa, Tempe, Chandler, and the Gila River Indian Community. While numerous trenches were dug and canals encountered during this project, I only studied two of them (T1 and T2).[1] The study of canal profiles so close to their headgates enabled simultaneous study of fluvial archives, irrigation canal fills, and signatures of maintenance. The socioenvironmental signatures obtained also indirectly provide information about the amount of water supplied to the secondary and tertiary canals downstream. Unfortunately, these samples do not allow investigation of connected and well-identified agricultural soils. While some bioturbated fluviosols were encountered adjacent to the canals, we do not think they were cultivated in this central area of canal construction. As a result, while some agricultural fields were studied further downstream in this system (Purdue 2011), they are not presented here, as they cannot be clearly connected to the irrigation canals studied here.

Chronostratigraphy and Micromorphological Sampling

Two overlapping canals were observed in T1 (Figure 3.1(b)) (Canal 1a and 1b). The lower canal (Canal 1a) is built on top of finely prismatic brown silts (T1, S VI)[2] and measures 2.6 m deep and 1.4 m wide. Based on the identification of discontinuous stratigraphic limits between the strata, which correspond to traces of past cleaning episodes, we distinguished 6 cleaning events within this structure filled with sand, silt, and clay and eroded in its upper part by a Salt River channel and flood deposits (T1, S VII and S VIII). Canal 1b was excavated into T1, S VIII. Much smaller in size than its predecessor, Canal 1b (75 cm deep, 75 cm wide) seems quickly filled by clayey deposits alternating with coarse sands. Charcoals recovered from the base of these fill deposits (T1, S24) provided a construction date in the interval

A.D. 1030–1189[3] and we obtained a date of A.D. 1489–1651 in the upper part of the canal fills (T1, S30)[4] (possible post-abandonment date) (Figure 3.1). Thirteen micromorphological samples were taken: 9 in the canal and 4 in flood deposits. Unfortunately, one went missing during transportation (T1, S VII).

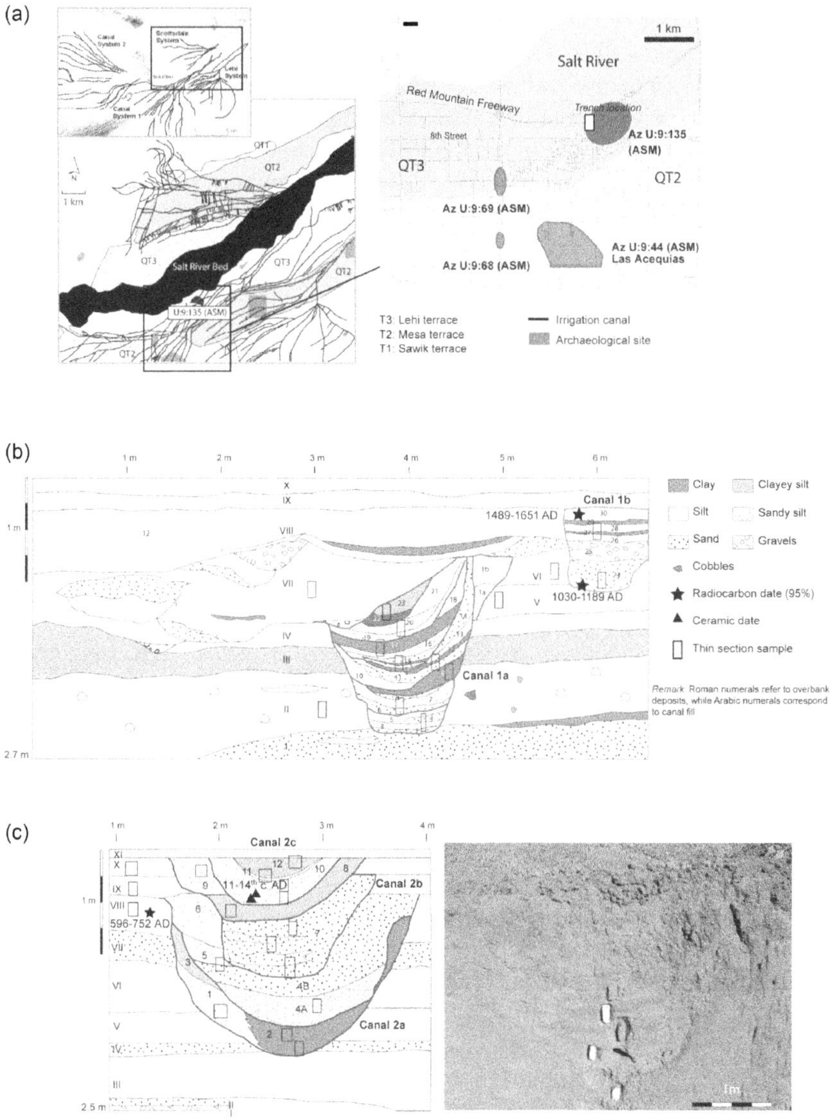

Figure 3.1 (a) Geographic and geomorphic location of AZ U:9:135 (ASM); (b) stratigraphy and chronology in Trench 1 (West profile), AZ U:9:135 (ASM), with Canal 1a and 1b identified; (c) stratigraphy and chronology in Trench 2 (North-East profile), AZ U:9:135 (ASM), with Canal 2a, 2b, and 2c identified.

Three superposed canals were uncovered in Trench 2 (Figure 3.1(c)) (Canal 2a, 2b, and 2c). The lower one, Canal 2a (1.75 m deep, 2m wide), was dug into sandy silts rich in charcoals (T2, S VIII) dated to A.D. 596–752[5] (Preclassic Period), and is filled with fine deposits. After a first cleaning episode, this structure was filled with cross-stratified sand and buried under two flood deposits (T2, S IXa and b). Canal 2b (1.25 m deep, 1.5 m wide) was then dug, but is also filled with cross-stratified sand and buried under new flood deposits (T2, S X). Canal 2c, wider and much smaller in size (2 m wide, 85 cm deep), was excavated into these deposits and is filled with alternating silty clay and laminated sands. One cleaning event was noticed. Numerous sherds were discovered in its fill (T2, S10) dating its use between A.D. 1000 and 1300 (Classic Period). Nineteen micromorphological samples were taken, three of which were in the flood deposits.

Marker Selection

As mentioned in the methodological section, numerous markers can be selected depending on the initial question. In the Riverview at Dobson Project, the fill of the canals directly reflects changes in the Salt River flow. We therefore selected and systematically described 11 markers (Table 3.1). Their interpretation is based on reference materials from historic canals (Huckleberry 1999; Purdue 2011) and geological formations (Purdue 2011), as well as pedological references (Soil Survey Staff 1975) and research on charcoal (Whitlock and Millspaugh 1996; Enache and Cumming 2006). However, we present here only the most relevant ones:

1. Flow intensity (texture obtained from image analysis; see Purdue 2011 for more details)
2. Sediment origin (mineral assemblage obtained by presence/absence, description, and counting of basalt grains followed by a multivariate statistical analysis integrating reference samples from the Salt River Basin (sedimentary basin) and Verde River Basin (volcanic basin)) (see Figures 3.2(b) and 3.2(d) and Purdue 2011 for more details)
3. Sediment transport (sedimentary structure, qualitative description)
4. Rhythmicity and rate of sedimentation (soil structure, qualitative description)
5. Fire regime inferred from charcoal content (+, +++, +++) and size (< or > 125 μm)
6. Organic matter content (+, +++, +++)

Results and Interpretation

EVOLUTION OF THE ALLUVIAL PLAIN

Out of the five flood events chronologically framed, four of them occurred between A.D. 596–752 and A.D. 1030–1189, and while modern agricultural

Table 3.1 Topical enquiry, micromorphological markers, and significance (Riverview at Dobson Project)

	Topic	*Marker*	*Marker Differentiation*	*Possible Interpretation*
SEDIMENTOLOGY	*Flow intensity*	Texture	Sand	High fluvial discharge
			Silt	Average fluvial discharge
			Clay	Decantation and low water circulation
	Sediment transport	Sorting and composition of the sediments	Positively graded	Rhythmic sedimentation, gradual rise and fall of the water level (anthropic or climatic); does not imply flooding (Fig. 3.2a)
			Homogeneous: uniform grain size and fabric	Balance between discharge and sediment supply (Fig. 3.2d)
			Heterogeneous: weak sorting + soil aggregates	Flooding episodes with turbulent flow, or sudden increase/decrease in water level or bioturbation
			Local	Berm breakage or local water diversion (Fig. 3.2c)
	Source to sink	Petrography	Basalt, microcline, etc.	Sediment origin, possible geomorphic and climatic extrapolation (Fig. 3.2b, 3.2d)
PEDOLOGY	*Biological activity and sedimentation rate*	Microstructure	Simple grain	Fast sedimentation (Fig. 3.2d)
			Massive to vughy	Fast sedimentation to hydromorphy (Fig. 3.2f)
			Blocky	Wetting and drying cycles (Fig. 3.2e)
			Channel/granular	High bioturbation and soil moisture (Fig. 3.2g)
	Water presence or absence	Pedological crusts		Scarce vegetation, canal drying out, increased sensitivity to erosion

(*Continued*)

Table 3.1 (Continued)

	Topic	*Marker*	*Marker Differentiation*	*Possible Interpretation*
ECOLOGY	*Fire regime*	Charcoal	%	Intensity of the fire regime (Fig. 3.2j, 3.2k)
		Charcoal size	Microcharcoal	Regional fire (Fig. 2j), effect of bioturbation or granulometric sorting
			Macrocharcoal (> 125 µm)	Local fire (Fig. 3.2l)
		Charcoal shape	Elongated	Possible gramineae vegetation (Fig. 3.2k)
	Vegetation cover	Organic matter	%	Vegetation cover
		Organic matter shape	Fresh	Local vegetation
			Humic	Vegetation cover (local to regional) (Fig. 3.2i)
			Aggregates / microparticules	Soil erosion
			Roots, phytoliths	Type of vegetation (Fig. 3.2h)

activity might have eroded other signatures, only 20 cm of arable land was identified on top of the last flood event dated after the sixteenth to seventeenth century. Of the earlier four flood deposits, three deposited sediments with similar texture and sorting: T1, SV / T2, SVIII; T1, SVII / T2, SIXb; T1, SVIII / T2, SX (Figure 3.3(a)). These flood deposits are composed of weakly sorted coarse silts to sands (Figure 3.3(b) and 3.3(c)), rich in quartz grains, feldspar, pyroxene, sandstone, plagioclase, amphibole, and most of all basalt grains, whose origin and type has been traced to the Verde River Basin, tributary of the Salt River (Purdue 2011; reference sample shown in Figure 3.2(a)). The use of micromorphology in source-to-sink analysis allowed link age of mineral assemblages to precipitation regimes. Indeed, both the Salt and Verde basins are sensitive to winter and summer precipitation, as part of the bimodal regional precipitation pattern. However, the Verde Basin, as a result of its topography and high altitude, as well as its north-south orientation, is supplied mainly by the input of winter precipitation and snow. Higher discharges and larger floods occur during the winter along the Verde River (Graf 1983; Hirschboeck 1985; Ely et al. 1993; House and Hirschboeck 1997). Therefore, the occurrence of sediments from both basins suggests that flow occurred probably dominantly during the winter. Over time, the absence of sediments from the Verde River could also

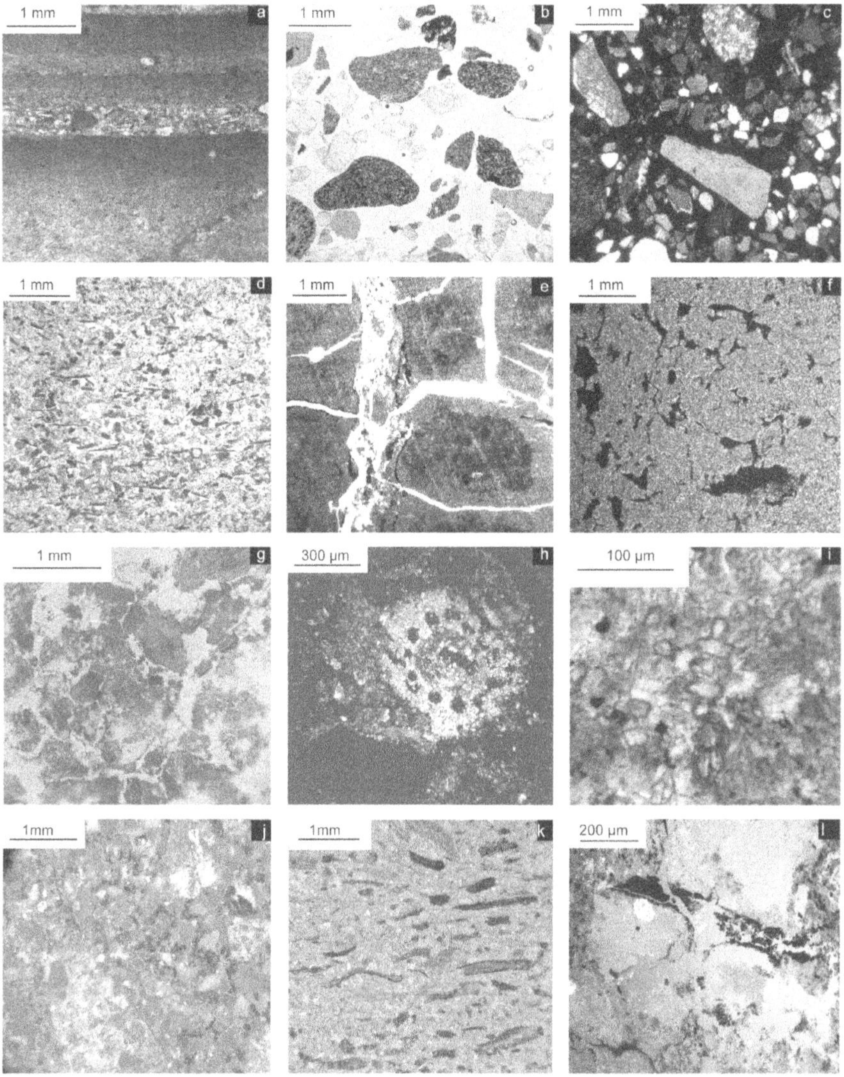

Figure 3.2 Microphotographs of features commonly observed in irrigation canals, flood deposits, and agricultural fields. PPL: Plane Polarized Light; XPL: Crossed Polarized Light; IL: Incident Light. (a) Positively graded sediments; (b) Subrounded basaltic grains (origin: Verde River) (PPL); (c) Heterogeneous subangular sands (local); (d) Well-sorted and homogeneous fine sand with horizontal mica particles, massive soil structure (PPL); (e) Water stagnation deposits and subangular blocky soil structure (PPL); (f) Vughy microstructure due to structural collapse under hydromorphic conditions (XPL); (g) Granular and subrounded blocky structure due to bioturbation (PPL); (h) In situ transversal cross-cut of a characeae stem (XPL); (i) Phytolith of the cyperaceae family (PPL); (j) Rounded to angular microcharcoal (IL); (k) Elongated charcoals and organic matter in flood deposits (IL); (l) In situ burnt organic matter and charcoals (IL).

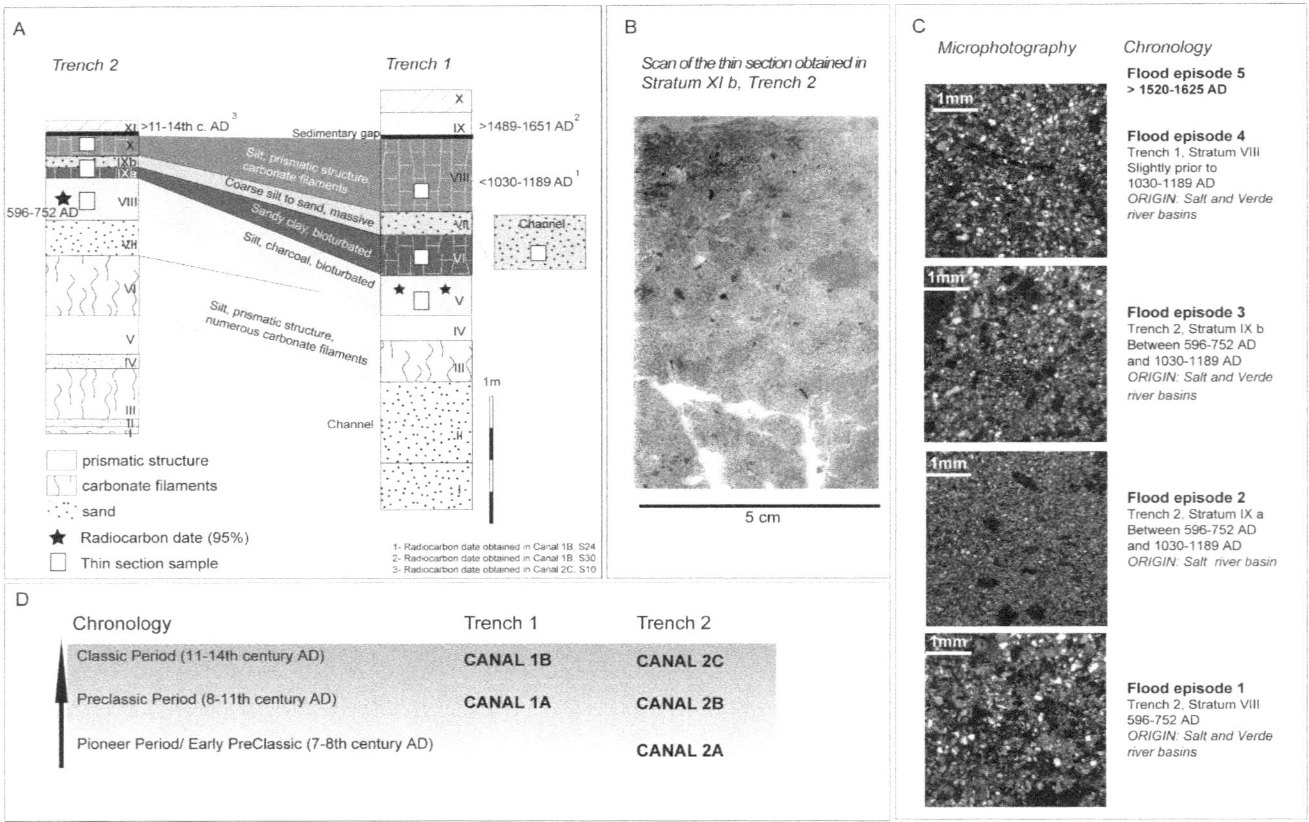

Figure 3.3 Stratigraphic correlation between T1 and T2, flood episodes, pedogenesis, and micromorphological characteristics.

(a) Stratigraphic correlation between Trenches 1 and 2 based on radiocarbon and ceramic dating obtained in both flood deposits and canal fills. In Trench 2, Stratum VIII was radiocarbon dated (A.D. 596–752), while Stratum XI buries Canal 2c, dated by ceramics to the eleventh to fourteenth century in its upper part (Figure 3.2(b)). In Trench 1, Stratum VIII corresponds to a flood deposit into which Canal 1b was excavated between A.D. 1049–1169 (Figure 3.1(c)). The last deposits filling Canal 1b were dated to A.D. 1520–1625. Therefore, Stratum IX, which covers Canal 1b, was deposited after that date. There is a sedimentary gap on the terrace between the twelfth and fifteenth century.

(b) Scan and binocular observation of the thin section sampled in Trench 2, Stratum IXb. The aim of the thin section samples was to pursue a source-to-sink analysis. First observations under the binocular highlight weakly sorted silts and sands, partly bioturbated, numerous soil aggregates, traces of organic matter, and rounded macrocharcoal.

(c) Microphotographs and chronology of the major flood deposits that occurred between the seventh and the sixteenth century. Description of the mineral and organic assemblage under the microscope allows identification of deposits originating from both the Verde and Salt River basins. Coarse and weakly sorted deposits rich in quartz, basalt, and amphibole originate from the Verde and Salt River basins, while finer deposits lacking basalt grains originate mainly from the Salt River Basin.

(d) Chronology of use of the irrigation canals based on stratigraphic correlations.

indicate reduced winter precipitation and shifts in precipitation patterns (summer precipitation) (Purdue 2011). The last flood deposit studied was composed of fine silt, with a homogeneous assemblage (T1, S VI; T2, S IXa), originating mainly from the Salt River Basin, which could point towards changing precipitation patterns.

The study of mineral assemblages also confirmed the first chronostratigraphic correlations established in the field. Based on radiocarbon dates obtained in the canals fills and overbank deposits, it was possible to date the use of the canals (Fig 3.3(d)). Of those canals sectioned by these two trenches, only one was extant during the early Preclassic Period, two during the Preclassic Period, and two much smaller ones were in operation at the beginning of the Classic Period.

MICROMORPHOLOGICAL DESCRIPTIONS

Based on the chronology established (Fig. 3.3(d)) and micromorphological observations, four phases were identified. The description of the deposits encountered in canals in T1 and the ones in T2 are presented in Table 3.2. Detailed micromorphological description of each stratum and microstratum are available in Purdue (2011). Strata identified in the field are numbered while microstrata identified in thin sections are identified by appending alphabetical letters to the stratum designation.

The first phase of canal construction and use in the area occurred around the seventh century. The silts rich in charcoal (T2, S VIII) into which T2, Canal 2a was excavated indicate probable slash and burn to clear vegetation. The deposits studied in the canal (T2, S 1 to 6) are composed of weakly sorted and cross-stratified coarse silts and sands, originating from both the Verde and Salt River basins (indicating regional erosion). Their configuration mimics natural channels and reveals high intensity turbulent flow, which probably favored berm breakage. The high content of organic matter throughout this structure and the presence of macrocharcoals suggest the presence of riparian vegetation at the scale of the watersheds as well as local fire events. Last, the dominant massive, vughy, and subangular structures indicate humid conditions, with well-expressed wetting and drying cycles. The thickness and homogeneity of the canal fill indicate a high sediment load but also a high discharge, as confirmed by the flood deposits (originating in the Salt River Basin), which plugged and buried the canal (T1, S VI; T2, S IXa).

The second phase corresponds to Canal 2a observed in T2, dug into S VIII. At the bottom of the canal (S 2 to 5), sediments are composed of microlaminae of very fine sand and sandy silt, from both the Salt and Verde basins. These deposits are homogeneous and very well-sorted; some are positively graded, while others are heterogeneous suggesting irregular flow. The single grain and massive to slightly bioturbated microstructure indicates fast sedimentation with occasional phases of evaporation and biological activity.

Table 3.2 Selected micromorphological results

Trench / Canal	*Phase*	*Stratum / Microstratum*	*Texture*	*Sediment Origin*	*Sediment Structure*	*Soil Microstructure*	*Organic Matter Content*	*Organic Matter Shape*	*Charcoal Content*	*Charcoal Size*
T1/1A	2	2a	ss	S	Het	c	+++	MP	+	mi
T1/1A	2	2b	vfs	V	Het	sg	X	X	+++	mi
T1/1A	2	2c	vfs	U	PG	sg	X	X	+	mi
T1/1A	2	2d	vfs	S	PG	m	X	X	+++	mi
T1/1A	2	3a	a	S	Hom	sg	+	MP	+	mi
T1/1A	2	3b	ss	U	Hom	m	+	MP	++	mi
T1/1A	2	3c	vfs	V	Het	c	+	MP	+++	mi
T1/1A	2	4a	vfs	V	Het	sg	X	X	+++	mi
T1/1A	2	4b	ss	U	Het	c	++	MP	+	mi
T1/1A	2	4c	ca	S	Het	m	++++	MP	++++	ma
T1/1A	2	4d	vfs	V	Het	mix	+	MP	++	mi
T1/1A	2	5	/	/	/	/	/	/	/	/
T1/1A	3	6	vfs	U		sg	X	X	+++	ma
T1/1A	3	7a	ca	US	Het	m	+	Ag	+++	mi
T1/1A	3	7b	ca	S	Het	sb	+	H	++++	mi
T1/1A	3	8	/	/	/	/	/	/	/	/
T1/1A	3	9a	ca	S	Het	c	++++	H	++++	ma
T1/1A	3	9b	ss	U	Het	sb	++++	H	+++	mi
T1/1A	3	9c	ss	S	PG	sb	+	MP	+	mi
T1/1A	3	9d	vfs	U	Het	sg	+	MP	+	mi
T1/1A	3	10	/	/	/	/	/	/	/	/
T1/1A	3	11	cs	U	Het	sb	+	F	++	ma
T1/1A	3	12a	ca	S	PG	sb	X	X	+	mi
T1/1A	3	12b	ca	V	Het	sb	++	H	++	ma
T1/1A	3	13	vfs	V	Het	sg	+	H	++	mi
T1/1A	3	14	cs	V	Het	sb	+	MP	++	ma

(*Continued*)

Table 3.2 (Continued)

Trench / Canal	*Phase*	*Stratum / Microstratum*	*Texture*	*Sediment Origin*	*Sediment Structure*	*Soil Microstructure*	*Organic Matter Content*	*Organic Matter Shape*	*Charcoal Content*	*Charcoal Size*
T1/1A	3	15a	vfs	U	Het	sb	+	MP	++	ma
T1/1A	3	15b	a	S	B	sb	+	MP	+	mi
T1/1A	3	15c	ss	S	Het	m	+	MP	+	mi
T1/1A	3	15d	vfs	V	Het	sg	+	MP	+	mi
T1/1A	3	15e	vfs	V	Het	v	+	MP	+++	mi
T1/1A	3	16	a	S	B	sb	+	MP	+	mi
T1/1A	3	17a	a	S	B	sb	+	MP	+	mi
T1/1A	3	17b	a	S	B	sb	+	Ag	+	mi
T1/1A	3	18	X	X	X	X	X	X	X	X
T1/1A	3	19	ca	S	B	sg	+++	H	++	mi
T1/1A	3	20	ca	S	Hom	m	++++	H	+++	mi
T1/1A	3	21	/	/	/	/	/	/	/	/
T1/1A	3	22	/	/	/	/	/	/	/	/
T1/1A	3	23	/	/	/	/	/	/	/	/
T1/1A	3	LOST	/	/	/	/	/	/	/	/
T1/1B	4	24a	vfs	V	Het	v	++++	H	++	ma
T1/1B	4	24b	cs	V	Het	v	+++	MP	+++	mi
T1/1B	4	26	cs	S	Het	m	++	MP	+++	mi
T1/1B	4	27	ca	S	PG	sb	+	MP	++	mi
T2/2A	1	1	vfs	U	Eolian	v	+++	H	++	mi
T2/2A	1	2a	cs	V	Het	sb	++	H	++	mi
T2/2A	1	2b	vfs	V	Het	m	+++	H	++++	mi
T2/2A	1	2c	ca	V	Het	sb	+++	F	+++	mi
T2/2A	1	3	/	/	/	/	/	/	/	/
T2/2A	1	4a	cs	US	Eolian	sg	+	MP	+++	mi
T2/2A	1	4b	ss	US	Het	v	+++	H	++++	ma
T2/2A	1	5	ss	S	PG	m	++++	H	++	mi

T2/2A	1	6	vfs	V	B	v to c	+++	H	++	mi
T2/2B	3	7a	vfs	S	PG	sg	++	H	++	mi
T2/2B	3	7b	vfs	US	PG	sg	X	X	+	mi
T2/2B	3	7c	vfs	V	Het	sg	X	X	+	mi
T2/2B	3	7d	vfs	US	Hom	sg	X	X	+	mi
T2/2B	3	7e	vfs	V	PG	sg	X	X	+	mi
T2/2B	3	7f	vfs	S	PG	sg	X	X	+	mi
T2/2B	3	7g	vfs	U	Het	sg	X	X	+	mi
T2/2B	3	7h	a	S	Het	c	+++	Ag	++	mi
T2/2B	3	7i	vfs	US	Het	sg	X	X	++++	ma
T2/2B	3	7j	ss	S	PG	sg	++	Ag	++	mi
T2/2B	3	7k	vfs	V	Het	sg	++	H	+++	mi
T2/2B	3	7l	vfs	V	PG	sg	X	X	++++	ma
T2/2B	3	7m	vfs	V	Het	sg	X	X	++++	ma
T2/2B	3	7n	vfs	V	Het	sg	+	MP	++++	mi
T2/2B	3	7o	vfs	US	Het	sg	+	Ag	++++	mi
T2/2B	3	7p	ss	S	Het	c	+++	H	+++	mi
T2/2C	4	8a	ca	S	B	sb	+++	Ag	++	mi
T2/2C	4	8b	ca	US	PG	sb	+++	H	+++	mi
T2/2C	4	8c	cs	V	Het	sg	+	MP	++++	ma
T2/2C	4	9	ss	V	Het	c	+	MP	+++	mi
T2/2C	4	10a	ca	S	Hom	sb	+++	H	+++	mi
T2/2C	4	10b	cs	US	Het	sg	++	H	+++	ma
T2/2C	4	10c	ca	U	Het	v	++	H	++	mi
T2/2C	4	10d	cs	U	Het	sg	+	H	++	mi
T2/2C	4	10e	ca	S	Hom	v	+++	H	+++	mi
T2/2C	4	10f	ca	U	Het	v	+++	H	++++	mi
T2/2C	4	11	ss	V	Hom	sb	+++	H	+++	ma
T2/2C	4	12a	ss	S	Hom	sb	++	MP	+	mi
T2/2C	4	12b	ss	V	Het	sb	+	H	++++	ma
T2/2C	4	12c	ca	V	Het	sb	+++	F	+	mi

(*Continued*)

Table 3.2 (Continued)

List of abbreviations	
Sedimentology	**Texture:** vfs: very fine sand; cs: coarse silt; ss: sandy silt; ca: clayey silt; a: clay
	Sediment origin: US: upstream Salt River; S: Salt River Basin; V: Verde and Salt River basins; U: undetermined (no basalt grains)
	Sediment structure: Het: heterogeneous deposits; PG: positively graded; Hom: homogeneous deposits; B: bioturbated
Organic components	**Type of organic matter:** F: fresh; H: humic; MP: microparticles; Ag: organic soil aggregates
	Content in organic matter: +: 0–2%; ++: 3–4%; +++: 5–8%; ++++: > 8%
	Content in charcoal particles: +: 0–2%; ++: 3–4%; +++: 5–8%; ++++: > 8%
	Size of charcoal: ma: macrocharcoal (> 125µm); mi: microcharcoal (< 125 µm)
Pedology	**Soil microstructure:** sb: subangular blocky; sg: simple grain; v: vughy; c: channel; mix: complex cleaning event

This is a typical signature of seasonal sedimentation (Courty 1990). Only a few organic microparticles have been observed in association with numerous rounded microcharcoal specks (possibly suggesting diminution of the vegetation cover and riparian vegetation upstream). The discrete change in canal shape suggests that this structure was possibly cleaned following that phase.

The third phase visible in T1 (Canal 1a, S 6 to 23) corresponds to a period of decreasing flow intensity but regular canal maintenance (5 cleaning events). Well-sorted positively graded silty clays and clays were initially deposited (T1, S 6 to 8) indicating rhythmic sedimentation in the canal but very low flow. Most of the sediments originate in the Salt Basin. Their blocky structure suggests episodes of complete drying out, and the abundance of charcoal and macrocharcoal points towards possible in situ burning events to clear vegetation. Coarser and weakly sorted sediments as well as numerous angular soil aggregates suggest episodes of berm breakage (T1, S9). Stratigraphic unconformities suggest that the canal was subsequently cleaned. Despite a temporary phase of turbulent flow characterized by deposits (T1, S13 to 14) composed of weakly sorted sands, silty sands, and coarse silts from the Verde and Salt Basin, maybe resulting from a sudden increase in water level in the river, massive well-sorted positively graded clayey silts (T1, S15 to 20) suggest low- to average-intensity flow as well as a rhythmic and fast sedimentation. Despite this sediment structure suggesting flows favorable to the transport and deposition of elongated particles of charcoals and organic matter, these two markers were only observed as isolates in low overall frequency (possibly suggesting regional decrease in the vegetation cover and/or drier conditions). The environment linked to the final use of the canal is unknown as the thin section sample was lost during shipping, but field data highlights a major in situ burning event in T1, S22. During its final use, Canal 1a was eroded and buried under flood deposits and channel sediments (T1, SVII; T2, SIXb) indicating fluvial morphological change; this probably occurred slightly prior to A.D. 1030–1189, based on the date obtained in T1, Canal 1b.

Before or during this flood event, Canal 2b was built in T2. This structure was filled with massive laminated or weakly sorted fine sands originating from the Verde Basin and various areas of the Salt River Basin (T2, S7). The mineralogical assemblage is much more diversified than the other sediments studied. We progressively observe increasing quantities of micro- and macrocharcoal in the upper part of T2, S7, suggesting local and regional fire events; landscape denudation may have contributed to an increasing sediment load. Canal 2b in T2, as well as the channel deposits in T1, was then buried under new flood deposits (T1, SVIII; T2, SX).

The fourth and last phase corresponds to two small Classic Period canals (T1, Canal 1b, S24 to 30; T2, Canal 2c, S8 to 12). Canal 1b in T1 was probably built prior to the Canal 2c in T2. Indeed, the massive weakly sorted sands rich in macrocharcoal, sands (from both the Verde and Salt rivers) and gravels (T1, S24 and 25) observed at the bottom of Canal 1b are very similar to the deposits into which this structure was excavated (T1, SVIII), suggesting a high bedload. Their heterogeneity, as well as vughy structure, implies uncontrolled flow and humid conditions (structural collapse due to humidity). Above, the graded and homogeneous coarse silts observed in both structures (T1, Canal 1b, S26 to 29; T2, Canal 2c, S8 to 12) indicate rhythmic flow alternating with temporary drying out as suggested by the massive to sub-angular soil structure (seasonal sedimentation, Courty 1990). Only small charcoals and altered fragments of organic matter have been identified. The upper deposits are weakly sorted and rich in angular soil aggregates possibly resulting from post-abandonment slopewash events and pedogenesis (T1, S30; T2, S12).

Discussion

The chronostratigraphic correlation between canal fills and flood deposits made it possible to distinguish four hydrosedimentary trends from the seventh to the thirteenth century, three of which took place during the Preclassic Period (A.D. 750–1150), and the last one during the Classic Period (A.D. 1150–1450). Correlating these results with dendroclimatic, climatic, paleodemographic, and archaeological data makes it possible to situate the hydrosedimentary trends in relation to broader socioenvironmental dynamics and discuss the sociopolitical versus environmental factors that control human-environment interactions. Results in the Phoenix Basin suggest that this co-evolution was overall controlled by hydroclimatic factors during the Preclassic Period, mixed anthropic and climatic influences during the transition to the Classic Period, and social drivers during the Classic Period.

The Preclassic Period (Seventh–Eleventh Century)

This study has enabled us to show a first period of canal construction at the middle of the Preclassic Period. The Canal 2a in T2 is filled with laminated and cross-stratified coarse sands, conditions seem humid, and the high content of both charcoals and organic matter testifies to landscape evolving possibly at the regional scale. However, it would be necessary to multiply local studies to validate this hypothesis. This dynamic is well correlated to the (dendro) climatic records, which reveal optimal conditions for irrigation with a high groundwater level and a possible narrow Salt River channel from which water could have been easily diverted (Graybill et al. 1989). This period of fluvial accretion favored water and sediment supply in downstream canals and fields but required coordination and management closer to the headgates to maintain canals and rebuild them after flood events.

The second trend corresponds to Canal 1a built in T1, on top of the flood deposits that buried the earlier Canal 2a visible in T2. The initial flow in the canal was turbulent and rhythmic, with possible signatures of seasonal sedimentation and regional soil erosion (i.e., sediments derived from both the Salt and Verde basins, rich in charcoal). This suggests efficient water delivery, but the high sediment load probably favored fast siltation. However, flow progressively decreased, as shown by the occurrence of positively graded very fine sediments (originating from the Salt River Basin) and traces of regular wetting and drying cycles. Graded sediments imply a regular increase and decrease in water level in the river or the action of headgates. The numerous cleaning events and ashes indicate local vegetation growth, contributing to low water flow. During this second period, the agricultural areas downstream, dependent on that canal, may have suffered from this reduced water supply. No definite chronology is available for this period, but we do know that it occurred between the eighth and the eleventh century. How to explain this shift in fluvial dynamic, as well as changes in sediment origin and high maintenance? The first hypothesis, supported by climatic data, is a regional decrease in precipitation and discharge. Numerous researchers have noticed increasingly drier conditions at regional scale during this chronological interval, and related them to the Medieval Climate Anomaly (e.g., Cook et al. 2004; Jones and Mann 2004; Moberg et al. 2005; Graybill et al. 2006). The dominance of Salt River deposits could also suggest evolution towards a summer precipitation pattern. The second possibility is that increasing population along the Salt River Valley and in the upstream watersheds (Verde and Tonto basins, lower Salt River Valley) (e.g., Hackbarth 1992; Ciolek-Torrello et al. 1994; Doelle 1995; 2000; Deaver 1997) could have contributed to increasing erosion through land clearance (evidenced by abundant charcoals in the deposits) and could subsequently also have diverted most of the available flow in the river leading to decreasing flow in canals.

Canal 1a in T1 is buried under flood deposits and channel sediments originating from both the Salt and Verde basins. The lack of bioturbation indicates fast sedimentation and burial. The only chronology available is that these destructive flood events and associated fluvial morphological change occurred after A.D. 596–752 and prior to A.D. 1030–1189 (2 sigma calibrated date), with a most probable date range prior to A.D. 1049 and 1169 (1 sigma calibrated date), remarkably close to the transition from Preclassic to Classic Period, which witnessed a major cultural, social, economic, and agricultural change around A.D. 1050–1150. This phase of floodplain instability may have led to two issues: (1) lateral soil erosion with possible loss of land close to the river, as well as issues cultivating it; and (2) water supply issues associated with necessary cooperation and collaboration to repair canals and/or build new ones. Decreasing agricultural activity on the Lehi Terrace slightly after the Preclassic to Classic cultural transition (Cable and

Doyel 1984; Greenwals et al. 1994; Henderson and Clark 2004) could be a result of these events. In the framework of understanding the controlling factors of socioenvironmental interactions, one can discuss the origin of this floodplain instability: was it anthropic or natural? While we know that our chronological framework needs refinement, other paleoenvironmental data available in the Southwest do indicate (1) increasing discharges associated with the Salt River's morphological change to a braided-like stream between A.D. 1051–1196 (Graybill et al. 2006); (2) a wetter and colder period within the Medieval Climate Anomaly and more precisely between A.D. 1034–1150 (Cook et al. 2004) associated with regional flooding at the scale of the American Southwest (Ely et al. 1992; 1993; 1994; Hirschboeck 1987; 1988) as a result of El Niño-Southern Oscillation anomalies (ENSO) (Waters and Haynes 2001), which increase the frequency of winter frontal storms; (3) a change from monsoon-dominant precipitation to winter-dominant precipitation during the eleventh century (indicated by dendroclimatic data; Ni et al. 2002). These results seem to confirm that our source-to-sink analysis is a marker of precipitation patterns, showing a shift from summer-dominant to winter-dominant precipitation during this period. They also emphasize that (1) the Phoenix Basin is sensitive to short-term climatic change, (2) seasonal precipitation shifts impact flooding events, and (3) winter floods are the most damaging along the lower Salt River. Summer storms, by their intensity, increase soil sensitivity to erosion, but their short duration does not favor soil transport over long distances, unlike the longer but less intense winter rains (Etheredge et al. 2004).

Going back to the Riverview at Dobson Project, it is also interesting to notice that following these events indicative of fluvial morphological change, only 20 cm of sediments were apparently deposited (as recorded in Trench 1 and 2). From a purely hydrogeomorphic perspective, if the discharge is higher than the sediment supply, the river will downcut. While no direct traces of downcutting events are visible on the site, we strongly support the hypothesis that this occurred along the Salt River, similarly to the Gila River (Waters and Ravesloot 2001) and its tributaries (e.g., Waters 1998; Cook et al. 2010). The anthropic impact on both the fluvial morphological changes and associated downcutting should, however, not be neglected. Indeed, the dominance of deposits rich in charcoals is in accordance with an increasing human occupation and agriculture in the upstream basins and tributaries of the Salt River (e.g., Hackbarth 1992; Hegmon 2002; Swanson and Diehl 2003). I therefore suggest that the transition from the Preclassic to the Classic Period was controlled by both short-term climatic shifts and long-term anthropic effects on the environment.

The Classic Period (> Eleventh Century)

The final trend, observed in the two small Classic Period canals (T1, Canal 1a; T2, Canal 2c) is one of contrasting sedimentation. Coarse and laminated deposits from both the Salt and Verde basins suggest rhythmic sedimentation, but phases of scarce water supply have also been identified. This combined signature could indicate a well-expressed increase and decrease of the water level in the river. Dendroclimatic data at the scale of the Verde and Salt River basins indicate phases of above average flow, possibly from winter-dominant precipitation, alternating with periods of very low discharge (Graybill et al. 2006). In both trenches, canal management decreased during this period. A plausible explanation for the evolution in canal size and maintenance could be a change in the organization of the agrosystem during the Classic Period. As a result of fluvial morphological change and a probable downcutting event, new systems had to be built upstream to meet water demands, such as the Lehi System, built upstream of Canal System 1 (Howard 1987). The potential issue is that water supply could have generated individualization within agricultural groups more than their collaboration, which could explain the evolution in canal size and management. The reuse over the long term of these irrigation canals does, however, imply a durability of traditions and techniques, with a desire for efficiency and maintenance of agricultural practices and plot locations even in geomorphologically sensitive areas. Finally, canals were abandoned, although no environmental constraint has been identified.

Conclusion

When studying human-environmental co-evolution, it is necessary to conduct systemic and multi-scalar studies that span both temporal scales (from single events to multi-century dynamics) and spatial scales (local to the scale of the watershed). Combined systematic geoarchaeological and micromorphological studies make it possible to examine these interactions as precisely and accurately as possible in past agricultural societies in semi-arid environments, as they exploited water and soils using gravity-fed irrigation systems. This direct linkage of agricultural fields to rivers through irrigation systems makes it possible to assess long-term socioenvironmental dynamics by examining these three elements. I have used the case of Hohokam prehistoric irrigation systems to illustrate the potential of a field geoarchaeological approach combined with the micromorphological study of irrigation canal fill and flood deposits from sedimentological, pedological, ecological, and anthropic perspectives in enabling reconstruction of these socioenvironmental interactions. In order to discuss which of climate or anthropic activities controlled these interactions, it is also necessary to incorporate archaeological, climatic, and demographic data, which I have done here by drawing on the rich record of published work for the area. Results highlight that when alluvial plains are stable (as evidenced by soil development or low intensity overbank deposition), adaptation to shifts in water supply (from, e.g., high flow leading to berm breakage and erosion or low flow favoring siltation) is visible through regular canal maintenance (cleaning and in situ burning events) and canal reconstruction. However, low frequency events (sudden floodplain accretion, repeated fluvial morphological change, and downcutting as a result of an abrupt climate change and shifting precipitation patterns, in parallel with the indirect impact of increasing population) can lead to social reorganization (e.g., the transition from Preclassic to Classic Period in our case study). On the one hand, the maintenance and reconstruction of water structures in similar locations highlights the resilience of past Hohokam communities. On the other, the smaller size and weak maintenance of these structures in the Classic Period, despite an existent labor force, followed by their abandonment a few centuries later in the absence of environmental constraints, suggest that the restructuration of the agrosystem was not durable. In this case, then, it seems that low-frequency environmental shifts were a triggering factor in agricultural reorganization, but the complete abandonment can be traced also to other factors.

Acknowledgments

I would like to thank Daniel Contreras, both as the organizer of the session "Correlation Is Not Enough: Building Better Arguments in the Archaeology of Human-Environment Interactions" held during the SAA 2014 conference in Austin, Texas, and for his insightful comments and suggestions on this chapter. I would also like to thank the School of Human Evolution and Social Change at Arizona State University (Tempe, Arizona) as well as Jerry Howard (Arizona Museum of National History), David Abbott (Arizona State University), and Jean François Berger for the support they provided in collecting the field data. I would also like to thank the GEOPHEN laboratory at the University of Caen Basse Normandie for help and support preparing the thin sections.

Methodological Vignette: Micromorphology and Agrosystems

In arid environments, gravity-fed irrigation systems have allowed agricultural societies to cope with social imperatives (water needs) and environmental constraints (water availability). The resulting intertwined social, natural, and/or artificial system, referred to as an anthroposystem (Lévêque et al., 2003), exists at multiple embedded spatial and temporal scales (micro-local to regional; past, present, or future). Simultaneously pursuing geomorphic, geoarchaeological, and agronomic studies to connect river dynamics to water networks and adjacent fields makes possible the relatively precise reconstruction of human-environment co-evolution, structured around phases of stability, thresholds, and shifts. The systematic analysis of alluvial archives, irrigation canal design, maintenance and sedimentary fill, as well as agricultural fields, can productively be combined with laboratory soil analyses such as micromorphology (Berger 2000; Purdue et al. 2010). Micromorphology is the study of sediments under the microscope (Bullock et al. 1985; Stoops 2003) and has proven extremely useful in the understanding of sedimentary processes (e.g., environments of sedimentation, dynamics of flow, identification of flood and soil erosion events), pedological and ecological dynamics (e.g., rhythm in water supply, soil development, vegetation cover, fire regime), human activities (slash and burn, cleaning events, manuring, irrigation), and dynamics related to soil climate (water stagnation/evaporation) (cf. Fédoroff et al. 1987; Courty et al. 1990; Kapur and Stoops 2008; Stoops et al. 2010). The systematic description of markers of these processes allows for their statistical classification and the grouping of synchronous events. Their interpretation, based on current or historical references and bibliography (e.g., Stoops et al. 2010), enables formulation of dynamic socioenvironmental models, such as reconstructions of changing irrigation regulars or impacts of climate changes. The multiplication of local studies is key for enlarging the scale of study to water networks or even entire catchments and discussing the controlling factors of socioenvironmental interactions.

Notes

1 T1 refers to Riverview at Dobson (RAD) project conducted in 2005, T 206, Feature 121; T2 refers to RAD 2005, T 204, Feature 119.
2 Overbank flood deposits are indicated by Roman numerals and canal fills by Arabic numerals. Stratigraphic units are designated with an S.
3 T1, S24: Microcharcoal, Lyon 45–15, 915 ± 30 B.P., 2 σ calibrated date (95.4%): A.D. 1030–1189 (Reimer et al. 2013).
4 T1, S30: Microcharcoal, Lyon 49–43, 305 ± 30 B.P., 2 σ calibrated date (95.4%): A.D. 1489–1651 (Reimer, 2013).
5 T2, S VIII: Microcharcoal, Lyon 49–42, 1380 ± 35 B.P., 2 σ calibrated date (95.4%): A.D. 596–752 (Reimer et al. 2013).

References Cited

Ackerly, N. W. and T. K. Henderson (editors) 1989 *Prehistoric Agricultural Activities on the Lehi-Mesa Terrace. Perspectives on Hohokam Irrigation Cycles*. Northland Research Inc, Tempe.

Alonso, P., C. Dorronsoro and J. A. Egido 2004 Carbonatation in palaeosols formed on terraces in the Tormes River Basin, Salamanca, Spain. *Geoderma* 118:261–276.
Amonette, J. 1994 *Quantitative Methods in Soil Mineralogy*. Soil Science Society of America. SSSA Miscellaneous Publication, American Society of Agronomy, Madison, WI.
Bal, L. 1973 *Micromorphological Analysis of Soil*. The Netherlands Soil Survey Institute, Soil Survey Papers 6, Wageningen.
Beach, Timothy, S. Luzzadder-Beach, N. Dunning, J. Jones, J. Lohse, T. Guderjan, S. Bozarth, S. Millspaugh and T. Bhattacharya 2009 A review of human and natural changes in Maya Lowland wetlands over the Holocene. *Quaternary Science Reviews* 28:1710–1724
Berger G. W. 1986 Dating Quaternary deposits by luminescence—recent advances. *Geoscience Canada* 13:15–21.
Berger, J.-F. 1996 Climat et Dynamique des Agrosystèmes dans la Moyenne Vallée du Rhône. In *Le III ème Siècle en Gaule Narbonnaise. Données Régionales sur la crise de l'Empire*, edited by Jean-Luc Fiches, pp. 299–332. Apdca, Sophia Antipolis.
——— 2000 Les Fossés Bordiers Historiques et l'histoire Agraire Rhodanienne. *Etudes Rurales: 153–154 La Très Longue Durée*:59–90.
Berger, J.-F. and C. Jung 1996 Fonction, evolution et "Taphonomie" des Parcellaires en Moyenne Vallée du Rhône. Un Exemple d'approche Intégrée en Archéomorphologie et en Géoarchéologie. In *Les Formes Du Paysage, Tome 2*, edited by G. Chouquer, pp. 95–112. Errance, Paris.
Bernigaud, N. 2012 Dynamique des Interactions Socio-Environnementales autour des Marais de Bourgoin-la Verpillière et du Grand Plan (Isère), depuis le Néolithique jusqu'au XIXe Siècle. Unpublished Doctorate in Archaeology, University of Nice Sophia Antipolis.
Bertrand, G. 1975 Pour une Histoire Ecologique de la France Rurale. In *Histoire de la France Rurale*, edited by G. Duby and A. Wallon, pp. 34–113. T.1. Editions du Seuil, Paris.
Bresson, L. and C. Valentin 1994 Soil surface crust formation: Contribution of micromorphology. In *Soil Micromorphology: Studies in Management and Genesis*, edited by A. J. Ringrose and G. S. Voase, pp. 737–762. Humphreys, Elsevier, Amsterdam.
Brewer, R. and J. R. Sleeman 1960 Soil structure and fabric. Their definition and description. *Journal of Soil Science* 11:172–185.
Bruins, H. J., M. Evenari and U. Nessler 1986 Rainwater-harvesting agriculture for food production in arid zones: The challenge of the African famine. *Applied Geography* 6:13–32.
Bullock, P., N. Fédoroff, A. Jongerius, G. Stoops, T. Tursina and U. Babel 1985 *Handbook for Soil Thin Section Description*. Waine Research Publications, Wolverhampton.
Cable, J. S. and D. E. Doyel 1984 The implications of field houses for modeling Hohokam agricultural systems. In *Archaeology of the Original Townsite: The Murphy's Addition*, edited by J. S. Cable, S. L. Henry, D. E. Doyel, pp. 259–280. City Of Phoenix, Soil Systems Publications in Archaeology No. 8, Phoenix, AZ.
Chen, Y., J. Tarchitzky, J. Brouwer, J. Morin and A. Banin 1980 Scanning electron microscope observations on soil crusts and their formation. *Soil Science* 130:49–55.

Chouquer, G. 2000 *L'étude des Paysages. Essai sur leurs Formes et leur Histoire.* Editions Errance, Paris.
Chouquer, G. and F. Favory 1991 *Les Paysages de l'antiquité, Terres et Cadastres de l'occident Romain.* Editions Errance, Paris.
Ciolek-Torrello R., S. M. Whittlesey and J. R. Welch 1994 A synthetic model of prehistoric land use. In *The Roosevelt Rural Sites Study. Vol. 3: Changing Land Use in the Tonto Basin*, edited by R. Ciolek-Torrello and J. R. Welch, pp. 437–472. J.R. Statistical Research Technical Series 28, Tucson.
Clarke, D., R. Sala, J.-M. Deom and E. Meseth 2005 Reconstructing irrigation at Otrar Oasis, Kazakhstan, AD 800–1700y. *Irrigation and Drainage* 54:375–388.
Cook, E. R., C. A. Woodhouse, C. M. Eakin, D. M. Meko and D. W. Stahle 2004 Long-term aridity changes in the western United States. *Science* 306:1015–1018.
Cook, J. P., P. A. Pearthree, J. Onken, A. Youberg and E. R. Bigio 2010 *Mapping of Holocene River Alluvium along Oak Creek, Wet Beaver Creek, West Clear Creek, Fossil Creek, and the East Verde River, Central Arizona.* Arizona Geological Survey, 40, Tucson, AZ.
Courty, M.-A. 1990 *Environnements Géologiques dans le Nord-Ouest de l'inde. Contraintes Géodynamiques au Peuplement Protohistorique*, Thèse d'état, Université de Bordeaux.
——— 1994 Le Cadre Paléogéographique des occupations humaines dans le Bassin du Haut-Khabour, Syrie Du Nord-Est. Premiers Résultats. *Paléorient* 20:21–59.
Courty, M. A. and N. Fédoroff 1985 Micromorphology of recent and buried soils in a semi-arid region of Northwestern India. *Geoderma* 35:287–332.
——— 1999 L'apport de la Micromorphologie des sols à la Reconstitution des Paléopaysages (Application au Bassin Méditerranéen pour la période 3000 av. J.C.—1800 ap. J.C.). In *Environmental Reconstruction in Mediterranean Landscape Archaeology*, edited by P. Leveau et al., pp. 55–65. The Archaeology of Mediterranean Landscapes 2, Oxford.
Courty, M. A., P. Goldberg and R. Macphail 1990 *Soils and Micromorphology in Archaeology.* Cambridge Manuals in Archaeology, Cambridge University Press, Cambridge.
Courty, M. A. and P. Nornberg 1985 Comparison between buried uncultivated and cultivated Iron Age soils of the west coast of Jutland, Denmark. *Isksos* 5:57–69.
Crook, D. C., R. J. Jones, M. Elvin, J. Shen and J. A. Dearing 2008 The history of irrigation and water control in China's Erhai catchment: Mitigation and adaptation to environmental change. In *Mountains: Sources of Water, Sources of Knowledge*, edited by E. Wiegandt, pp. 21–42. Springer, Dordrecht.
Davis, A. P. 1902 *Water Storage on the Salt River, Arizona.* US Geological Survey Water Supply Paper 73.
Deaver, W. L. 1997 Chronological issues of the LVAP. In *Vanishing River. Landscapes and Lives of the Lower Verde Valley*, edited by S. M. Whittlesey, R. S. Ciolek-Torrello and J. H. Altschul, pp. 447–490. Statistical Research Inc., Tucson.
Doelle, W. H. 1995 Tonto basin demography in a regional perspective. In *The Roosevelt Community Development Study. New Perspectives on Tonto Basin Prehistory*, edited by M. D. Elson, M. T. Stark and D. A. Gregory, pp. 201–226. Center for Desert Archaeology Anthropological Papers 15, Tucson.
——— 2000 Tonto basin demography in a regional perspective. In *Salado*, edited by J. S. Dean, pp. 81–105. Amerind Foundation New World Studies Series 4, Dragoon, AZ.

Dorronsoro, C. 1994 Micromorphological index for the evaluation of soil evolution in central Spain. *Geoderma* 61:237–250.

Downing, T. and G. Mcguire 1974 *Irrigation's Impact on Society.* University of Arizona Press, Tucson.

Doyel, D. E. 1979 The prehistoric Hohokam of the Arizona desert. *American Scientist* 67:544–554.

——— 1991 Hohokam cultural evolution in the Phoenix basin. In *Exploring the Hohokam: Prehistoric Desert Peoples of the American Southwest*, edited by G. J. Gumerman, pp. 231–278. University of New Mexico Press, Albuquerque.

Doyel, D. E., A. Black and B. Macnider 1995 *Archaeological Excavations at Pueblo Blanco. The MCDOT Alma School Road Project.* Archaeological Consulting Services Cultural Resources Report 90, Tempe.

Ely, L. 1992 *Large Floods in the Southwestern United States in Relation to Late-Holocene Climate Variations.* Ph.D. Dissertation, University of Arizona, Tucson.

Ely, L., Y. Enzel, V. R. Baker and D. R. Cayan 1993 A 5000 year record of extreme floods and climate change in the southwestern United States. *Science* 262:410–412.

Ely, L., Y. Enzel and D. R. Cayan 1994 Anomalous North Pacific atmospheric circulation and large winter floods in the southwestern United States. *Journal of Climate* 7:977–987.

Enache, M. D. and B. F. Cumming 2006 Tracking recorded fires using charcoal morphology from the sedimentary sequence of prosser lake. British Columbia (Canada). *Quaternary Research* 65:282–292.

Ensor, B. E., M. Ensor and G. V. De Vries 2003 Hohokam political ecology and vulnerability: Comments on waters and ravesloot. *American Antiquity* 68:169–181.

Eswaran, H. 1968 Point-count analysis as applied to soil micromorphology. *Pédologie* 18:238–252.

Etheredge, D., D. S. Gutzler and F. J. Pazzaglia 2004 Geomorphic response to seasonal variations in rainfall in the southwest United States. *GSA Bulletin* 116(5–6):606–618.

Evans, D. D. and S. W. Buol 1968 Micromorphological study of soil crusts. *Soil Science Society America Proceeding* 32:19–22.

Fédoroff, N., L. M. Bresson and C. A. Courty (editors) 1987 *Micromorphology des sols, Soil Micromorphology.* AFES, Paris.

Fédoroff, N. and C. A. Courty 1989 Indicateurs Pédologiques d'Aridification: Exemples du Sahara. *Bulletin de la Société Géologique Française* 5:43–53.

——— 1994 Organisation du Sol aux Echelles Microscopiques. In *Pédologie: Constituants et Propriétés Du Sol, 2ème Edition*, edited by M. Bonneau and S. Souchier, pp. 349–375. Edition Masson, Paris.

——— 1999 Soil and soil forming processes under increasing aridity. In *Paleoenvironmental Reconstructions in Arid Lands*, edited by A. K. Singhvi and E. Derbyshire, pp. 72–108. Oxford Et Ibh Publishing, Calcutta.

Fenneman, N. M. 1931 *Physiography of Western United States.* Mcgraw-Hill, New York.

Fitzpatrick, E. A. 1980 *The Micromorphology of Soils, a Manual for the Preparation and Description of Thin Sections of Soils.* University of Aberdeen, Aberdeen, UK.

Fowler, M. J. F. 2002 Satellite remote sensing and archaeology: A Comparative study of satellite imagery of the environs of Figsbury Ring, Wiltshire. *Archaeological Prospection* 8(2):55–69.

Francfort, H.-P. and O. Lecomte 2002 Irrigation et Société en Asie Centrale des Origines à l'époque Achéménide. In *Annales. Histoire, Sciences Sociales* 57(3):625–663.

Gebhardt, A. 1988 Evolution du paysage bocager sub-atlantique en morbihan. Apport de la micromorphologie. *Bulletin de l'association Française pour l'étude du Quaternaire* 4:197–203.

——— 1993 Soil micromorphological evidence of soil deterioration since the Mid-Holocene in Brittany, France. *The Holocene* 3(3):333–341.

——— 2000 Le Rôle de la micromorphologie des Sols dans la formation des Paysages. *Etudes Rurales, La Très Longue Durée* 153–154:139–149.

Gentelle, P. 1980 Le croissant fertile. *Le Monde de la Bible et Terre Sainte* 15:4–8.

Gilbert, G. K. 1875 Report of the geology of portions of Nevada, Utah, California and Arizona Examined in the years 1871 and 1872. *U.S. Geographic Surveys West Of the One-Hundredth Meridian* 3:17–187.

Glick, T. 1996 *Irrigation and Hydraulic Technology: Medieval Spain and its Legacy.* Aldershot, Variorum.

Goldberg Paul and Richard I. Macphail 2006 *Practical and Theoretical Geoarchaeology*. Blackwell Science, Oxford.

Goodman-Elgar Melissa 2007 Evaluating soil resilience in long-term cultivation: A study of pre-columbian terraces from the Paca Valley, Peru. *Journal of Archaeological Science* 35:3072–3086.

——— 2008 Agricultural fields. Identification and study. *Encyclopedia of Archaeology* 1:110–115.

Graf, W. 1983 The arroyo problem: Paleohydrology and paleohydraulics in the short term. In *Background to Paleohydrology*, edited by K. E. Gregory, pp. 279–302. Wiley, New York.

Graybill, D. A. 1989 The reconstruction of prehistoric salt river streamflow. In *The 1982–1984 Excavations At Las Colinas: Environment and Subsistence*, edited by D. A. Graybill, D. A. Gregory, F. L. Nials, S. K. Fish, R. E. Gasser, C. H. Miksicek and C. R. Szuter, pp. 25–38. Arizona State Museum Archaeological Series162, Tucson.

Graybill, D. A., D. A. Gregory, G. S. Funkhauser and F. L. Nials 2006 Long-term streamflow reconstructions, river channel morphology, and aboriginal irrigation systems along the salt and Gila Rivers. In *Environmental Change and Human Adaptation in the Ancient Southwest*, edited by J. S. Dean and D. E. Doyel, pp. 69–123. University of Utah Press, Salt Lake City.

Greenwals, D. H., M. Zyniecki and D. M. Greenwals (editors) 1994 *Early Desert Farming and Irrigation Settlements: Archaeological Investigations and the Phoenix Sky Harbor Center, Volume 2, Dutch Canal Ruin.* Anthropological Research Papers No. 4. Swca Environmental Consultants, Flagstaff and Tucson.

Gregory, D. A. 1991 Form and variation in Hohokam settlement pattern. In *Chaco and Hohokam: Prehistoric Regional Systems in the American Southwest*, edited by P. L. Crown and W. J. Judge, pp. 159–193. School of American Research Press, Santa Fe.

Gregory, K. J. and C. Park 1974 Adjustment of river channel capacity downstream from a reservoir. *Water Ressources Research* 10:363–374.

Guilloré, P. 1983 *Méthode de Fabrication Mécanique et en Série de lames Minces.* Institut National Agronomique Paris Grignon, Département Des Sols.

Gumerman George 1991 Understanding the Hohokam. In *Exploring the Hohokam: Prehistoric Desert Peoples of the American Southwest*, edited by G. J. Gumerman, pp. 1–27. University of New Mexico Press, Albuquerque.

Hackbarth, M. R. 1992 *Archaeological testing at Verde Park, City of Phoenix.* Northland Research Archaeology Series 3, Phoenix, 127 pp.

Hackbarth, M. R., T. K. Henderson, D. Craig, G. Huckleberry, C. Miksicek, M. Palacios-Fest, S. Smith 1995. *Archaeology at the Head of the Scottsdale Canal System. Vol. 3 : Canal and Synthetic Studies*. Northland Research Anthropological papers of the University of Arizona, No. 95–1.

Hallaire, V., M. Hachicha and C. Cheverry 1998 Evolution structurale d'un horizon de surface argileux sous irrigation. *Etude et Gestion des Sols* 5(2):107–116.

Harden, J. 1982 A Quantitative index of soil development from field descriptions: Examples from a chronosequence in central California. *Geoderma* 28:1–28.

Harrower, Michael J. 2006 *Environmental versus Social Parameters, Landscape, and the Origins of Irrigation in Southwest Arabia (Yemen)*. Unpublished Ph.D., Ohio State University.

Haury, Emil W. 1976 *The Hohokam, Desert Farmers and Craftsmen*. University of Arizona Press, Tucson.

Hegmon M. 2002 Recent issues in the archaeology of the Mimbres region of the North American Southwest. *Journal of Archaeological Research* 10(4):307–357.

Henderson, T. K. 1989 The chronology of the Las Acequias Canals. In *Prehistoric Agricultural Activities on the Lehi-Mesa Terrace: Perspectives on Hohokam Irrigation Cycles*, edited by N. W. Ackerly and T. K. Henderson, pp. 184–199. Northland Research Inc, Flagstaff, AZ.

Henderson, T. K. and T. C. Clark 2004 Changing perspectives: Considerations of past agricultural use of the Salt River floodplain. In *Hohokam Farming on the Salt River Floodplain: Refining Models and Analytical Methods*, edited by T. K. Henderson. Anthropological Papers No. 43. Tucson, AZ: Center for Desert Archaeology.

Hirschboeck, K. K. 1985 *Hydroclimatology of Flow Events in the Gila River Basin, Central and Southern Arizona*. Ph.D. Dissertation, University of Arizona.

——— 1987 Catastrophic flooding and atmospheric circulation anomalies. In *Catastrophic Flooding*, edited by L. Mayer and D. Nash, pp. 23–56. Allen and Unwin, Concord, Massachusetts.

——— 1988 Flood hydroclimatolgy. In *Flood Geomorphology*, edited by V. R. Baker, R. C. Kochel and P. C. Patton, pp. 27–49. John Wiley & Sons, New York.

House, P. K. and K. K. Hirschboeck 1997 Hydroclimatological and paleohydrological context of extreme winter flooding in Arizona, 1993. Storm-Induced Geologic Hazards. In *Case Histories From The 1992–1993 Winter in Southern California and Arizona*, edited by E. R. Larson and J. E. Slosson, pp.1–25. Geological Society of America Reviews in Engineering Geology.

Howard J. B. 1987 The Lehi Canal System: Organization of a Classic Period community. In *The Hohokam Village: Site Structure and Organization*, edited by D. E. Doyel, pp. 211–255. Southwestern and Rocky Mountain Division of the American Association for the Advancement of Science, Glenwood Springs, CO.

Howard, J. B. and Gary Huckleberry 1991 *The Operation and Evolution of an Irrigation System: The East Papago Canal Study*. Soil Systems Publication in Archaeology 18, Phoenix.

Huckleberry, Gary 1999 Stratigraphic identification of destructive floods in relict canals: A case study from the middle Gila River, Arizona. *The Kiva* 65:7–33.

Huckleberry, Gary, J. Onken, W. Graves and R. Wegener 2013 Climatic, geomorphic, and archaeological implications of a late quaternary alluvial chronology in the lower Salt River, Arizona, USA. *Geomorphology* 185:39–53.

Hunt, E. and R. C. Hunt 1973 Irrigation, conflict, and politics: A Mexican case. In *Irrigation's Impact on Society*, edited by T. Downing and G. Macguire, pp. 129–157. University of Arizona Press, Tucson.

Hunt, R. C. and E. Hunt 1976 Canal irrigation and local social organization. *Current Anthropology* 17(3):389–411.

Jones, P. D. and M. E. Mann 2004 Climate over past millennia. *Reviews of Geophysics* 42(2):1–42.

Jones, R. E., B. S. J. Isserlin, V. Karastathis, S. P. Paramarinopoulos, G. E. Syrides, J. Uren, I. Balatsas, C. Kapopoulos, Y. Maniatis and G. Facorellis 2000 Exploration of the Canal of Xerxes, Northern Greece: The role of geophysical and other techniques. *Archaeological Prospection* 7:147–170.

Kapur, S. and G. Stoops (editors) 2008 *New Trends in Micromorphology*. Springer-Verlag, Berlin.

Keay, S., G. Earl, S. Kay, J. Ogden and K. Strutt 2009 The role of integrated geophysical survey methods on the assessment of archaeological landscapes: The case of portus. *Archaeological Prospection* 16(3):154–166.

Kemp, R. A., C. A. Whiteman and J. Rose 1993 Palaeoenvironmental and stratigraphic significance of the valley farm and barham soils in Eastern England. *Quaternary Science Reviews* 12:833–848.

Kubiëna, W. L. 1938 *Micropedology*. Collegiate Press, Ames, IA.

Lebedeva-Verba, M. P. and M. I. Gerasimova 2009 Micromorphology of diagnostic Horizons in aridic soils (complementary to the new classification system of soils of Russia). *Eurasian Soil Science* 42(13):1427–1434.

Leblanc, S. A. 1999 *Prehistoric Warfare in the American Southwest*. University of Utah Press, Salt Lake City.

Lees, S. 1973. *Sociopolitical aspects of canal irrigation in the valley of Oaxaca*. Memoirs of the Museum of Anthropology, University of Michigan, Ann Arbor.

Lees, S. and K. V. Flannery (editors) 1973 *Sociopolitical Aspects of Canal Irrigation in the Valley of Oaxaca*. University of Michigan, Memoirs of the Museum of Anthropology, Ann Harbor.

Leroyer, C. and V. Krier 1991 Etudes Environnementales des Remplissages de Fossés: Premiers Résultats Micromorphologiques et Palynologiques Expérimentaux. In *Archéologie Expérimentale, Tome 2 —La terre: l'os et la Pierre, la Maison et les Champs*, Actes du Colloque International Expérimentation en Archéologie: Bilan et Perspectives, pp. 228–235. Editions Errance, Paris.

Lévêque, C., T. Muxart, L. Abbadie, A. Weil and S. van der Leeuw 2003 L'anthroposystème: entité structurelle et fonctionnelle des interactions sociétés-milieux. In *Quelles natures soulons-nous?*, edited by C. Lévêque and S. van der Leeuw, pp. 110–129. Elsevier, Paris.

Macphail, Richard I. 1986 Paleosols in archaeology: Their role in understanding flandrian pedogenesis. In *Paleosols, Their Recognition and Interpretation*, edited by P. Wright, pp. 263–290. Blackwell Scientific Publication, Oxford.

——— 1990 *Soil Report on the Deansway Archaeology Project, Worcester*. Ancient Monuments Laboratory Report 82/91, English Heritage.

——— 1992 Soil micromorphological evidence of ancient soil erosion. In *Past and Present Soil Erosion*, edited by M. Bell and J. Boardman, pp. 197–2015. Oxbow Monograph 22, Oxford.

Macphail, Richard I., M. A. Courty and A. Gebhardt 1990 Soil micromorphological evidence of early agriculture in North-West Europe. *World Archaeology* 22: 53–69.

Macphail, Richard I., J. C. C. Romans and L. Robertson 1987 The application of micromorphology to the understanding of Holocene soil development in the British Isles, with special reference to early cultivation. In *Micromorphologie des*

Sols, edited by N. Fédoroff, L. M. Bresson and M. A. Courty, pp. 647–656. Afes, Plaisir.
Marcolongo, B. and D. M. Morandi Bonacossi 1997 L'abandon du Système d'irrigation Qatabanite dans la Vallée du Wadi Bayhan (Yémen): Analyse Géo-Archéologique. *Comptes Rendus de lacadémie des Sciences—Series Iia—Earth and Planetary Science* 325 (1):79–86.
Masse, W. B. 1987 *Archaeological Investigations of Portions of the Las Acequias-Los Muertos Irrigation System.* Arizona State Museum Archaeological Series 176, University Of Arizona, Tucson.
Mathieu, C. and A. Ruellan 1987 Evolution Morphologique des Sols Irrigués en Région Méditerranéenne Semi-Aride: l'exemple de la Basse Moulouya (Maroc). *Cahiers Orstom Pédologie* 13:3–25.
Mitchell, D. R. and T. Motsinger (editors) 1998 *Life at the River's Edge: Hohokam Irrigation and Settlement along the Red Mountain Freeway between Price Road and Mckellips Road.* Swca Archaeological Report 97–23, Scottsdale.
Mitchell, W. 1973 The hydraulic hypothesis. A reappraisal. *Current Anthropology* 14:532–534.
Moberg, A., D. M. Sonechkin, K. Holmgren, N. M. Datsenko and W. Karlen 2005 Highly variable northern hemisphere temperatures from low- and high-resolution proxy data. *Nature* 433:613–617.
Monger, H. C., L. A. Daugherty, W. C. Lindemann and C. M. Liddell 1991 Microbial precipitation of pedogenic calcite. *Geology* 19:997–1000.
Mouton, M. 2004 Irrigation et Formation de la Société Antique dans les Basses-Terres du Yémen: un Essai de Modèle, *Syria* 81:81–104.
Murphy, C. and R. Kemp 1987 Micromorphology and the argillic Horizon-a reappraisal. In *Micromorphologie des Sols*, edited by N. Fédoroff, L. M. Bresson and M. A. Courty, pp. 257–261. Afes, Plaisir.
Nabhan, G. P. 1986 Papago indian desert agriculture and water control in the sonoran desert, 1697–1934. *Applied Geography* 6:43–59.
Nanson, G. C. and J. C. Croke 1992 A genetic classification of floodplains. *Geomorphology* 4:459–486.
Nelson, M.C., K. Kintigh, D. R., Abbott and J. M. Anderies 2010 The cross-scale interplay between social and biophysical context and the vulnerability of irrigation-dependent societies: Archaeology's long-term perspective. *Ecology and Society* 15(3):31.
Netting, R. 1974 The system nobody knows: Village irrigation in the swiss alps. In *Irrigation's Impact on Society*, edited by T. E. Downing and G. Mcguire, pp. 67–75. Anthropological Papers of the University of Arizona 25, Tucson.
Ni, F., T. Cavazos, M. K. Hughes, A. C. Comrie and G. Funkhauser 2002 Cool-season precipitation in the Southwestern USA Since AD 1000: Comparison of linear and nonlinear techniques for reconstruction. *International Journal of Climatology* 22(13):1645–1662.
Nials, F. L., D. A. Gregory and D. A. Graybill 1986 Salt River streamflow and the Hohokam irrigation system. In *The 1982–1984 Excavations at Las Colinas. Vol. 5: Environment and Subsistence*, edited by D. A. Graybill, D. A. Gregory, F. L. Nials, S. K. Fish, R. E. Gasser, C. H. Miksicek and C. R. Szuter, pp. 59–78. Arizona State Museum Archaeological Series 162, Tucson.
Onken, J., M. R. Waters and J. A. Homburg 2004 *Geoarchaeological Assessment for the Tres Rios Project, Maricopa County, Arizona.* Statistical Research Inc., Technical Report 03–68, Redlands, California.

Orengo, H. and C. Miro i Alaix 2013 Reconsidering the water system of Roman Barcino (Barcelona) from supply to discharge. *Water History* 5:243–266.

Ortloff, Charles E. 2009 *Water Engineering in the Ancient World*. Oxford University Press, Oxford.

Pal, D. K., P. Srivastava and T. Bhattacharyya 2003 Clay illuviation in calcareous soils of the semi-arid part of the indo-gangetic plains, India. *Catena* 115:177–192.

Palacios-Fest, M. R. 1994 Nonmarine ostracode shell chemistry from ancient Hohokam irrigation canals in central Arizona: A paleohydrochemical tool for the interpretation of prehistoric human occupation of the western United States. *Geoarchaeology: An International Journal* 9:1–29.

Palacios-Fest, M. R., J. B. Mabry, F. Nials, J. P. Holmlund, E. Miksa and O. K. Davis 2001 Early irrigation systems in southeastern Arizona: The ostracode perspective. *Journal of South American Earth Sciences* 14(5):541–555.

Pawluk, S. 1978 The pedogenic profile in the stratigraphic section. In *Quaternary Soils*, edited by W. C. Mahaney, pp. 61–75. Geo Abstracts, Norwich.

Pearthree, P. A. and G. Huckleberry 1994 *Surficial Geologic Map of the Mesa 30'× 60' Quadrangle, Arizona*. Arizona Geological Survey Open-File Report 94–23, Scale 1:100,000.

Péwé, T. 1978 Terraces of the lower salt river valley in relation to the late cenozoic history of the phoenix basin, Arizona. In *Guidebook to the Geology of Central Arizona, State of Arizona*, edited by D. M., Burt and T. L. Péwé, pp. 1–46. Bureau of Geology and Mineral Technology, No. 2, Tucson, AZ.

Phillips, B. G., A. Gibbson, A. Miller, M. S. Droz and L. A. Walker 2004 Paleoenvironnement. In *Life on the Lehi Terrace: The Archaeology of the Red Mountain Freeway between State Route 87 and Gilbert Road*, edited by S. Powell, R. L. Boston, pp 11–47. Archaeological Consulting Services Cultural Resources Report 135, Tempe.

Pournelle, J. R 2003 The littoral foundation of the Uruk State: Using satellite photography toward a new understanding of the 5th/4th millennium bce landscapes in the warka survey area, Iraq. In *Chalcolithic and Early Bronze Age Hydrostrategies*, edited by D. Gheorghiu, pp. 5–23. Bar International Series 1123, Oxford Archeopress, Oxford.

Powlesland, D. J., J. Lyall, D. Hopkinson, D. N. M. Donoghue, M., Beck, A. Harte and D. Stott 2006 Beneath the sand—remote sensing, archaeology, aggregates and sustainability: A case study from Heslerton, the Vale of Pickering, North Yorkshire, UK. *Archaeological Prospection* 13:291–299.

Presley, D. R., M. D. Ransom, G. J. Kluitenberg and P. R. Finnell 2004 Effects of thirty years of irrigation on the genesis and morphology of two semiarid soils. *Soil Science Society of America Journal* 68:1916–1926.

Purdue, Louise 2011 *Dynamique des Paysages Agraires et Gestion de l'eau dans le Bassin Semi-Désertique de Phoenix, Arizona, de la Préhistoire à l'époque moderne*. Unpublished Ph.D. Dissertation, Université of Nice.

——— 2013 Dynamiques Alluviales et Adaptations aux Contraintes Hydro-Climatiques dans le Sud-Ouest Américain au cours des deux Derniers Millénaires (Phoenix, Arizona). *Géomorphologie: Relief, Processus, Environnement* 19(3):243–264.

Purdue, Louise E., W. Miles, K. Woodson, A. Darling and J.-F. Berger 2010 Micromorphological study of irrigation canal sediments: Landscape evolution and hydraulic management in the Middle Gila River Valley (Phoenix Basin, Arizona) during the Hohokam occupation. *Quaternary International* 216:129–144.

Reimer, P. J., E. Bard, A. Bayliss, J. W. Beck, P. G. Blackwell, C. Bronk Ramsey, C. E. Buck, H. Cheng, R. L. Edwards, M. Friedrich, P. M. Grootes, T. P. Guilderson,

H. Haflidason, I. Hajdas, C. Hatté, T. J. Heaton, D. L. Hoffmann, A. G. Hogg, K. A. Hughen, K. F. Kaiser, B. Kromer, S. W. Manning, M. Niu, R. W. Reimer, D. A. Richards, E. M. Scott, J. R. Southon, R. A. Staff, C. S. M. Turney and J. van der Plicht 2013 IntCal13 and Marine13 radiocarbon age calibration curves 0–50,000 years cal BP. *Radiocarbon* 55(4):1869–1887.

Richards, G. E. 1982 *Rivers: Form and Process in Alluvial Channels*. Methuen, London.

Sanders, William T. 1982 *Introduction, a Reconstruction of a Classic Period Landscape in the Teotihuacan Valley: Report to the National Science Foundation*, 1–19. Manuscript on File, Department of Anthropology, Pennsylvania State University.

Sandor, A. S., J. B. Norton, J. A. Humburg, D. A. Muenchrath, S. E. Williams, C. I. Havener and P. D. Stahl 2007 Biochemical studies of a North American runoff. *Geoarchaeology* 22(3):359–386.

Sehgal, J. L. and G. Stoops 1972 Pedogenic calcite accumulations in arid and semiarid regions of the indo-gangetic alluvial plain of Erswhile Punjab (India)—their morphology and origin. *Geoderma* 8:59–72.

Sellers, W. D. 1965 *Physical climatology*. University of Chicago Press, Chicago.

Service, Elman R. 1975 *Origins of the State and Civilization: The Process of Cultural Evolution*. W. W. Norton, New York.

Singhvi, A. K. and E. Derbyshire (editors) 1999 *Paleoenvironmental Reconstruction in Arid Lands*. A. A. Balkema. Rotterdam, Netherlands.

Soil Survey Staff 1975 *Soil Taxonomy, a Basic System of Soil Classification for Making and Interpreting Soil Surveys*. US Department Of Agriculture Handbook.

Stoops, G. 2003 *Guidelines for Analysis and Description of Soil and Regolith Thin Sections*. Soil Science Society of America, Madison, Wisconsin.

Stoops, G., V. Marcelino and F. Mees 2010 *Interpretation of Micromorphological Features of Soils and Regoliths*. Elsevier Science, Amsterdam.

Swanson S. and M. W. Diehl 2003 Mimbres pithouse dwellers. *Archaeology Southwest* 17(4):3

Valentine, K. W. G. and J. B. Dalrymple 1976 Quaternary buried paleosols: A critical review. *Quaternary Research* 6:208–222.

Verba, M., L. Aidabekova and I. Yamnova 2002 *Micromorphological and Mineralogical Peculiarities of old-Irrigated Soils on Aral Sea Basin*. Proceedings of the 17th World Congress of Soil Science, Bangkok, Thailand, Poster.

Verba, M. P., A. S. Al-Kasiri, N. A. Goncharova and N. P. Chizhikova 1995 Impact of irrigation on micromorphological features of desert soils of Hadramout Valley (Yemen). *Eurasian Soil Science* 27:108–124.

Waters Michael, R. 1998 The effect of landscape and hydrologic variables on the prehistoric salado: Geoarchaeological investigations of the Tonto basin, Arizona. *Geoarchaeology* 13:105–160.

——— 2000 Alluvial stratigraphy and geoarchaeology in the American southwest. *Geoarchaeology* 15:537–557.

Waters, Michael R. and C. Vance Haynes 2001. Late quaternary arroyo formation and climate change in the American southwest. *Geology* 29(5):399–402.

Waters, Michael R. and J. Ravesloot 2001 Landscape change and the cultural evolution of the Hohokam along the Middle Gila River and other river valleys in south central Arizona. *American Antiquity* 66(2):285–299.

Western Regional Climate Center 2015 Historical climate information. http://www.wrcc.dri.edu/

Whitlock, C. and S. H. Millspaugh 1996 Testing assumptions of fire history studies: An examination of modern charcoal accumulation in Yellowstone National Park. *The Holocene* 6:7–15.

Wilkinson, Tony J. 1998 Water and human settlement in the Balikh Valley, Syria: Investigations from 1992–1995. *Journal of Field Archaeology* 25(1):63–87

——— 2003 Hydraulic landscapes and irrigation systems of sumer. In *The Sumerian World*, edited by H. Crawford, pp. 33–54. Routledge, London, New York.

Wilson, C., I. A. Simpson and E. J. Currie 2002 Soil management in pre-Hispanic raised field systems: Micromorphological evidence from Hacienda Zuleta, Ecuador. *Geoarchaeology* 17(3):261–283.

Wittfogel, Karl A. 1957 *Oriental Despotism. A Study of Total Power*. Yale University Press, New Haven.

4 Regional Climate, Local Paleoenvironment, and Early Cultivation in the Middle Wadi el-Hasa, Jordan

Daniel A. Contreras and Cheryl Makarewicz

Abstract

For several decades, researchers have sought to define a relationship between environmental change and initial human experimentation with food production in southwest Asia. Many models emphasize the role of the Younger Dryas, a Northern Hemisphere–wide cold and arid climatic period that lasted from approximately 12,900 B.P.—11,700 B.P., in providing the impetus for first human experimentation with cultivation in southwest Asia. In general, these models focus on macrolevel environmental patterns, discounting variability in the effects of the Younger Dryas at local scales. In so doing, these models elide the question of whether or not regional paleoenvironmental data provide spatial and chronological resolution adequate for the purposes of establishing and examining relationships between environmental change, resource stress, and nascent cultivation in southwest Asia. Here, we consider this problem of scale in the southern Levant through geoarchaeological analyses of stratified in-stream wetland deposits dating to the Late Pleistocene and Early Holocene, including the crucial Younger Dryas period, in the Wadi el-Hasa, Jordan. The evidence of local landscape morphology and conditions that this provides, in close proximity to the Pre-Pottery Neolithic A settlement of el-Hemmeh (where pre-domestication cultivation was practiced alongside gathering of wild food plants), provides an opportunity to closely examine the environmental contexts of early experimentation with agriculture in the southern Levant, particularly their change over time and the relationship of local to regional evidence.

Introduction

The transition from hunting and gathering to food production not only changed how people acquired food, but also revolutionized the cultural practices and social structures of human communities. Because this transition was so fundamental, much debate revolves around precisely what contributed to the emergence of food production over 10,000 years ago in southwest Asia. Demographic thresholds, resource stress, cognitive changes, and competitive feasting are frequently suggested as influential in the origins of agriculture (Cauvin 2000; Watkins 2005; Bocquet-Appel and Bar-Yosef 2008; Hayden 2009; 2011), but these factors are generally overshadowed by the explanatory role attributed to environmental change (e.g., Bar-Yosef 2011). From the early days of Pumpelly's (1908) "oasis theory," which posited that increasing aridity compelled hunter-gatherers, plants, and animals to concentrate in well-watered oases, stimulating plant and animal domestication, to more recent archaeological models invoking "environmental deterioration" as a prompt for the development of wholly new subsistence practices, environmental change has long been viewed as a potential driver of human subsistence innovations.

The idea that the Younger Dryas, generally understood as a period of intensely cold and arid conditions throughout the Northern Hemisphere that began abruptly at approximately 12,900 cal B.P. and lasted until approximately 11,700 cal B.P., was the primary forcing mechanism that prompted the "origins of agriculture" in southwest Asia still holds considerable sway (e.g., Moore and Hillman 1992; Wright 1993; Bar-Yosef 1998; 2001a; 2001b; 2009; 2011; Hillman et al. 2001; Mithen 2006; Cordova 2008). This model specifies that, under the cold and dry conditions of the Younger Dryas, the productivity of wild cereal stands declined and their distribution contracted significantly (Bar-Yosef 2001b); variants also suggest that the (hypothesized) natural decline in cereal availability may have been further exacerbated by resource overexploitation caused by hunter-gatherer groups (Bar-Yosef 1998). In response to this climate-induced resource stress, the model suggests, hunter-gatherers increased their mobility in some cases and, in others, developed a new subsistence strategy—plant cultivation—to compensate for the decreased availability of wild cereals (Bar-Yosef 2001a; Grosman and Belfer-Cohen 2002).

At the same time, while the idea that the Younger Dryas had a profound and irreversible impact on the hunter-gatherer populations remains pervasive, the role of this climatic perturbation as the impetus for cereal cultivation by foragers has been recently challenged. Buoyed by the recent increase in the number of well-contextualized archaeobotanical datasets from the Early Holocene, there has been a proliferation of new models detailing the complexity and regional variability in nascent plant cultivation strategies in the Near East (e.g., Willcox 1999; 2005; Savard et al. 2006; Allaby et al. 2008; Gustafson et al. 2008; Abbo et al. 2010; Fuller et al. 2011; Rosen and Rivera-Collazo 2012). These newer models focus on local developmental trajectories that minimize the particular role of the Younger Dryas in driving changes in human subsistence. For instance, addressing the potential "push" of climatic deterioration, Willcox (2005) has suggested that the effects of the Younger Dryas on the shift to cultivation have been overemphasized, arguing that it would have had only a minor impact on the availability and distribution of wild cereals. Complementary arguments suggest that Late Pleistocene hunter-gatherer subsistence strategies were highly adaptive, so much so that any changes in plant resource distribution that may have resulted from marked climate change, such as that hypothesized for the Younger Dryas, would have entailed shifts in the focus of resource exploitation rather than the development of entirely novel subsistence strategies (Rosen 2012; Rosen and Rivera-Collazo 2012). The chronological correlation between environmental change and cereal cultivation has also been challenged by critics who suggest that archaeological chronologies are at best too uncertain and inconclusive to suggest causality, and may in fact contradict models relating the Younger Dryas to emergent cultivation (e.g., Willcox 2005; Blockley and Pinhasi 2011; Maher et al. 2011). Moreover, alternative relationships have been posited based on slightly different chronologies, linking the emergence of cultivation to the climatic amelioration *following* the Younger Dryas rather than the deterioration putatively associated with it (e.g., Robinson et al. 2006; Stein et al. 2010).

In sum, there are now a number of macroevolutionary models that link climate change to emergent agriculture, looking either to the supposed stress of the Younger Dryas or the supposed new possibilities afforded by the climatic changes that constituted the end of the Younger Dryas and the beginnings of the Holocene. The profusion of different suggested correlations is testament to chronological uncertainties, which combine with the small sample of known sites to create a situation in which even small shifts in the evidence can suggest or enable major reworkings of climate-culture hypotheses. The ubiquity and variety of these *kinds* of arguments is evidence that they are hypotheses at best, rather than empirically driven explanations, while the persistence of climate-culture hypotheses regardless of how specifics are rearranged is evidence that the models are underdetermined. In the absence of a unitary and mechanistic relationship between environment and human behavior (i.e., climate x necessarily produces cultural response y), arguments about the interplay of culture and climate at the dawn of agriculture are limited in large part by the spatial scale and chronological resolution of the paleoclimatic and archaeological data mobilized to construct them. As a result, selection between these competing hypotheses cannot be made by simply adding *more* coarse (temporally, spatially, or both) data. Rather, what we need to understand are *local* conditions, ideally over spans of time relevant to human decision-making.

Both deterioration and amelioration models share an understanding of the effects of the Younger Dryas that is dependent on paleoenvironmental data derived from regional records, namely, that the cooler and dryer period observable at a hemispheric scale resulted in declines in resource abundance in the foraging catchments of Levantine communities. Disagreements are primarily over whether these effects were of magnitude sufficient to compromise subsistence systems, over what precisely that magnitude might be, and over the chronological relationship of the Younger Dryas to early experiments with cultivation—but both the deterioration model and most of its critics embrace the *potential* significance of the regional-local relationship. In other words, generalizations about climate that are regional in scale (e.g., a hemispheric cooling and drying) are matched to mechanisms of explanation (e.g., foraging returns and behavioral responses to their variability) that are local in scale. One question left largely unaddressed is the character of the regional-local link; i.e., how were such regional climatic shifts manifest at the local scales at which they would have intersected the subsistence activities of inhabitants of southwest Asia?

Relating Models to Evidence: Scalar and Chronological Challenges

The questions that archaeologists have generally asked of paleoclimate data are implicitly local and short term: e.g., did the environments available to foragers change with regard to the key parameters of resource availability (water and animal and plant foods), and if so, how much, how fast, and for how long? Faced with spatially coarse-grained paleoclimate data with which to address these questions, it has been common to rely on an implicit downscaling that has tacitly posited a 1:1 relationship between regional averages and the local conditions under discussion.

Research investigating the specifics of any relationship between the Younger Dryas and emergent cultivation has been hampered by several factors: (1) the relative scarcity of archaeological sites that span the period of the development of the first food production strategies, (2) imprecise and/or underdetermined archaeological and paleoenvironmental chronologies, (3) ambiguity in the interpretation of palaeoclimatic proxies for the period (cf. Rambeau 2010:5230–5233), and (4) a scalar mismatch between the relatively coarse temporal and spatial resolution offered by paleoenvironmental records available for the southern Levant and the spatial coverage and temporal precision of archaeological evidence. We summarize these challenges before turning to the local evidence from the Wadi el-Hasa.

Archaeological Patterns and Evidence

Sites dating to the Terminal Pleistocene and Early Holocene are widely distributed throughout the southern Levant, in many different ecozones, with their only common factor being an association with water sources. Major landscape changes since the Early Holocene, including both natural and anthropogenic, have likely both destroyed some sites and masked others. In addition, archaeological research has been very uneven over the region. These two factors combine to make it difficult to know whether spatial patterning of sites from the Early Holocene or earlier can be considered reliable.

Nevertheless, as no single occupation spanning the entire archaeological sequence from Early Natufian to Pre-Pottery Neolithic A (see Table 4.1 for a sketch of the relevant cultural chronology of the region) is known, it is necessary to examine temporal changes in aggregate. As individual sites

Table 4.1 Approximate cultural chronology of southern Jordan

Southern Levantine Cultural Chronology	
cal B.P.	*Period*
post-9000	Later PPNB and subsequent periods
10000	Middle Pre-Pottery Neolithic B (MPPNB)* (ca. 10,300–9,000)
11000	Late PPNA (ca. 10,800–10,300)
12000	Pre-Pottery Neolithic A (PPNA) (ca. 12,000–10,800)
	Final Natufian (ca. 12,500–12,000)
13000	Late Natufian (ca. 13,500–12,750)
14000	Early Natufian (ca. 14,900–13,700)
pre-15000	Epipaleolithic and earlier periods

Source: After Goring-Morris and Belfer-Cohen 2011 (pre-PPNA) and Finlayson and Makarewicz in press (PPNA and later)

*No EPPNB evident in southern Jordan.

represent different seasons of occupation, scales, and functions, such a process doubtless conflates some spatial and environmental diversity.

Intensive exploitation of cereals was not a phenomenon restricted to the Neolithic. Wild wheat and barley were staple food grasses during the Upper Paleolithic (ca. 23,000 cal B.P.) at Ohalo II, and the exploitation of low-ranked (i.e., having high collection and processing costs and low yields) small-grained grasses alongside larger-grained wild cereals suggests use of a broad spectrum gathering strategy that may have been instigated by low-level resource stress (Weiss et al. 2004). Such "broad spectrum" strategies, in which foragers incorporate a more diverse array of foodstuffs into their diets as high-ranked resources become more difficult to acquire (see summary in Codding and Bird 2015), are frequently invoked as forerunners of cultivation.

Against a background of such Late Pleistocene subsistence activity, the Early Natufian (15,000 to 13,000 cal B.P.) is key to understanding hunter-gatherer subsistence strategies on the eve of the Younger Dryas. This period is defined by substantial settlements that are largely limited in their distribution to the Galilee region (but see Richter 2014 and Jones et al., this volume) and noted for sizeable circular stone architecture, human interments, and the use of large groundstone mortars and sickle blades. Wild barley, goat-faced grass, lentils, lupines, and peas, as well as other plants and fruits, were collected during the Early Natufian (Hopf and Bar-Yosef 1987; Edwards et al. 2002), but the general absence of carbonized plant remains at most Early Natufian sites has challenged attempts to document the precise plant exploitation strategies in use during this critical period. The Late Natufian, which roughly coincides with the onset of the Younger Dryas, is generally understood as a mobile adaptation to the hypothesized colder and drier environmental conditions of the Younger Dryas. Late Natufian settlements were considerably smaller and occupied at lower intensities relative to their Early Natufian counterparts, and contained dwellings much reduced in scale. Interestingly, there is little evidence for faunal resource stress during the Late Natufian. Low-ranked small game were heavily exploited during the Early Natufian, but capture of higher-ranked small game actually increased during the Late Natufian, suggestive of a relaxation of hunting pressure due to increased mobility of forager groups. At the same time, the continued exploitation of juvenile gazelle suggests large game prey availability and hunting pressure remained unchanged during the Younger Dryas (Munro 2003).

First experimentation with pre-domestication cultivation is one of the defining characteristics of the subsequent Pre-Pottery Neolithic A (PPNA) period, beginning ca. 12000 cal B.P. The presence of large-sized barley (*Hordeum* sp.) grains typical for cultivars, barley rachis internodes exhibiting a smooth abscission scar associated with wild types, and "weeds of cultivation" at PPNA settlements strongly suggests that humans actively cultivated barley, although gathering wild plant foods was still intensively pursued (Edwards et al. 2004; Meadows 2005; White and Makarewicz 2012). The emergence of the PPNA is generally thought to coincide with the onset of Early Holocene pluvial conditions, but this chronological relationship is as much based on the hypothesized relationship as constitutive of it (cf. Blockley and Pinhasi 2011; Maher et al. 2011). Uncertainties in archaeological chronologies (stemming from the scarcity of deep and extensive excavations in Natufian and PPNA sites, the potential problems of old wood effects on ^{14}C dating in arid environments, and the inherent measurement uncertainty in ^{14}C dates) are thus small enough to suggest broad contemporaneity and relationships where it is not actually possible with available data to assess their reality (e.g., Grosman 2013), and large enough to leave questions of particular causal relationships frustratingly difficult to address (if not impossible to assess on purely chronological grounds). In the southern Levant, for instance, the PPNA may have commenced before the end of the Younger Dryas (Finlayson et al. 2011).

Paleoclimatic Data in the Southern Levant

While the broad parameters of climate in the Northern Hemisphere following the Last Glacial Maximum are well understood, the reconstruction of specific local climates for particular regions remains complex. In the case of the southern Levant, the use of climate modeling (e.g., Robinson et al. 2006; Brayshaw et al. 2011) is clarifying the relationships between the North Atlantic forcing factors of the Younger Dryas and the climate of the Eastern Mediterranean, and improving understandings of changes in (for instance) seasonality and inter-annual variability, but remains at a regional resolution (e.g., 50 km cells in the higher-resolution of the two examples cited above). The paleoclimate data available for the southern Levant specifically, primarily derived from the carbon and oxygen isotopic composition of speleothems (e.g., Verheyden et al. 2008; Bar-Matthews and Ayalon 2011), pollen spectra in sapropels and lake cores (e.g., Rossignol-Strick 1999; Bottema 2002; Yasuda et al. 2000), and Dead Sea and Lake Lisan lake level records (e.g., Frumkin and Elitzur 2002; Stein et al. 2010; Torfstein et al. 2013) are interpreted as regional records representative of spatially averaged southern Levantine environmental conditions. It is important to note, however, that there is sufficient variability in these records to suggest that impact of the Younger Dryas was not uniform across the southern Levant (Rosen 2007; Makarewicz 2012), and these records do not capture local expressions of climatic variability in the southern Levant, which is characterized by pronounced topographic diversity and, consequently, high variability in precipitation and temperature even over short distances. For example, although it involves an area only approximately 700 km on its long axis, the modern southern Levantine precipitation regime is characterized overall by decreasing precipitation levels along west-east and north-south gradients (Ben-Gai et al. 1998), and varying also according to altitude. This spatial variability, in conjunction with seasonal and interannual variability in temperature and precipitation levels, would have produced a mosaic of environments and attendant plant and animal resources available at any given moment in time in the southern Levant. As a result, climatic or environmental archives from one locale may not be reliable indicators of even nearby past climates and environments. Such spatial diversity, with its potentially differential responses to regional climatic patterns, underlies the error that Maher and colleagues describe as, "a tendency to assume that interstadials were universally welcome and that cooler periods, especially the Younger Dryas, were universally cold, dry, resource-poor and unpleasant" (2011:20).

Scale and Resolution

As discussed above, any model of cultural impacts of climatic and environmental change is faced with the challenges of articulating paleoenvironmental generalizations that are generally regional and local archaeological datasets that address landscape scales. As we use the terms here, *landscapes* are anthropocentric units of analysis defined by combinations of landforms, hydrology, and biotic communities, which determine the type and distribution of plant, animal, and water resources available for human exploitation, and whose extent is commonly a function of behavior (i.e., a human resource catchment, defined from the bottom-up; they are also freighted with various layers of meaning and social significance, and their extents and characteristics are dynamic). *Regions*, conversely, are top-down, defined without reference to inhabitants, and as such constitute mosaics of multiple landscapes. The Younger Dryas climate stress model is based on paleoenvironmental generalizations about the southern Levant, a geographical region rather than a landscape, and one characterized by sharp changes in topography, precipitation levels, and temperature that together contribute to wide variation in environmental conditions and biotic diversity over short distances (Rambeau 2010:5232).

In addition, relating patterns of archaeological data to paleoclimatic records remains a chronological challenge as well as a spatial one. Both archaeological and paleoclimatic datasets incorporate significant uncertainty (if not outright error; cf. Meadows 2005). Chronological data underpin arguments linking the Younger Dryas and early agriculture, but even setting aside uncertainties in paleoclimatic chronologies, the archaeological chronological data tends not to stand up to rigorous reexamination. For example, although it is still commonly referenced (e.g., Goring-Morris and Belfer-Cohen 2011; Grosman 2013), the model of the Late Natufian as a specialized adaptive response to the Younger Dryas is not particularly well supported by the chronological data, and rigorous approaches to ^{14}C data highlight the imprecision in dating and vulnerability to sampling effects that make establishing correlation difficult. Specifically, reanalysis of radiocarbon dates from archaeological sites across the region and the Soreq Cave speleothem record suggests that the emergence of highly mobile Late Natufian lifeways did not coincide with the Younger Dryas, as previously believed, but appeared earlier during the latest portion of the Bølling-Allerød and persisted into the Younger Dryas (Blockley and Pinhasi 2011), while Maher and colleagues (2011) point out that the various correlations of particular cultural changes in southwest Asia with climatic periods remain speculative at best, given the limited precision and quality of the chronological data on which they are based. Moreover, the sample on which such hypotheses are based is inevitably small, and the observed patterns are very sensitive to the addition of new data: in the Negev and southern Jordan, for instance, both material cultural and chronological evidence argue for a continuous local development from the Harifian variant of the Late Natufian to the early PPNA (Finlayson et al. 2011; Finlayson and Makarewicz, in press), undermining the chronological gap proposed by Blockley and Pinhasi (2011:106) between the onset of the PPNA at the beginning of the Holocene and end of the Late Natufian.

The sensitivity of Blockley and Pinhasi's conclusions to the addition of small amounts of additional data suggests that hypotheses of broad climate-culture relationships are driven more by general chronological correlation than by particular data about specific links. Even as Blockley and

Pinhasi dismiss the association of agriculture and the Younger Dryas, they suggest (2011:106) that the emergence of the PPNA largely coincides with the onset of pluvial conditions during the Early Holocene, writing that "the hypothesis to be tested is that climate change was the direct driver of the cultural innovation of the adoption of agriculture in the Southern Levant." The replacement of the Younger Dryas with the onset of the Holocene as a proposed environmental driver of early agriculture reflects the fact that these proposed relationships are based more on conviction that some relationship *should* exist than on specific evidence arguing for such a relationship.

One way to address these problems of explanation and scale is through the identification and exploration of paleoenvironmental archives that can be directly linked to human activity. Ultimately such an approach can enable finer-grained modeling of subsistence possibilities and local experiences of climatic change. The investigation of such an archive in the Middle Wadi el-Hasa, we argue here, challenges some of the basic assumptions about how the Younger Dryas impacted the lives of those inhabitants of the Hasa who first began experimenting with agriculture. The local character of environmental archives is often taken as a limitation, as researchers struggle to define and advocate the regional relevance and implications of their data. It can also, we suggest, be an advantage, capturing vital aspects of landscape-scale change that are otherwise lost to regional averaging.

El-Hemmeh and the Middle Wadi el-Hasa

Here, we investigate the impact of the Younger Dryas at a local scale through an analysis of Late Pleistocene and Early Holocene paludal sediments located in the Middle Wadi el-Hasa, Jordan, and closely associated with the Early Neolithic (Pre-Pottery Neolithic A) site of el-Hemmeh (see Figures 4.1 and 4.2). The settlement of el-Hemmeh, currently the subject of ongoing archaeological investigation directed by one of us (Makarewicz), consists of a series of free-standing and semi-subterranean circular, oval, and irregular structures (see Figure 4.2, inset) constructed of stone and *pisé* (a mix of clay, water, and plant temper). To date, 15 structures have been identified; stratigraphic relationships indicate that not all structures were contemporaneous in construction or occupation. Occasionally, architectural features contained within structures were remodeled and renewed, and earlier structures were often partially dismantled to allow subsequent construction. Floors within individual structures were regularly renewed, with new surfaces installed after periods of brief, perhaps seasonal or annual, abandonment. After falling into disuse, some structures were used as middens while others were allowed to simply collapse. These structures were also associated with extramural spaces, which served as both middens and informal activity areas (Makarewicz and Rose 2011).

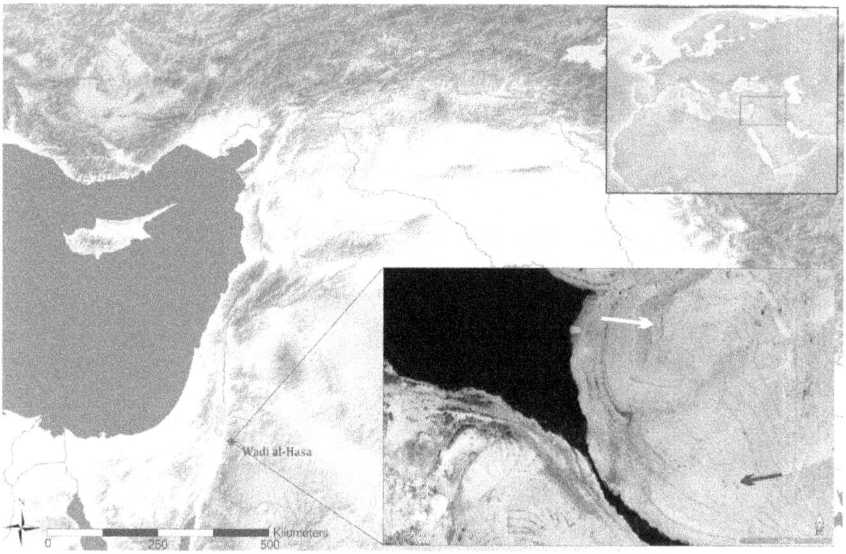

Figure 4.1 Location of the Wadi el-Hasa on the Jordan Plateau. The modern channel where it enters the Tannur Reservoir, along with the excavations of the PPNA area (white arrow) and paludal deposits (black arrow), are visible in the inset 2015 Google Earth image.

Figure 4.2 View east across the Tannur Reservoir, up the Wadi el-Hasa. Black and white arrows mark the locations of PPNA Hemmeh and the paludal deposits described here, respectively. The inset shows an excavated portion of the PPNA site.

Source: Photo by Zak Alnaimat

Animal exploitation strategies pursued at el-Hemmeh appear to have been broadly similar to those at many other southern Levantine PPNA settlements (e.g., Tchernov 1994; Horwitz 2010). Preliminary zooarchaeological analyses of the el-Hemmeh faunal assemblage indicate that a broad spectrum of animal resources native to a variety of biomes were exploited including wild goat, gazelle, aurochsen, boar, and waterfowl. Initial paleobotanical analyses indicate wild plants including lentils, vetch, and fig were commonly exploited. Predomestication cultivation was also practiced at el-Hemmeh, evidenced by the presence of large barley grains, predominance of smooth abcision scars on barley rachis internodes, and a high abundance of weedy taxa commonly associated with cultivation plots (White and Makarewicz 2012).

A single radiocarbon determination obtained from a hearth located inside one of the circular stone structures suggests that el-Hemmeh was occupied during the later portion of the PPNA (ca. 8800–8600 cal B.P.) (Makarewicz et al. 2006); additional dates are forthcoming. Lithic techno-typological analyses and architectural style further support a late PPNA occupation. The absence of el-Khiam points and Hagdud truncations, two tool types commonly found at occupations dating to the early half of the PPNA, and the presence of stone-built, above-ground architecture at el-Hemmeh both parallel developments observed in the occupations of Wadi Faynan 16 (Trench 3) and Zaharat adh-Dhra 2 (ZAD 2) to form what appears to be a distinct Late PPNA phase (Finlayson and Makarewicz in press; Smith et al. in press).

El-Hemmeh is located on a dissected alluvial fan above the modern-day floodplain of the Wadi el-Hasa, which today is a deeply incised watercourse draining the central Jordanian Plateau into the Dead Sea Basin. Modern precipitation levels throughout the Hasa are quite low and on average reach only 80 mm per annum (Attewill and Humphreys 1996). As a result, water levels and flow within the drainage vary markedly interannually and are strongly dependent on winter precipitation levels. The steep slopes of the Wadi el-Hasa are largely devoid of vegetation (see Figure 4.2), but this may owe something to overgrazing as well as to generally arid conditions; in the absence of exclosures protecting wild vegetation from grazing sheep and goats, it is difficult to evaluate the precise impact of either.

Small in-stream wetlands supporting abundant *Phragmites* growth are scattered throughout the wadi bottom where sufficient perennial surface- or groundwater is available. The modern hydrologic regime is significantly altered by diversion of surface water and groundwater mining for small-scale agriculture throughout the Hasa, and by the Tannur Dam, completed in 2001. The former certainly contributes to the present aridity of even the floodplain above the dam, while dam release has created stable and perennial (if low) flows below the dam, promoting formation of in-stream wetlands.

During the Late Pleistocene and Early Holocene, the topography of the Wadi el-Hasa was significantly different from the modern deeply incised drainage. A series of studies have addressed the question of exactly how different, and investigated the alluvial history of the drainage, confronting the basic problem of defining the number of aggradation and incision events and determining their chronology. Vita-Finzi and Copeland (Vita-Finzi 1966; Copeland and Vita-Finzi 1978), in pioneering work on Late Quaternary alluvial history in the Mediterranean generally, proposed a chronosequence of alluvial fills in the Wadi el-Hasa dated on the basis of associations of alluvial fills with chronologically diagnostic material culture. Their scheme (summarized in Copeland and Vita-Finzi 1978:23 and Fig. 1) suggested the existence of four post-LGM fills, with three of them dating to the Holocene, and guided subsequent archaeological interpretations of the drainage (e.g., Schuldenrein and Clark 2003; Hill 2006). Schuldenrein (2007) subsequently suggested some revision of the sequence, arguing that the latter two fills (III and IV, in Copeland and Vita-Finzi's scheme) both dated to the last ~4,500 years (rather than the last 6,000), and that Fill IV had accumulated over the last 1,000 years (rather than the last 2,000). As we have argued elsewhere (Contreras et al. 2014:41), evidence from the Middle Hasa shows that this significantly oversimplifies Late Holocene alluvial history. In addition, our recent work, along with that of Jason Rech and Emily Winer (Winer 2010; Contreras et al. 2014), is demonstrating that the Late Pleistocene and Holocene history of incision and aggradation was significantly more complicated than previously reported (e.g., Vita-Finzi 1966; Copeland and Vita-Finzi 1978; Hill 2006; Schuldenrein 2007). With regard to the Terminal Pleistocene and Early Holocene, Schuldenrein (2007:564) points out that "there is no evidence of Early Holocene or even Middle Holocene landforms in the alluvial bottoms of the Wadi Hasa." As we discuss below, the description of an alluvial landform dating to this period demonstrates that Holocene erosion has effaced evidence of an additional cycle of aggradation and incision, described below.

The date of the initial incision of Fill I (Unit C in Winer's scheme; i.e., the massive alluvial fill deposited in the Pleistocene), and the subsequent aggradation/incision dynamics of the Middle Hasa, are vital to understanding not only the ages and locations of eroded deposits (which structure site survival and inferred settlement patterns) and the ages of surfaces accessible to archaeological survey and research (which structure survey results), but also the environments available to past inhabitants of the Hasa. The paludal deposits described below, in addition to providing an archive of the local environment during and immediately following the Younger Dryas, provide critical evidence about the post-Unit C (Later Pleistocene) cut-and-fill activity of the Hasa.

Paludal Deposits in the Middle Wadi el-Hasa

A geoarchaeological survey beginning in 2010 identified an in-stream wetland deposit in the Wadi el-Hasa, located only 160 m SE of el-Hemmeh (see Figure 4.1, inset). Three 1 × 2 m exposures were excavated on what was determined to be an eroded fragment of a Late Pleistocene/Early Holocene alluvial terrace; the uppermost surface of these is ~10 m above the current watercourse, which is today incised approximately 6 m into limestone (the Late Cretaceous Karak Limestone formation, according to Tarawneh 1988) bedrock approximately 60 m to the south. These descend down the eastern (eroded) flank of the deposit, overlapping to capture a combined ~5.5 m of stratigraphy (Figure 4.3). In spite of the repeated aggradation and incision within the Wadi el-Hasa, which has effaced much of the Late Pleistocene and Early Holocene stratigraphic record (see above), this preserved alluvial terrace fragment provides a stratigraphic record of the Hasa floodplain.

The alluvial sands and gravels characteristic of most of the modern, episodically inundated, floodplain are absent from these deposits, which instead consist of paludal material: greyish silts incorporated into organic-rich black mat deposits (cf. Quade et al. 1998; Pigati et al. 2014), tufa deposits accreted around organic matter (e.g., Phragmites), and marls, interspersed in strata ranging from sub-cm to ~20 cm thickness (see Figure 4.3). Throughout ~5.5 m of stratigraphy exposed in the three excavations, spanning an estimated 2,000–4,000 years, these strata are uninterrupted by alluvial, colluvial, or aeolian sediments, with the exception of one small colluvial wedge intruding into the easternmost, upslope exposure.

In addition to the black mats—stratigraphic indicators of organic matter deposited in anoxic conditions—macroscopic evidence of Phragmites in tufa casts and carbonized vegetation suggests that these deposits formed under wet and marshy conditions. Ostracod analyses of a selection of strata confirm that the depositional environment was a shallow wetland (see Contreras et al. in prep). A modern analogue of such depositional conditions can be found in stands of Phragmites situated in perennially wet and often subaqueous organic-rich sediments, located approximately 3 km downstream from the sample site.

The middle portion of this deposit is now securely dated to 13000–12000 cal B.P. (based on four ^{14}C dates on carbonized organic matter from black mats or sealed in tufa accretions in intact deposits; see Figure 4.3). The extrapolated age-depth curve (see Figure 4.3 and Contreras et al. in preparation) suggests that deposition began ~14000 cal B.P. and continued until ~10500 cal B.P., suggesting the presence of in-stream wetlands before, during, and after the Younger Dryas—and likely overlapping the period of PPNA occupation at el-Hemmeh (as noted above, a series of ^{14}C dates suggest that the site was occupied from ~10800 cal B.P. until ~10600 cal B.P.). The recovery of four lithic artifacts from strata estimated to date between 13000 and 11000 cal B.P. (see Figure 4.3) provides further evidence of the interaction of the Hasa's human inhabitants with this wetland environment, and the presence of waterfowl in the faunal assemblage from el-Hemmeh indicates exploitation of this environment as well.

These paludal deposits also provide a dated post-LGM sequence of aggradation in the middle Hasa, the key ingredient to describing a significantly more complex sequence of aggradation and incision than those previously proposed. This added complexity has important implications for understanding the environments and resources available to Terminal Pleistocene and Early Holocene inhabitants of the Middle Hasa and changes therein.

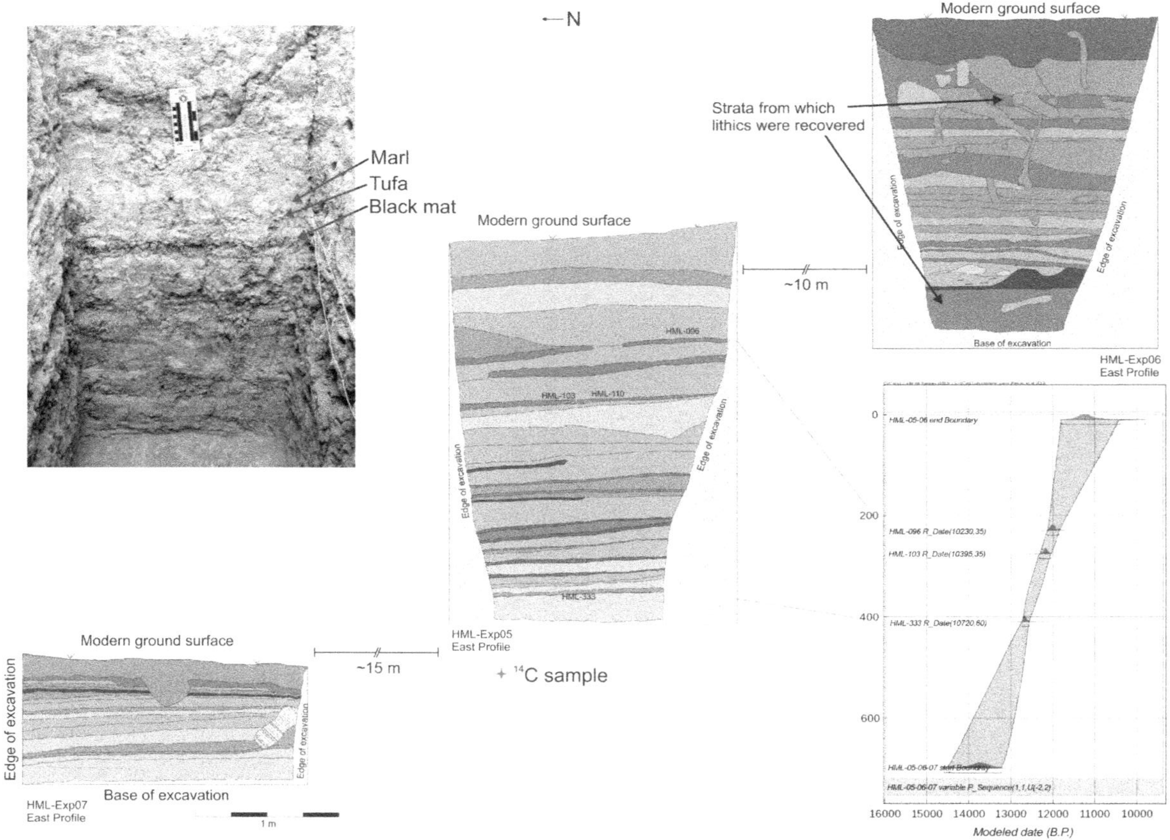

Figure 4.3 Composite stratigraphic profile of the paludal deposits; note that excavation did not reach the base of these, estimated (based on bedrock exposed in the channel cut to the south) to be ~1.5 m deeper. The age-depth curve (lower right) includes four ^{14}C dates from carbonized vegetation sealed in tufa accretions (see Contreras et al. in preparation for detail).

The consensus view (Copeland and Vita-Finzi 1978; Schuldenrein 2007; Winer 2010) is that Fill I/Unit C is the product of Pleistocene aggradation, while differences coalesce over the timing of the post-Unit C incision, and the number, magnitude, and timing of subsequent cut-and-fill cycles. While the broad parameters of the prior scenarios remain intact, three archaeologically significant revisions are salient: (1) the massive Late Pleistocene terrace that is preserved in select areas (Fill I according to Copeland and Vita-Finzi, and later Schuldenrein, and Unit C in Winer's scheme) had finished aggrading by something closer to 27 kya than 15 kya (Winer 2010:21), (2) Terminal Pleistocene and Early Holocene dynamics (post-Unit C) included incision to near-modern levels, approximately 10 m of aggradation, and renewed incision to near-modern levels, all by the Early Holocene, and (3) the generalized chronosequence masks considerable complexity in the later Holocene (see Contreras et al. 2014:41).

The most recent chronological information (Winer 2010:18–22), coupled with the dates described here (and in more detail in Contreras et al. in preparation), suggest aggradation (forming Unit C) until ~27 kya, followed by incision to near-modern levels before ~15 kya, and then aggradation of the paludal deposits described here, characterized by ~6–10 m of accumulation until incision began at an estimated 10500 B.P. (the least well-constrained of these dates). This incision beginning ~10500 B.P. may be contemporary with Fill II/Unit D or may predate it; *all* of the alluvial activity encompassed by Fills III and IV postdates this incision. This is reasonably compatible with Winer's (2010:22) implication that Holocene aggradation (of what she terms Unit E) began ~9200 B.P. (particularly as that date, on bulk sediment, is a *terminus post quem*).

Most important for archaeological interpretation of el-Hemmeh and the early Neolithic, this datum signaling an aggrading alluvial plain during the Terminal Pleistocene/Early Holocene in the Middle Hasa also testifies to the height of the valley floor when Hemmeh was occupied. Situated approximately 10 m above the modern valley floor, these paludal deposits imply a much broader watered floodplain during the Younger Dryas and Early Holocene relative to that seen today. They demonstrate that the modern landscape (see Figure 4.2), even discounting the reservoir, is actually a very poor guide to the Late Pleistocene/Early Holocene landscape: its aridity is the product in part of modern land-use practices, and its morphology has been significantly altered by Holocene incision. Using the elevation (relative to the modern watercourse) of the deposits described here as an indicator of the elevation of the Late Pleistocene/Early Holocene floodplain and a simple nearest neighbor interpolation to generate a surface at this elevation, reconstruction of a paleosurface DEM shows that within a 2-km radius of Hemmeh, such a change in valley topography is the difference between ~22 and ~130 ha of floodplain (see Figure 4.4).

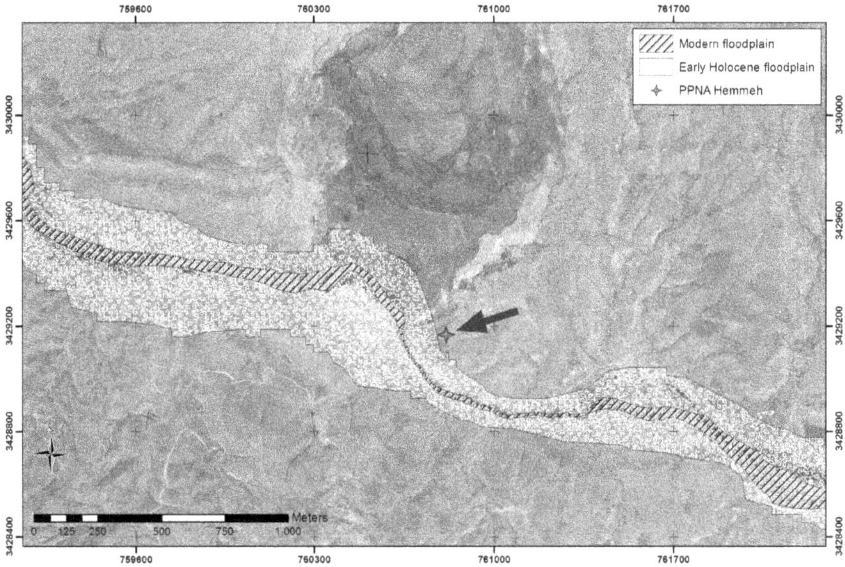

Figure 4.4 Contrasting modern (black) and reconstructed Early Holocene (white) floodplains. Hemmeh's location is indicated with the black arrow. The underlying aerial photograph is from the 1953 Hunting Aerial Survey, predating the flooding of the area by the Tannur Reservoir.

Discussion

The data presented here strongly suggest that either the aridity of the Younger Dryas in the southern Levant has been previously over-emphasized or, perhaps more likely, that aridity associated with the Younger Dryas was not necessarily uniform across the region (in fact the model of a cold and dry Younger Dryas itself has recently been questioned; cf. Enzel et al. 2008; Stein et al. 2010; Torfstein et al. 2013). The presence of a continuously wet floodplain in the Wadi el-Hasa throughout the Younger Dryas, as well as before and possibly after, strongly suggests consistent levels of at least some precipitation within the Hasa catchment throughout the Younger Dryas and into the early Holocene. More to the point with regard to archaeological interpretation, whatever increase in aridity was associated with the Younger Dryas, wetland environments remained accessible to the inhabitants of the Hasa throughout.

If, in the case of the Middle Hasa, wetland resources remained accessible, it may no longer be tenable to consider the emergence of cultivation as a subsistence strategy as a response to resource stress—at least at the local scale. If the Younger Dryas was a wet period (or at least wetter than regional averages suggest) within certain areas such as the Wadi el-Hasa, the distinctions between the Bølling-Allerød, Younger Dryas, and the Early Holocene in those areas are severely eroded. The blurring of these environmental distinctions severely undermines models linking environmental deterioration with resource scarcity and subsequent development of plant cultivation. The evidence from the Middle Hasa argues that if environmental changes had an effect on subsistence strategies of particular human communities, it would have been through changes in regional patterning—e.g., increasingly spatially circumscribed resources, heightened patchiness—rather than through direct impacts on local resource returns. *Whether* the Younger Dryas had such effects on regional patterning and diversity remains to be explored and will require environmental reconstructions with high spatial and temporal resolution, as well as regional data on well-dated settlement pattern dynamics.

One vital and historically under-appreciated element may be the floodplains of incised wadi systems like the Hasa. As discussed above, the Hasa was not only wetter than expected; its topography was also significantly different. As subsequent cut-and-fill cycles have largely removed floodplain deposits from drainage systems like the Hasa, the remaining landforms with surfaces dating to the Terminal Pleistocene and Early Holocene are, in order of frequency, wadi slopes (above the range of cut/fill cycles), apices of alluvial fans, and (rarely) remnant alluvial terraces. That this preservation bias may skew reconstructions of settlement patterns has long been recognized (e.g., Maher 2011). We would suggest that, in tandem with an over-reliance on regional generalizations about paleoenvironments, this preservation bias has also structured reconstructions of the environments and resources available to inhabitants.

A broader and wetter Hasa floodplain would not only have provided fertile, well-watered land eminently suitable for experimentation with cultivation, but also an area that was less susceptible to arid conditions caused by either annual variation in precipitation or regional climatic shifts. In arid and semi-arid environments, floodplains can be vital resources for foragers and cultivators alike. The instability of floodplains is more than offset by their productivity, and they provide “high values for resource diversity, productivity, and reliability” (Nicholas 1998:720). For cultivators, floodplains offer locations in which viable and intensifiable agriculture may be carried out without infrastructural investments in irrigation or landscape modification (Doolittle 2006). Indeed, previous work has suggested the importance of floodplains as locations of early cultivation (e.g., Sherratt 1980; Bintliff et al. 2006; Roberts and Rosen 2009).

The prominence of in-stream (floodplain) wetlands easily accessible from el-Hemmeh throughout the Younger-Dryas and into the period in which the site’s inhabitants began to experiment with cultivation demonstrates that there was no dramatic change in the local environment associated with the Younger Dryas, and shows that these experiments took place in the context of a floodplain approximately six times larger than the modern one.

These local scale developments have important implications for understanding the adoption of cultivation in the Hasa. The spatial variability within such floodplains, provided by shifting channels and water levels that varied seasonally and interannually, may have served to buffer against climate shifts: while particular locations might change, a mosaic of microenvironments would have remained available. Even absent the stability that Willcox (2005) argues was necessary for the development of agriculture, particular microenvironments would have been predictably available (except in case of heightened competition for a reduced number of valuable patches, but such overall reduction remains to be evaluated). In such a context, foraging returns would have remained relatively stable and nascent cultivation strategies would have been optional rather than desperate; in any case it is the Younger Dryas' impact at these local scales that would have affected foraging returns and influenced subsistence strategies.

Conclusion

The hypothesized relationship between the onset of the Younger Dryas and the origins of plant agriculture has long served as an example of how major climate change can contribute to, indeed even force, human adaptive responses. In the case of the Near East, older models draw heavily from regional-scale palaeoenvironmental records to emphasize the Younger Dryas as a cold, arid event that evenly impacted the diverse environments and topographies of the Near East to create conditions of resource stress that provided the impetus for the shift to agriculture. Although critiques of models linking the Younger Dryas to emergent agriculture have generally focused on chronology (i.e., whether in fact the onset or end of the Younger Dryas correlates chronologically with the emergence of agriculture at given sites or in the region generally) or on broad theories of forager behavior (i.e., whether the effects of the Younger Dryas were sufficient to force behavioral change), the local paleoenvironmental evidence from the Middle Wadi el-Hasa suggests that *scale* and *spatial variability* may be challenges as important as chronology.

Resource stress models are implicitly grounded in a logic of *local* environmental effects: their explanatory power is drawn from the implied consequences for everyday human activity in particular places. The mismatch between the local logic of such models and the coarse-grained regional paleoenvironmental data on which they draw means that the very data upon which resource stress models are built cannot provide the requisite spatial or temporal resolution to flesh out the mechanisms suggested by the models. What such models demand is an understanding of climate-driven environmental changes, and related shifts in food resources within the particular niches inhabited by foragers. In the Levant, such niches of course reflect the region's environmental variability, and as others (e.g., Maher et al. 2011) have pointed out, we should not expect any less diversity in the region's past than can be observed today. Climatic shifts were experienced at local scales, and links between climatic drivers and human behavior must have been articulated through very local experiences of such change—i.e., shifts in available water sources, distributions of edible plants, availability of game, etc.

Continued collaborative work on sedimentology, ^{18}O isotopes in tufas as climate proxies, and microfauna will refine our picture of climate and environment, letting us address this question of local, human experience of environment and environmental change. Even while these paleoenvironmental studies are still in progress, this evidence of Early Holocene landscape morphology in the Middle Hasa has already redefined our understanding of the environmental context of early experimentation with cultivation at Hemmeh. Examining such a local scale in the Middle Wadi el-Hasa suggests new ways of thinking about early experiments with plant cultivation. The continuous presence of in-stream wetlands argues that in experimenting with cultivation, hunter-gatherers intentionally manipulated a stable resource base that was readily available to them throughout the periods in question. This highlights that risk-reduction in the strict sense may not have been a factor in the development of cereal cultivation. Moreover, the sustained presence of wetlands and associated resources meant that hunter-gatherers accumulated knowledge of the local biotic community over a long period of time and, consequently, nascent cultivation was not necessarily a high-risk endeavor that entailed manipulation of a totally unknown resource. Instead, hunter-gatherers already familiar with the exploitation of local plant resources were able to modify those resources in a relatively stable environment that provided a secure foundation for experimentation.

In broad terms, then, looking more closely at human-environment interactions at the dawn of agriculture is suggesting that we need to model paleoenvironments in greater complexity and greater detail, at more local scales. Addressing environmentally driven changes in such parameters requires paleoenvironmental/paleoclimatic proxies local to the archaeological sites that would have experienced such effects—potentially also in order to establish a relationship between local and regional climates so that regional records may be employed to consider local conditions. In order to link environmental shifts to cultural changes by characterizing particular mechanisms rather than simply through hopeful juxtaposition of broadly contemporaneous trends, both spatial specificity and chronological resolution need to be sufficient—or at least somehow downscalable—to address human experience. Paleoenvironmental archives local to archaeological sites from the key periods of the Late Pleistocene and Early Holocene, like those we have described here, offer a promising means of exploring such links.

Acknowledgements

The authors gratefully acknowledge the contributions of Jason Rech, Steffen Mischke, and Regina Gonda to the research presented here, and thank them for the productive intellectual exchanges that formed part of the process. We would also like to thank the Jordanian Department of Antiquities and the members of the 2010–2014 el-Hemmeh excavation teams for making the fieldwork possible, and the Alexander von Humboldt Foundation for supporting Contreras' stay in Germany during which this work was carried out. The aerial photograph used as a background image in Figure 4.4 was kindly provided by the Aerial Photographic Archive for Archaeology in the Middle East (APAAME).

Methodological Vignette—Paleolandscape Reconstruction in Archaeology

Reconstruction of landscape morphology is a useful tool for examining past human-environment interactions generally, and a vital one in contexts of significant post-occupation landscape change, whether climatogenic or anthropogenic. With the computational capacity for interpolating surfaces from discontinuous point data now well within the reach of laptop computers, a variety of geostatistical tools make the data processing possible (surface interpolation is a standard tool in ArcGIS, GRASS, QGIS, and Surfer, for example, and may be managed also in R). Data can be managed in either local or global coordinate systems, and various interpolation algorithms suited to the character of the data available (e.g., constraining the resultant surface more or less closely to the observed datapoints and specifying varying degrees of spatial autocorrelation (Wheatley and Gillings 2002; Bivand et al. 2008; Hengl 2009)).

Such geospatial tools allow integration of disparate data (from, e.g., exposures documented in field geoarchaeological and geomorphic survey, remote-sensing data, and previously published geological and archaeological observations) and offer a specifiable and replicable means of overcoming the interpretive limitations of discontinuous data. This can enable, for instance, assessment of past subsistence potential (as in the floodplain modeling employed in this chapter) or estimates of the scale of anthropogenic and/or geomorphic landscape modification (e.g., Katsianis 2004; Contreras 2009). Taking advantage of them, however, requires spatially explicit field data that can identify contemporaneous landscape features on-, near-, and off-site, and relating such features to human activity generally requires some absolute chronologies and dating programs as well.

References Cited

Abbo, Shahal, Simcha Lev-Yadun and Avi Gopher 2010 Yield stability: An agronomic perspective on the origin of Near Eastern agriculture. *Vegetation History and Archaeobotany* 19(2):143–150.

Allaby, Robin G, Dorian Q. Fuller and Terence A. Brown 2008 The genetic expectations of a protracted model for the origins of domesticated crops. *Proceedings of the National Academy of Sciences* 105(37):13982–13986.

Attewill, L. J. S. and Howard Humphreys 1996 Wala, Mujib and Tannur Dams, Jordan. In *The Reservoir as an Assset: The British Dam Society*, pp. 128–136. Thomas Telford Publishing, London.

Bar-Matthews, M. and A. Ayalon 2011 Mid-Holocene climate variations revealed by high-resolution speleothem records from Soreq Cave, Israel and their correlation with cultural changes. *The Holocene* 21(1):163–171.

Bar-Yosef, Ofer 1998 The Natufian culture in the Levant, threshold to the origins of agriculture. *Evolutionary Anthropology: Issues, News, and Reviews* 6(5):159–177.

——— 2001a From sedentary foragers to village hierarchies: The emergence of social institutions. *Proceedings of the British Academy* 110:1–38.

——— 2001b The world around Cyprus: From Epi-Paleolithic foragers to the collapse of the PPNB civilization. In *The Earliest Prehistory of Cyprus: From Colonization to Exploitation*, edited by Stuart Swiny, pp. 129–164. Cyprus American Archaeological Research Institute Monograph Series. American Schools of Oriental Research, Boston.

——— 2009 Social changes triggered by the Younger Dryas and the early Holocene climatic fluctuations in the Near East. In *The Archaeology of Environmental Change: Socionatural Legacies of Degradation and Resilience*, edited by Christopher T. Fisher, J. Brett Hill and Gary M. Feinman, pp. 192–208. University of Arizona Press, Tucson.

——— 2011 Climatic fluctuations and early farming in west and east Asia. *Current Anthropology* 52(S4):S175–S193.

Ben-Gai, T., Arie Bitan, A. Manes, P. Alpert and S. Rubin 1998 Spatial and temporal changes in rainfall frequency distribution patterns in Israel. *Theoretical and Applied Climatology* 61(3–4):177–190.

Bintliff, John, Emeri Farinetti, Kalliope Sarri and Renato Sebastiani 2006 Landscape and early farming settlement dynamics in central Greece. *Geoarchaeology* 21(7):665–674.

Bivand, Roger S, Edzer J. Pebesma and Virgilio Gómez-Rubio 2008 Interpolation and geostatistics. In *Applied Spatial Data Analysis with R*, edited by Roger S. Bivand, Edzer J. Pebesma and Virgilio Gómez-Rubio, pp. 213–261. Springer-Verlag, New York.

Blockley, S. P. E. and R. Pinhasi 2011 A revised chronology for the adoption of agriculture in the Southern Levant and the role of Lateglacial climatic change. *Quaternary Science Reviews* 30(1–2):98–108.

Bocquet-Appel, Jean-Pierre and Ofer Bar-Yosef 2008 Explaining the neolithic demographic transition. In *The Neolithic Demographic Transition and Its Consequences*, edited by Jean-Pierre Bocquet-Appel and Ofer Bar-Yosef, pp. 35–55. Springer, Dordrecht.

Bottema, Sytze 2002 The use of palynology in tracing early agriculture. In *The Dawn of Farming in the Near East: Studies in Early Near Eastern Production,*

Subsistence, and Environment, edited by R. T. J. Cappers and Sytze Bottema, pp. 27–38. Ex oriente, Berlin.

Brayshaw, D. J., Claire M. C. Rambeau and S. J. Smith 2011 Changes in Mediterranean climate during the Holocene: Insights from global and regional climate modelling. *The Holocene* 21(1):15–31.

Cauvin, Jacques 2000 *The Birth of the Gods and the Origins of Agriculture*. Cambridge University Press, Cambridge.

Codding, Brian F. and Douglas W. Bird. 2015 Behavioral ecology and the future of archaeological science. *Journal of Archaeological Science* 56:9–20.

Contreras, Daniel A. 2009 Reconstructing landscape at Chavín de Huántar, Perú: A GIS-based approach. *Journal of Archaeological Science* 36(4):1006–1017.

Contreras, Daniel A., Cheryl Makarewicz and Jason Rech. In preparation. "What Do We Mean by Dry? The Younger Dryas in the Middle Wadi el-Hasa."

Contreras, Daniel A., Vincent Robin, Regina Gonda, Rachel Hodara, Marta Dal Corso and Cheryl A. Makarewicz 2014 (Before and) After the flood: A multiproxy approach to past floodplain usage in the middle Wadi el-Hasa, Jordan. *Journal of Arid Environments* 110(C):30–43.

Copeland, L. and C. Vita-Finzi 1978 Archaeological dating of geological deposits in Jordan. *Levant* 10:10–25.

Cordova, Carlos E. 2008 Floodplain degradation and settlement history in Wadi al-Wala and Wadi ash-Shallalah, Jordan. *Geomorphology* 101(3):443–457.

Doolittle, William E. 2006 Agricultural manipulation of floodplains in the southern Basin and Range Province. *CATENA* 65(2):179–199.

Edwards, Phillip C., John Meadows, Ghattas Sayej and Mary C. Metzger 2002 Zahrat Adh-Dhra' 2: A new pre-pottery neolithic a site on the Dead Sea plain in Jordan. *Bulletin of the American Schools of Oriental Research* 327:1–15.

Edwards, Phillip C, John Meadows, Ghattas Sayej and Michael Westaway 2004 From the PPNA to the PPNB: New views from the Southern Levant after excavations at Zahrat adh-Dhra'2 in Jordan. *Paléorient*:21–60.

Enzel, Yehouda, Rivka Amit, Uri Dayan, Onn Crouvi, Ron Kahana, Baruch Ziv and David Sharon 2008 The climatic and physiographic controls of the eastern Mediterranean over the late Pleistocene climates in the southern Levant and its neighboring deserts. *Global and Planetary Change* 60(3–4):165–192.

Finlayson, Bill and Cheryl Makarewicz In press The neolithic of Southern Jordan. In *Quaternary of the Levant*, edited by Ofer Bar-Yosef and Yehouda Enzel. Cambridge University Press, Cambridge.

Finlayson, Bill, Steven Mithen and Samuel J. Smith 2011 On the edge: Southern Levantine Epipaleolithic-Neolithic chronological succession. *Levant* 43:127–138.

Frumkin, Amos and Yoel Elitzur 2002 Historic dead sea level fluctuations calibrated with geological and archaeological evidence. *Quaternary Research* 57(3):334–342.

Fuller, Dorian Q, George Willcox and Robin G Allaby 2011 Cultivation and domestication had multiple origins: Arguments against the core area hypothesis for the origins of agriculture in the Near East. *World Archaeology* 43(4):628–652.

Goring-Morris, A Nigel and Anna Belfer-Cohen 2011 Neolithization processes in the Levant. *Current Anthropology* 52(S4): S195–S208.

Grosman, Leore 2013 The Natufian chronological scheme—new insights and their implications. In *Natufian Foragers in the Levant: Terminal Pleistocene Social Changes in Western Asia*, edited by Ofer Bar-Yosef and François R Valla, pp. 1–22. International Monographs in Prehistory, Ann Arbor, MI.

Grosman, Leore and Anna Belfer-Cohen 2002 Zooming onto the "Younger Dryas." In *The Dawn of Farming in the Near East: Studies in Early Near Eastern Production, Subsistence, and Environment*, edited by R. T. J. Cappers and Sytze Bottema, pp. 49–54. ex oriente, Berlin.

Gustafson, Perry, Manfred Heun, Sylvi Haldorsen and Kari Vollan 2008 Reassessing domestication events in the Near East: Einkorn and Triticum urartu. *Genome* 51(6):444–451.

Hayden, Brian 2009 The proof is in the pudding: Feasting and the origins of domestication. *Current Anthropology* 50(5):597–601.

——— 2011 Feasting and social dynamics in the epipaleolithic of the Fertile Crescent. In *Guess Who's Coming to Dinner: Feasting Rituals in the Prehistoric Societies of Europe and the Near East*, edited by G. Aranda, S. Monton-Subias, and M. Sanchez, pp. 30–63. Oxbow Books, Oxford.

Hengl, Tomislav 2009 *A Practical Guide to Geostatistical Mapping*. Office for Official Publications of the European Communities, Luxembourg.

Hill, J. Brett 2006 *Human Ecology in the Wadi el-Hasa: Land Use and Abandonment through the Holocene*. University of Arizona Press, Tucson.

Hillman, Gordon, Robert Hedges, Andrew Moore, Susan Colledge and Paul Pettitt 2001 New evidence of Lateglacial cereal cultivation at Abu Hureyra on the Euphrates. *The Holocene* 11(4):383–393.

Hopf, Maria and Ofer Bar-Yosef 1987 Plant remains from Hayonim cave, western Galilee. *Paléorient* 13(1):117–120.

Horwitz, Leora 2010 Fauna from the sites of Gilgal I, II and III. In *Gilgal: Early Neolithic Occupations in the Lower Jordan Valley*, edited by Ofer Bar-Yosef, Nigel Goring-Morris, and Avi Gopher, pp. 263–295. Oxbow Books, Oxford.

Katsianis, M. 2004 Stratigraphic modelling of multi-period sites using GIS: The case of neolithic and early Bronze age Knossos. *BAR International Series* 1227:304–307.

Maher, Lisa A. 2011 Reconstructing paleolandscapes and prehistoric occupation of Wadi Ziqlab, Northern Jordan. *Geoarchaeology* 26(5):1–44.

Maher, Lisa A., Edward Bruce Banning and Michael Chazan 2011 Oasis or Mirage? Assessing the role of abrupt climate change in the prehistory of the Southern Levant. *Cambridge Archaeological Journal* 21(1):1–30.

Makarewicz, Cheryl A. 2012 The Younger Dryas and hunter-gatherer transitions to food production in the Near East. In *Hunter Gatherer Behavior: Human Response During the Younger Dryas*, edited by Metin I. Eren, pp. 195–230. Left Coast Press, Walnut Creek, CA.

Makarewicz, Cheryl A. and Kate Rose 2011 Early pre-pottery neolithic settlement at el-Hemmeh: A survey of the architecture. *Neo-Lithics* 11(1):19–25.

Makarewicz, Cheryl, Nathan Goodale, P. Rassman, Chantel White, H. Miller, J. Haroun, E. Carlson, A. Pantos, M. Kroot, S. Kadowaki, A. Casson, J. T. Williams, A. E. Austin, and B. Fabre 2006 El-Hemmeh: A multi-period pre-pottery neolithic site in the Wadi el-Hasa. Jordan. *Eurasian Prehistory* 4(1–2):183–220.

Meadows, John 2005 The Younger Dryas episode and the radiocarbon chronologies of the Lake Huleh and Ghab Valley pollen diagrams, Israel and Syria. *The Holocene* 15(4):631–636.

Mithen, Steven 2006 *After the Ice: A Global Human History, 20,000–5000 BC*. Harvard University Press, Boston.

Moore, A. M. T. and Gordon C. Hillman 1992 The pleistocene to Holocene transition and human economy in Southwest Asia: The impact of the Younger Dryas. *American Antiquity* 57(3):482–494.

Munro, Natalie D. 2003 Small game, the Younger Dryas, and the transition to agriculture in the southern Levant. *Mitteilungen der Gesellschaft für Urgeschichte* 12(4):47–71.

Nicholas, George P. 1998 Wetlands and hunter-gatherers: A global perspective. *Current Anthropology* 39(5):720–731.

Pigati, Jeffrey S., Jason A. Rech, Jay Quade and Jordon Bright 2014 Desert wetlands in the geologic record. *Earth Science Reviews* 132(C):67–81.

Pumpelly, Raphael 1908 *Explorations of Turkestan-Prehistoric Civilizations of Anau. Origins, Growth, and Influence of Environment.* Carnegie Institution of Washington, Washington, DC.

Quade, Jay, Richard M. Forester, William L. Pratt and Claire Carter 1998 Black mats, spring-fed streams, and late-glacial-age recharge in the southern Great Basin. *Quaternary Research* 49(2):129–148.

Rambeau, Claire M. C. 2010 Palaeoenvironmental reconstruction in the Southern Levant: synthesis, challenges, recent developments and perspectives. *Transactions of the Royal Society A* 368:5225–5248.

Richter, Tobias 2014 Margin or centre? The epipalaeolithic in the Azraq Oasis and the Qa' Shubayqa. In *Settlement, Survey and Stone: Essays on Near Eastern Prehistory in Honour of Gary Rollefson*, edited by Bill Finlayson and Cheryl A. Makarewicz, pp. 27–36. ex oriente, Berlin.

Roberts, Neil and Arlene Miller Rosen 2009 Diversity and complexity in early farming communities of southwest Asia: New insights into the economic and environmental basis of Neolithic Çatalhöyük. *Current Anthropology* 50(3):393–402.

Robinson, Stuart A., Stuart Black, Bruce W. Sellwood and Paul J. Valdes 2006 A review of palaeoclimates and palaeoenvironments in the Levant and Eastern Mediterranean from 25,000 to 5000 years BP: Setting the environmental background for the evolution of human civilisation. *Quaternary Science Reviews* 25(13–14):1517–1541.

Rosen, Arlene 2012 Change and stability in an uncertain environment: Foraging strategies in the Levant from the Early Natufian through the beginning of the Pre-Pottery Neolithic B. In *Sustainable Lifeways: Cultural Persistence in an Ever-Changing Environment*, edited by Miller, Naomi F. Miller, Katherine M. Moore, and Kathleen Ryan, pp. 128–149. University of Pennsylvania Press, Philadelphia.

Rosen, Arlene Miller 2007 *Civilizing Climate: Social Responses to Climate Change in the Ancient Near East.* Rowman Altamira, Lanham.

Rosen, Arlene Miller and Isabel Rivera-Collazo 2012 Climate change, adaptive cycles, and the persistence of foraging economies during the late Pleistocene/Holocene transition in the Levant. *Proceedings of the National Academy of Sciences of the United States of America* 109(10):3640–3645.

Rossignol-Strick, Martine 1999 The Holocene climatic optimum and pollen records of sapropel 1 in the eastern Mediterranean, 9000–6000BP. *Quaternary Science Reviews* 18(4):515–530.

Savard, Manon, Mark Nesbitt and Martin K Jones 2006 The role of wild grasses in subsistence and sedentism: New evidence from the northern Fertile Crescent. *World Archaeology* 38(2):179–196.

Schuldenrein, Joseph 2007 A Reassessment of the Holocene Stratigraphy of the Wadi Hasa Terrace and Hasa Formation, Jordan. *Geoarchaeology* 22(6):559–588.

Schuldenrein, Joseph and Geoffrey A. Clark 2003 Prehistoric landscapes and settlement geography along the Wadi Hasa, west-central Jordan. Part II: Towards a model of palaeoecological settlement for the Wadi Hasa. *Environmental Archaeology* 8:1–16.

Sherratt, Andrew 1980 Water, soil and seasonality in early cereal cultivation. *World Archaeology* 11(3):313–330.

Smith, Samuel J., Jonathan Paige and Cheryl A. Makarewicz In press A first look at the PPNA chipped stone tool assemblage from el-Hemmeh, southern Jordan: Implications for recognising Early Neolithic diversity. *Paleorient.*

Stein, Mordechai, Adi Torfstein, Ittai Gavrieli and Yoseph Yechieli 2010 Abrupt aridities and salt deposition in the post-glacial Dead Sea and their North Atlantic connection. *Quaternary Science Reviews* 29(3–4):567–575.

Tarawneh, Bassam 1988 The geology of at-tafila. *Geology Directorate Geological Mapping Division* Bulletin 12:1–34.

Tchernov, Eitan 1994 *An Early Neolithic Village in the Jordan Valley: The Fauna of Netiv Hagdud.* Peabody Museum of Archaeology, Cambridge, MA.

Torfstein, Adi, Steven L. Goldstein, Mordechai Stein and Yehouda Enzel 2013 Impacts of abrupt climate changes in the Levant from Last Glacial Dead Sea levels. *Quaternary Science Reviews* 69(C):1–7.

Verheyden, Sophie, Fadi H. Nader, Hai J. Cheng, Lawrence R. Edwards and Rudy Swennen 2008 Paleoclimate reconstruction in the Levant region from the geochemistry of a Holocene stalagmite from the Jeita cave, Lebanon. *Quaternary Research* 70(3):368–381.

Vita-Finzi, Claudio 1966 The Hasa Formation: An alluvial deposition in Jordan. *Man* 1(3). New Series:386–390.

Watkins, Trevor 2005 The Neolithic revolution and the emergence of humanity: A cognitive approach to the first comprehensive world-view. In *Archaeological Perspectives on the Transmission and Transformation of Culture in the Eastern Mediterranean*, edited by J. Clarke, p. 2: Levant Supplementary Series. Council for British Research in the Levant & Oxbow Books, Oxford.

Weiss, Ehud, Wilma Wetterstrom, Dani Nadel and Ofer Bar-Yosef 2004 The broad spectrum revisited: Evidence from plant remains. *Proceedings of the National Academy of Sciences of the United States of America* 101(26):9551–9555.

Wheatley, David and Mark Gillings 2002 *Spatial Technology and Archaeology. The Archaeological Applications of GIS.* Taylor and Francis, London.

White, Chantel E. and Cheryl A. Makarewicz 2012 Harvesting practices and early Neolithic barley cultivation at el-Hemmeh, Jordan. *Vegetation History and Archaeobotany* 21(2):85–94.

Willcox, George 1999 Geographical variation in major cereal components and evidence for independent domestication events in Western Asia. In *The Dawn of Farming in the Near East*, edited by R. T. J. Cappers and Sytze Bottema, pp. 133–140. ex oriente, Berlin.

——— 2005 The distribution, natural habitats and availability of wild cereals in relation to their domestication in the Near East: Multiple events, multiple centres. *Vegetation History and Archaeobotany* 14(4):534–541.

Winer, Emily Rachel 2010 *Interpretation and Climatic Significance of Late Quaternary Valley-Fill Deposits in Wadi Hasa, West-Central Jordan.* Unpublished M.S. Thesis, Miami University of Ohio, Oxford, OH.

Wright, H. E 1993 Environmental determinism in Near Eastern prehistory. *Current Anthropology* 34(4):458–469.

Yasuda, Yoshinori, Hiroyuki Kitagawa and Takeshi Nakagawa 2000 The earliest record of major anthropogenic deforestation in the Ghab Valley, northwest Syria: A palynological study. *Quaternary International* 73:127–136.

5 Human-Environment Interactions Through the Epipalaeolithic of Eastern Jordan

Matthew D. Jones, Lisa A. Maher, Tobias Richter, Danielle Macdonald, and Louise Martin

Abstract

On- and off-site environmental archives relating to archaeological sites in the Azraq Basin, Jordan, are used to relate Epipalaeolithic human behavior to local, and then regional, records of palaeoenvironmental change. We review recent work from three sites: Ayn Qasiyya, Kharaneh IV, and Shubayqa. Human occupation of the basin was more or less continuous throughout the Epipalaeolithic but shifted from locality to locality, varying in density and type. The environmental data suggest that this was, at least in part, due to a complex landscape of changing local environments. We use this example to discuss issues of scale in developing an understanding of human—climate—environment relationships.

Introduction

The Azraq Basin drains approximately 12,000 km^2 of northeast Jordan, southern Syria, and northwest Saudi Arabia (Figure 5.1). It has a long and complex history of human occupation, as outlined below, and over the last decade the multidisciplinary Epipalaeolithic Foragers in Azraq Project[1] (EFAP) has been reconstructing the prehistoric landscapes of what is today the eastern desert of Jordan. Working at three separate locations, we have attempted to integrate archaeological sites and the local off-site stratigraphy, evaluating the evidence of the environment in which the sites exist today and existed in the past. This chapter discusses using on- and off-site environmental archives to address the challenges of relating prehistoric human behavior to global, regional, and local records of palaeoenvironmental change.

The Azraq Basin is an area where the impact of environmental change, driven partly by the activities of people themselves, is clearly evident. The unsustainable use of natural resources, such as water, has been one reason for changing settlement patterns and agricultural practices in recent times (e.g., Fariz and Hatough-Bouran 1998). A point often made in prehistory is the need for hunter-gatherer groups to intimately know their environment and carefully schedule and manage their activities around particular resources and their availability (e.g., Lee and Devore 1968; Binford 1980;

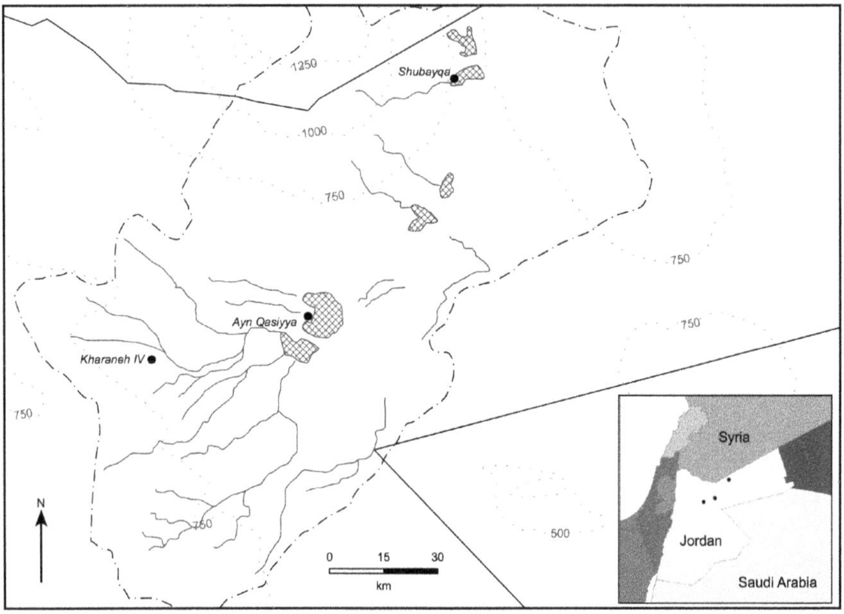

Figure 5.1 Map of the Azraq Basin (dashed black line) highlighting the three principal sites of the EFAP excavations to date. The inset shows these locations within the wider region.

Winterhalder and Smith 1981; Bettinger 1987; Lourandos 1997; Kelly 2013). Despite this, these activities may not have necessarily been sustainable, with evidence pointing to increasingly intensified use of gazelle and other resources throughout the Epipalaeolithic (Stiner et al. 2000; Munro 2003; 2009) and the potential that prehistoric groups manipulated their landscapes (e.g., Ramsey et al. 2015). It has likely been a long time since there was a "natural" environment in the Azraq Basin and, this, linked with today's continued concerns with sustainability, make clear the need to better understand people's relationships with the environment throughout time. This improves our understanding of past societies and societal change and informs how societies can mitigate or adapt to environments and environmental change now and in the future.

In delving into the palaeoenvironmental record on- and off-site at a number of locations in the Azraq Basin, we have discovered that existing conceptual models of people's relationship with their natural resources are sometimes very helpful tools for interpretation, but can also lead us to shape the interpretation of evidence collected to fit the model. People and environments vary widely in space and time, and this complicates the issue, making the notion of one single catchall human-environment model for a region unrealistic in terms of improving our understanding of these complex relationships and dynamics with any kind of nuance. Our work in the Azraq Basin is showing that we must be aware of the limitations of general models or hypotheses, and carefully consider issues of chronology, scale, and resolution, be they in time or space.

A good example of this in Levantine prehistory is the attempt to link or match cultural change at the end of the Pleistocene and into the early Holocene to environmental change. Whether or not one takes a causal approach to this human-environment dynamic recent work (e.g., Blockley and Pinhasi 2011; Maher et al. 2011) has highlighted the problems inherent in correlating large-scale regional, or hemispheric, palaeoenvironmental records with chronologies of occupation to create models of culture change. In addition to the difficulties of using rough correlation to argue for particular relationships, such broad patterns elide critical regional variation. For example, spatial climate-culture models of societal developments through this time period have historically placed our study region at the "margin" (Bar-Yosef and Belfer-Cohen 1989), and yet evidence from our work (reviewed here) suggests the Azraq Basin was far from marginal.

We argue here that the reason we can't always seem to make these different temporal or spatial datasets "correlate," with each other or with a model pattern, is because we try to correlate inappropriate scales of data.

Azraq Basin: A Fragile Ecosystem?

There is a long history of research into the prehistory of the Azraq Basin, starting with the early work of John Waechter and Veronica Seton-Williams in the early part of the twentieth century (Waechter et al. 1938), followed by Henry Field (1960) and intensifying in the 1980s with the initiation of several long-term prehistory projects. Fieldwork by Andrew Garrard and colleagues (e.g., Garrard and Stanley-Price 1975–77; Garrard 1998; Garrard and Byrd 1992; 2013; Garrard et al. 1988, 1994), Gary Rollefson (e.g., Rollefson 1982; Rollefson et al. 1997), Alison Betts (1998), and Lorraine Copeland and Francois Hours (1989) detailed the occupational history of the Azraq region from the earliest hominins to the emergence of the Neolithic. Garrard's excavations at 14 Palaeolithic and Neolithic sites made perhaps the most critical contribution to our understanding of the regional settlement chronology. In the 1990s and 2000s, Palaeolithic projects undertaken by Leslie Quintero, Philip Wilke, and Gary Rollefson (e.g., Rollefson et al. 1997; 2001; Cordova et al. 2008) continued to explore the activities of some of Jordan's earliest inhabitants. More recently, the Druze Marsh Archaeological and Palaeoecological Project continues to explore the early Palaeolithic occupation of the Azraq wetlands (Cordova et al. 2013; Ames et al. 2014). Given the wealth of data on the Palaeolithic and Neolithic periods provided by these projects, EFAP continues this tradition of prehistoric research by focusing on the intervening Epipalaeolithic periods, and in 2005 began excavations at the site of Ayn Qasiyya.

Understanding the palaeoenvironmental history of the Azraq Basin has been a key research aim of most of these inter-disciplinary projects, and this continues with EFAP. Eastern Jordan has often been considered to be somewhat peripheral to trajectories of cultural change in the Late Pleistocene, yet perhaps paradoxically contains a rich and continuous record of prehistoric occupation. This long record, as well as its environmental context, makes it a key geographical zone in which the relationship between cultural change and environmental change in prehistory can be studied.

Ranging from approximately 500 m to 1,200 m above sea level, the Azraq Basin today receives an average of between 50 and 150 mm of rainfall annually (Al-Kharabsheh 2000). Rainfall decreases notably from the northwest reaches of Jebel Druze to the headwaters of the Wadi Sirhan in the southeast. Although rainfall can be sparse in the western and eastern parts of the basin, surface water and groundwater drainage towards the Azraq Oasis create multiple seasonal pools and playas that are important habitats for wildlife. Underground drainage from the Jebel Druze has created a series of aquifers throughout the center of the basin. This water is discharged partly as a series of springs within the Azraq Oasis. The oasis was, until about 30 years ago when substantial pumping and drilling of water began, a unique and incredibly rich natural habitat in Jordan, providing a refuge for migratory birds and a range of other wildlife (Nelson 1973).

The waters of the main Azraq aquifers are thousands of years old with radiocarbon dating of the waters suggesting substantial recharge periods in the late glacial and early Holocene (Bajjali and Abu-Jaber 2001). This hydrological system was largely kept in equilibrium through natural rainfall and spring discharge, but this delicate balance has been completely disturbed by drilling and pumping of water in recent decades. The Azraq aquifers are one of Jordan's largest and principal fresh water sources, with water being pumped to Amman since the 1960s to supply the capital's population, as well as Azraq and its surrounding area, with fresh water. Within the last 10 years, several springs in the oasis have ceased to flow, and the once-rich marshland is rapidly disappearing. In a valiant effort to maintain a small proportion of the oasis, the Royal Society for the Conservation of Nature (RSCN) pumps water into the former wetland area to provide birds and other wildlife with a continuous refugium. The oasis has traditionally provided a rich livelihood for the local community in terms of farming, commercial salt production, tourism, and as a waypoint on long trips farther east (Nelson 1973; Fariz and Hatough-Bouran 1998). All but the last of these is now dwindling significantly.

Issues of resource availability, especially in relation to the availability of water, were likely key factors for the choice of settlement location at all three sites discussed here: Ayn Qasiyya, Kharaneh IV, and Shubayqa. At Kharaneh IV large numbers of people, based on the density of artifacts at the site and the relatively short period of occupation (ca. 1,200 years), were supported multi-seasonally (Richter et al. 2013). At Ayn Qasiyya relatively low population densities existed within the immediate oasis area over most of the span of the Epipalaeolithic. Initial work around the Qa' Shubayqa (*Qa'* is the regional word for dry lake or playa) has demonstrated a substantial settlement phase during the Late Epipalaeolithic along the margins of this mudflat (Richter et al. 2012; 2014). Thus, we are interested in the long-term patterning in the availability of water and how human societies in the region adjusted their ways of life to continually changing habitats. This has particular relevance to issues of habitat conservation and change in Azraq today, and this chapter shows that local, near-site, palaeoenvironmental records are vital for understanding human behavior and human-environment relationships.

The Sites

Wetlands are a key archive for palaeoclimatic and archaeological work, particularly in otherwise arid regions, as they provide a focus for human occupation and preserve environmental information. The sedimentary records from Ayn Qasiyya, a spring on the edge of the Azraq Qa, Kharaneh IV, an aggregation site within a well-vegetated grassland habitat, and Shubayqa 1, a village site on the edge of a seasonal water source, provide well-dated archaeological and environmental sequences through the last glacial–interglacial transition.

Ayn Qasiyya

Ayn Qasiyya was the second-largest spring in the southern Azraq marshlands. Originally described by Rollefson et al. (1997; 2001), the archaeological sediments at Ayn Qasiyya were explored through a series of excavations between 2005 and 2007 that aimed to understand in detail this early Epipalaeolithic site (e.g., Richter et al. 2007; 2010a). The site, dated to between 20,900–19,200 cal B.P. (Jones and Richter 2011; Richter et al. 2013), consisted of accumulations of chipped stone artifacts and faunal remains. Human remains were also found embedded in the palaeomarsh sediment (Richter et al. 2010b). Five sections through the sedimentary succession were described from the edge of the Ayn Qasiyya pool to place the excavation trenches in a wider environmental context (Figure 5.2). As well as field description, some sedimentological analysis, e.g., particle size, magnetic susceptibility, and loss on ignition, were undertaken. The chronology for the site was established using radiocarbon dating of the Epipalaeolithic occupation horizons (individual charcoal fragments were identified to genus level and samples chosen from short-lived species where possible to reduce chronological errors from long-lived species; Richter et al. 2013) and Optically Stimulated Luminescence (OSL) dating of archeologically sterile units which placed the human occupations in a sedimentary sequence spanning the last 60,000 years.

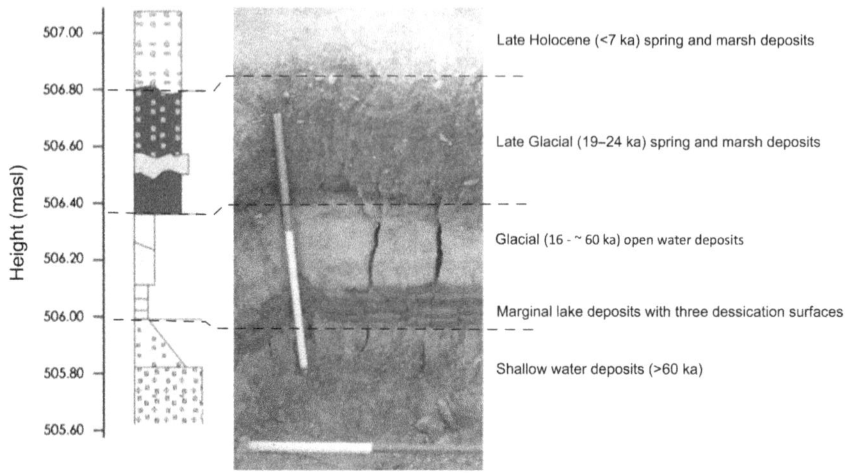

Figure 5.2 Summary of the main stratigraphic units from Ayn Qasiyya Section 1. The Epipalaeolithic occupation evidence is found in the upper section of the Late Glacial deposits. A further unit, of flood and marsh deposits, dating to the early Holocene, is observed elsewhere on the site.

Source: After Jones and Richter 2011

The strong relationship between groundwater level and spring discharge apparent from recent monitoring (Bajjali and Al-Hadidi 2005; El-Naqa et al. 2007) was used to help reconstruct potential environmental conditions in the past based on the sedimentary record from Ayn Qasiyya. Water levels fell, following a late Pleistocene highstand, during the Last Glacial Maximum (LGM) at Ayn Qasiyya, leading to the development of a marsh at the spring site, although open water was still accessible close by. This suggests that spring discharge was reduced or reducing through the LGM. Between 16 and 10.5 ka B.P., there is no net deposition at the site. One reason for this could be the complete drying of the springs during this time period, as in Azraq today. However, we would expect some deflation and removal of material in such a dryland, particularly over the course of 5,500 years, and so we cannot say for sure that there were consistently dry conditions through this time, just that net deposition was zero over this period. Even if the springs were dry for a considerable period during this interval, water may still have been available locally within the basin due to surface runoff into the Qa', at least on a seasonal basis, and particularly in the higher areas to the north of Azraq where rainfall is generally greater today. The early Holocene again shows deposition of material at the site, particularly large clasts and mixed sediments suggestive of flood deposits, rather than deposits from spring waters. During the later Holocene, spring conditions were re-established with carbonate concretions forming in the marshland due to the warmer conditions during the Holocene compared to the LGM.

Kharaneh IV

Kharaneh IV is located approximately 37 km west of Ayn Qasiyya, at the western edge of the Azraq Basin. The site is situated approximately 1 km south of Qasr Kharaneh, one of the more notable desert castles. Kharaneh IV is a multi-component Epipalaeolithic site, with Early and Middle Epipalaeolithic occupations. Wadi Kharaneh lies adjacent to the site and flows into the western side of the central Azraq Qa'.

The current EFAP excavations have been ongoing since 2008. Kharaneh IV is an exceptionally large Epipalaeolithic site, approximately 21,000 m^2 in surface area, making it one of the largest Terminal Pleistocene occupations in the Levant. The most abundant artifacts at the site are chipped stone tools, with an estimated 2 million lithics excavated to date in just over 100 m^2 of excavated area. The site sits as a low mound on the desert landscape; the accumulation of cultural deposits has created an Epipalaeolithic "tell" rising 2 m above the surrounding sediment surface (Figure 5.3). This mound has two peaks, with Early Epipalaeolithic deposits located at the highest point of the site to the east (Area B) and Middle Epipalaeolithic deposits on the western portion of the mound (Area A). Beyond our main excavation areas, we have opened several small test trenches to explore deposits underlying the occupational horizons and one larger geological trench 9 m in length (into the side of the mound on the southern side) to trace the relationships between the off-site and on-site deposits. Despite some surface deflation, the subsurface stratigraphy is very fine-grained across the entire site, with in situ cultural deposits extending to a depth of ~1.5 m below the surface (Figure 5.3).

Thirteen radiocarbon age estimates date the Early Epipalaeolithic occupation sequence between 19,830 to 18,800 cal B.P. and the Middle Epipalaeolithic occupation sequence to between 18,800 and 18,600 cal B.P. (Richter et al. 2013). This gives a very early date for the Middle Epipalaeolithic occupation at Kharaneh IV, overlapping slightly with the Early Epipalaeolithic occupation of the site. Despite these early dates, the geometric microliths present in the assemblage aligns the occupation with other Middle Epipalaeolithic sites in the southern Levant.

During the 2010 excavations evidence for two hut structures was uncovered in the Early Epipalaeolithic area. Radiocarbon dates from above and below the floor of Hut 1 date the structure between 19,400 and 18,800 cal B.P. (Maher et al. 2012). The structure is just over 2 × 3 meters in size and is covered by an organic-rich, black layer containing abundant charcoal fragments. This suggests that the huts were burned after abandonment. Situated beneath the burned layer, but on top of the hut floor, are groundstone fragments, red ochre, and articulated aurochs vertebrae. Near the center of the structure are three distinct concentrations of pierced marine shells accompanied by large chunks of red ochre. These concentrations contain over 1,000 shells from both the Mediterranean and Red Seas, imported from up to 270 kilometers away (Maher et al. 2012). Excavations of the structure in 2013 revealed a series of deposits, perhaps representing several floor deposits or

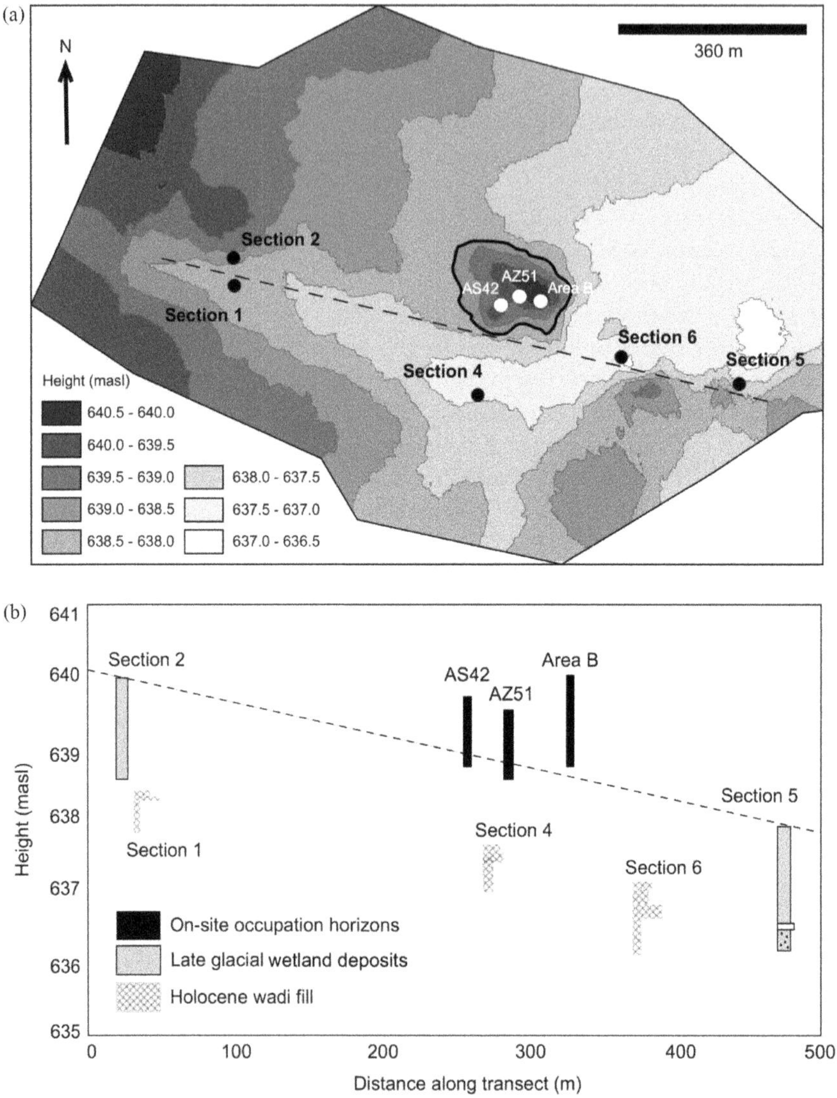

Figure 5.3 (a) The spatial distribution of sedimentary sections and the site of Kharaneh IV (within the thick black line). AS42 is an excavation square in Area A (as described in the text); AZ51 a deep sounding between the two main excavation areas. The dashed line represents a transect down the wadi from which the vertical distribution of the deposits can be shown. (b) The present-day surface of the wetland sediments (dashed line in b) runs parallel to the wadi gradient, sloping east towards the Azraq Qa'.

Source: After Jones et al. 2016

in-fill after abandonment. In situ artifacts including lithics, articulated bone, and pierced marine shell were found on these surfaces. A potential third structure was identified in the excavation section, suggesting a larger occupation at the site.

The Middle Epipalaeolithic deposits contain a series of compact occupation surfaces on which people were discarding lithics and remains of hunted animals. Cut into these surfaces are a number of hearths. Several small dark features are also cut into the compact surfaces and are interpreted as postholes, suggesting ephemeral structures placed over the fireplaces, perhaps as cooking structures or as drying racks for the processing of gazelle remains that are found in abundance here. Future excavations at the site will continue to expose the horizontal extent of the Middle Epipalaeolithic deposits at Kharaneh IV to identify potential structural features like the ones found in the Early Epipalaeolithic deposits.

Beyond exploring the wide range of activities conducted on-site, our main research questions have always revolved around trying to understand what drew people to the site and what types of occupation created such a large, artifactually dense site. Our on-site work makes it clear that substantial numbers of people congregated at the site, in some cases for prolonged periods of time—as evidenced by the density of material, as well as the faunal records that show multi-seasonal occupation throughout most phases of occupation (Martin et al. 2010; Jones 2012). The distribution of faunal material, particularly gazelle, as well as analyses of mortality profiles also suggests intensification, with a clear focus on extensive gazelle carcass processing, including probably large-scale drying and storage of gazelle meat and communal hunting efforts. Marine shells and technological analyses of the chipped stone assemblage suggests people were moving both objects and knowledge substantially long distances, involved in trade and exchange networks much greater than previously considered for the Early and Middle Epipalaeolithic (Richter et al. 2011). The construction of hut structures, specific outdoor food processing areas, flint knapping concentrations, and caches are just some examples of the well-structured use of space within the site. Our work has therefore looked to understand the resources required to support such a site.

In order to answer these questions we have paired our on-site excavations with an extensive program of on-site and off-site geological and geomorphological sampling to reconstruct the local environment. In addition, a program of 100% flotation provides a rich archaeobotanical record, which is currently under investigation. Soundings and geological trenches on-site indicate the presence of wetland deposits at the base of and underlying the occupational deposits in each sampled area. These deposits are ostracod-rich, carbonate-concreted marls interstratified with the earliest Early Epipalaeolithic occupation of the site (Jones et al. 2016; Maher et al. 2016).

Careful surveying of the off-site stratigraphy, and comparison to that on-site, suggests that a substantial wetland preceded the occupation site and still existed, at least to some degree, when the site was occupied (Figure 5.3).

This was established by (1) dating on- and off-site material; in this case using largely OSL and radiocarbon age estimates, respectively, and (2) surveying the off-site sedimentary units into the site grid (Jones et al. 2016). Work is ongoing to characterize this freshwater body, specifically its spatial and temporal extent. Current evidence suggests it was unlikely that the freshwater resource evident as the site was established continued throughout its occupation history. Our initial interpretation is that during low water levels, or seasonal episodes of drying around 20 ka B.P., Epipalaeolithic groups were taking advantage of this local water resource and occupying the margins of the wetland. Once the water levels retreated enough, the site was occupied on a "regular" basis for about 1,000 years.

Shubayqa 1

The Qa' Shubayqa is situated in the basaltic Harra desert, ca. 20 km north of the town of Safawi (Figure 5.1). The Qa' itself is a 12 km^2 large basin that is dominated by an extensive alluvial fan, with both features interrupting the extensive, low-rising basalt boulder fields characteristic of the Harra. Today the area is a semi-arid steppe zone that receives less than 200 mm of mean annual rainfall. However, the Qa' Shubayqa is fed by a series of seasonal wadis that transport extensive amounts of water from the Jebel Druze to the mudflat. The largest of these is the Wadi Rajil whose catchment area encompasses the high rainfall zone of the Jebel Druze (Whitehead et al. 2008). This causes seasonal flooding of the Qa', which can be extensive and rapid. Given sufficient rainfall, and our current understanding of regional geomorphological change, this situation is likely to have occurred for a long time, until the recent construction of dams upstream in Syria. It would have provided the area with a seasonally reliable source of fresh water and could have thus enabled periodic intensive settlement.

Evidence for the human occupation of the Qa' Shubayqa in northeastern Jordan was first reported by Alison Betts (1998). Her surveys and excavations in the area provided initial baseline evidence that Late Epipalaeolithic groups were present in this region at the end of the Pleistocene. Shubayqa 1 was first reported by Betts (1998) and briefly test-excavated in 1996. Since 2013 the site has been subject to renewed excavations. These excavations are accompanied by landscape survey and geomorphological sampling of the Qa'.

Recent field-walking survey around the Qa' Shubayqa (Richter et al. 2012) has produced evidence for nine Late Epipalaeolithic and Early Neolithic localities. Four of these represent substantial occupations with architecture, dense chipped stone artifacts scatters and large numbers of ground stone tools. These sites date (based on radiocarbon dates and lithic typology) from the Early Natufian to the PPNA (Richter et al. in prep), suggesting a more or less continuous presence of people in this landscape between ca. 14.8 ka B.P. to 9 ka B.P. This would have only been possible if ample water and other resources were available. Given the settlements' proximity to the

Qa' Shubayqa it is highly plausible that the basin provided a reliable source of water, which also attracted animals and allowed a wide range of plants to grow in the area.

Excavations at the Natufian site Shubayqa 1 have provided some evidence to support this idea. Shubayqa 1 is a ca. 5,000-m^2 large site centered on a ca. 3-m high mound. Large boulder mortars, ground stone, and chipped stone artifacts litter the site's surface. Excavations in a 60 m^2 area have revealed evidence for two major phases of occupation characterized by the construction and use of two round buildings. Rich chipped stone, ground stone, faunal, and botanical assemblages have been recovered, in addition to worked bone, stone and marine shell beads, and seven human burials. Preliminary analysis of the botanical data indicates the presence of tubers, cereals (*Hordeum Spontaneum*, *Triticum spp.*), charred wood (e.g., *Tamarix sp.*, *Chenopodiaceae* and *vitex agnus-castus*), and a range of other wild plants (e.g., *Heliotropium sp.*, *Buglossoides* sp.). Although gazelle dominates the fauna from the site, numerous bird bones, including waterfowl, form a significant part of the assemblage (Yeomans and Richter in press).

Unlike at Ayn Qasiyya and Kharaneh IV, the sedimentary record of environmental change in and around the Qa' Shubayqa cannot be directly linked to the occupation sites. The Qa' itself provides the main sedimentary environment locally, and lies topographically below the on-site, archaeological evidence. The Qa' contains approximately 4.5 m of sediment, based on augering at two locations in the basin, and preliminary dates from both OSL (5.6 ± 0.3 ka B.P. at a depth of 135 cm) and radiocarbon age estimates (6,400–6,220 cal B.P. at a depth of 195 cm) suggest this sediment is largely Holocene in age.

The "Big Picture" . . . *Writ Small*

The archaeological evidence described above and from previous work in the Azraq Basin (see "Azraq Basin: A Fragile Ecosystem?" above) indicates that occupation of the area was more or less continuous throughout the Epipalaeolithic but shifted from locality to locality, varying in density and type (Figure 5.4). The intensity and type of settlement is likely to have varied in accordance with fluctuations in surface water availability. It appears that environments that were sufficiently resource rich to support human settlement were available *somewhere* in the basin at most times, although subtle changes in site function and length of occupation as well as in the environmental evidence are noticeable and suggest a dynamic patchwork of local environments.

Through the LGM the Azraq Basin appears to have had good water availability, providing a potential refugium for human populations. The Ayn Qasiyya archaeological sequence shows a fairly continuous sequence of occupations, especially through the LGM, which had a residential, as opposed to a short-term opportunistic, character, as suggested by the

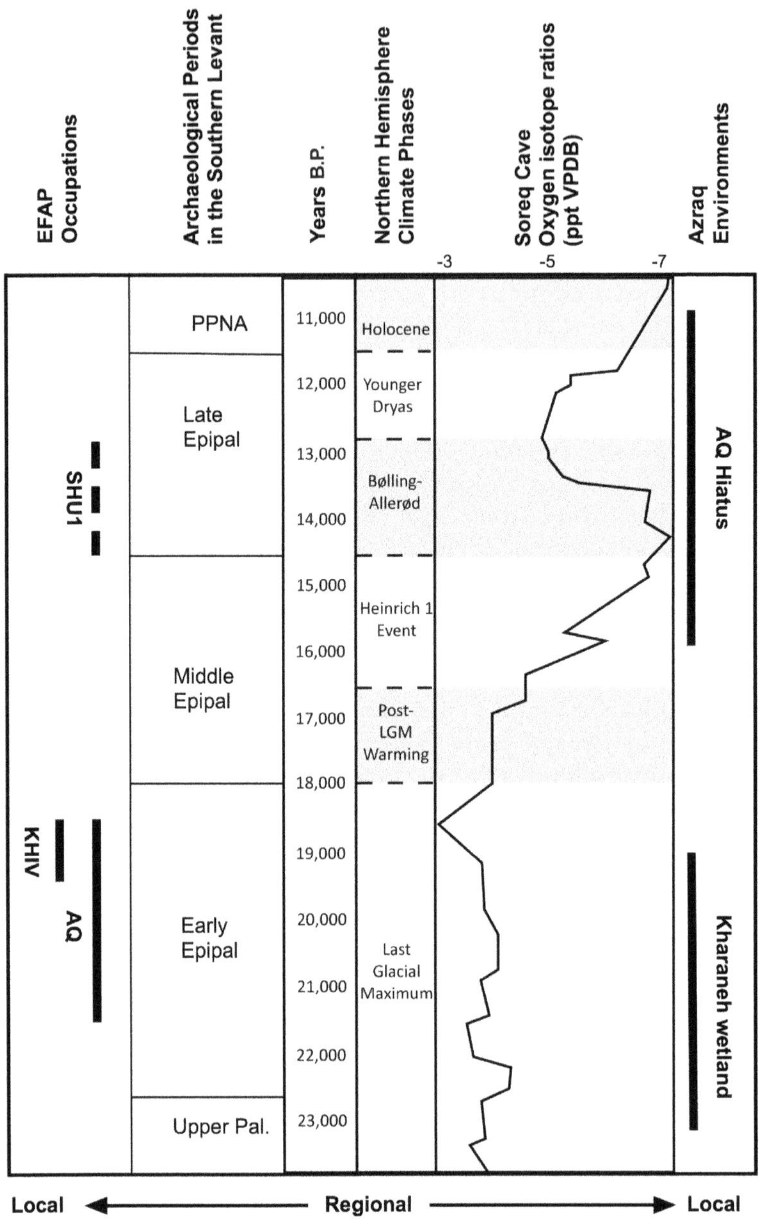

Figure 5.4 Azraq Basin occupation and palaeoenvironmental evidence from the EFAP work described in this chapter compared to more regional patterns of change. This figure makes no correlations or causal links between the local and regional data shown; see the main text for discussion. (Occupation dates for Ayn Qasiyya (AQ) and Kharaneh IV come from radiocarbon dating of the sites (Richter et al. 2013); the Shubayqa 1 (SHU1) dates are based on the evidence presented in this chapter. The Azraq environments column highlights the period of most positive water balance, as recorded by the wetland deposits around Kharaneh IV, and the period where our palaeoenvironmental evidence is still unclear, highlighted by the hiatus in the Ayn Qasiyya (AQ) sedimentary sequence.)

Source: After Bar-Matthews et al. 1997; Maher et al. 2011

evidence for on-site carcass butchery and a lithic tool kit that includes many scrapers and large retouched pieces, as well as evidence for the maintenance of composite microlithic tools (Richter et al. 2010a). These kinds of activities are typically associated with longer-term residential sites, as opposed to short-term knapping or butchering sites. The continuous deposition of archaeological material in a 40–60 cm thick deposit also reflects longer-term repeated use of the same locale by human groups. Faunal evidence, especially data from gazelle age profiles and seasonal migratory bird species, tentatively support repeated, multi-seasonal occupation (Edwards forthcoming). Elsewhere in the Azraq Basin, numerous sites were established between ca. 20–16.5 ka B.P., including the two very large sites at Kharaneh IV and Jilat 6 (Muheisen 1988; Garrard and Byrd 1992; Garrard et al. 1994; Garrard and Byrd 2013). Environmental evidence from Ayn Qasiyya and Kharahneh IV discussed above suggest maximum water availability immediately prior to the development of these sites.

This fits regional patterns of highest lake levels prior to the LGM, e.g., in Lake Lisan (Torfstein et al. 2013), Konya (Roberts 1983), and Van (Çağatay et al. 2014) which had begun to fall by 20 ka B.P. A lack of speleothem and stromatolite deposition in the Dead Sea Valley between 25 and 18 ka B.P. (Sorin et al. 2010) also suggests rainfall was significantly reduced, at least in that area, at this time. However, open water conditions continue, at least locally, at Ayn Qasiyya until the break in stratigraphy between 16 and 10.5 ka B.P., during which time the Lake Lisan levels rise again before falling rapidly towards the beginning of the Holocene (e.g., Torfstein et al. 2013).

This well-watered Azraq environmental situation apparently changed following the onset of climatic "amelioration" during the time of the North Atlantic Bølling-Allerød. Sedimentary evidence from Ayn Qasiyya shows that there was a lack of deposition between 16 and 10.5 ka B.P., and human occupation appears to have shifted to other areas of the oasis. About 1.5 km southeast of Ayn Qasiyya, multiple small Middle Epipalaeolithic (Geometric Kebaran) surface lithic scatters were identified within an approximately 3,000-m^2 area (Richter 2009). The analysis of the material suggests a highly specialized tool kit, dominated by geometric microliths. These sites, while indicating a certain degree of continuity in occupational history, also represent a more dispersed settlement pattern, which was not primarily focused on the spring localities. Given that these sites were short-term occupations that reflect tool preparation and hunting camps, the oasis was clearly still attractive to human groups although possibly no longer suitable for long-term occupation. This pattern would fit with continued seasonal wetting of the Qa' but a lack of a permanent spring-sourced water supply.

Site abandonment at Kharaneh IV, and a similar pattern documented at other nearby Early and Middle Epipalaeolithic sites, including the only other substantial aggregation site in the region, Jilat 6 (Garrard and Byrd 2013), suggests a substantial and negative change in local resources that previously sustained hunter-gatherer populations. Although further data is necessary to confirm this, it seems that settlement generally shifted from these large aggregation sites in the western portion of the basin in the first part of the Epipalaeolithic to the oasis and northern part of the basin such as Shubayqa in the later Middle Epipalaeolithic and Late Epipalaeolithic. The site of Jilat 22 is an exception to this pattern, with a number of Middle Epipalaeolithic horizons within marsh deposits (Garrard and Byrd 1992), although the deposits are cemented, suggesting frequent drying, including during the occupation horizons themselves. These deposits are unlikely to be groundwater fed, as the Wadi Jilat is 280 m above the Azraq Qa', and probably reflect seasonal wetting of the site, with possible damming of the wadi allowing the development of a localized marsh at this time and providing a local resource despite basin-wide patterns.

By the Late Epipalaeolithic (14.8–11.7 ka B.P.; Maher et al. 2011), the frequency of occupations in the oasis and elsewhere in the basin had once again increased (Richter and Maher 2013). A substantial site dating to this period has been documented to the south of the marshes at Azraq 18 (Garrard et al. 1994), and smaller, more ephemeral sites found at the eastern edge of the marsh and Qa' at Bawaab al-Ghazal (Rollefson et al. 1999) and at Ayn Qasiyya (Richter et al. 2007; 2010b). Numerous Late Epipalaeolithic sites are also found in the Badia to the north and east (Betts 1991; Richter and Maher 2013; Richter et al. 2013; 2014) including Shubayqa 1.

At the Qa' Shubayqa evidence is beginning to point to a more permanent body of open water through the Epipalaeolithic than is evident today. It is unclear at present how seasonal this water body may have been and whether its appearance matches the timescales of site occupation. It is highly probable that more water reached Shubayqa in the past, since modern dams hold back much of the water upstream and there would have been more space in the basin to hold water prior to Holocene sediment in-filling. It is also likely that the water body, if it was seasonal, lasted longer than it would today, but this is as yet difficult to demonstrate. The influence of any climatic changes during the time of the Younger Dryas on the local settlement sequence is as yet not well understood, since our environmental data for this time frame are still under investigation. Shubayqa 1's main phases of occupation predate the Younger Dryas, falling into the time period of the North Atlantic Bølling-Allerød interstadial. The more prominent and substantial sites in the Qa' Shubayqa also appear to either date to the Early Natufian or the PPNA, with smaller more ephemeral sites perhaps representing late Natufian occupations. This would seem to suggest that there might have been a reduction in settlement intensity during Younger Dryas times in the area. However, this hypothesis needs to be further tested through additional fieldwork that will hopefully produce some direct data for late Natufian settlement patterns and environments in the area. As shown from EFAP's work across the basin through the Epipalaeolithic, even though settlement intensity may have been reduced at the time of the Younger Dryas, people did not entirely disappear from the area, but likely adjusted settlement patterns and economies to the local resources available.

Acknowledging Our Known Unknowns . . .

The wider Levant has seen much interest in the Epipalaeolithic and the correlation of societal change (namely, the emergence of sedentary farming villages) with large-scale post-glacial climate transitions. But any arguments that these changes represent cause and effect remain only hypotheses to test. Archaeological and palaeoenvironmental work in the Azraq Basin and beyond continues to provide evidence with which to do this testing.

Our ability to "know" what was going on in the past becomes more attenuated as we go further back in time, and the Epipalaeolithic is certainly far back enough to be a pretty mysterious place. The current data, both archaeological and palaeoenvironmental, do not have the spatial or temporal coverage to give us the answer to a hypothesis of climatically driven societal change in Epipalaeolithic Azraq. The current data do suggest a complex landscape of changing environments and occupation types even within the Azraq Basin itself, urging caution for a one-size-fits-all model for an even wider region.

These data maybe incomplete, but are appropriate for answering research questions at the resolution necessary for getting at an (arguably) useful understanding of human-environment interactions—the local scale. We advocate a bottom-up, data-driven approach to the investigation of past human-environment interactions that focuses on the local scale of investigation, rather than large-scale contextual explanations of culture change. We need to build not one single regional model, but models scalable to particular research areas, and acknowledge the fact that microenvironments and localized human activity matters much more than has sometimes been allowed for.

Our work in Azraq continues to fill in pieces of the puzzle, using a combination of archaeological and geological techniques. Further work will focus on hydrological modeling and vegetation reconstruction as well as continuing with the work described here. But it is the reconstruction of local environments, experienced by the local inhabitants, that continue to be our basis for human—climate—environment discussions.

Acknowledgments

We thank the Department of Antiquities of Jordan and its Director Dr. Monther Jamhawi, as well as former Director Dr. Fawwaz al-Khreshah, for their continued support of our work. EFAP has been funded by the Arts and Humanities Research Council of Britain, Council for British Research in the Levant, Royal Geographical Society, the Danish Council for Independent Research, the Danish Institute in Damascus, the H.P. Hjerl Mindefondet for Dansk Palæstinaforskning, the University of Nottingham, University of Copenhagen, University of California, Berkeley, Fragmented Heritage (University of Bradford), and the Wenner-Gren Foundation. This chapter was written while M.J. was a Visiting Fellow in the School of Geography, Planning and Environmental Management at the University of Queensland. Importantly, we thank the numerous colleagues and students who have helped in our excavations and surveys and past and present colleagues working in Azraq with whom we have discussed various aspects of this research.

Methodological Vignette: Integrating Archaeological and Palaeoenvironmental Data Through On-Site and Off-Site Stratigraphy

The relationship between the archaeological deposits and the off-site paleoenvironmental stratigraphy is different at the three sites discussed in this chapter in that occupation at Ayn Qasiyya is found within the paleoenvironmental stratigraphy, Kharaneh IV is on top of and adjacent to it, and the occupation sites at Shubayqa are topographically above and spatially removed from the Qa' deposits. These require slightly different approaches to comparing evidence from on- and off-site.

However, in all cases the off-site stratigraphy provides a local archive of environment for some or all of the occupation phases. As we discuss in this chapter, we have tried to keep our off-site data local in this work, as paleoenvironments can be as spatially heterogeneous as those we observe today. It is difficult to answer the question of how local such records should be; this clearly depends on a number of factors, for example, what aspect of the environment or human activity are we interested in reconstructing and how continuous, temporally or spatially, our palaeoenvironmental information needs to be. A guide to the spatial relevance of a specific site or proxy can come from looking at the modern environment, e.g., pollen distributions (e.g., Djamali et al. 2009) or climatic zones (Jones 2013).

Useful palaeoenvironmental data can come from directly near the site (e.g., our work described here, or that at Çatalhöyük in Turkey (Boyer et al. 2006; Rosen and Roberts 2006)) or from slightly farther away if environmental proxies, for example, pollen and charcoal, can give a more regional picture of change (England et al. 2008; Turner et al. 2010). These regional reconstructions are useful to compare to local data and can be a useful link to larger-scale patterns of environmental and climatic change. However, a good understanding of the spatial limitations of such data is important; otherwise they can be misleading about the local environments of archaeological sites.

Note

1 http://epipalaeolithicforagers.wordpress.com/

References Cited

Al-Kharabsheh, A. 2000 Ground-water modelling and long-term management of the Azraq basin as an example of arid area conditions (Jordan). *Journal of Arid Environments* 44:143–153.

Ames, Chris, April Nowell, Carlos Cordova, James Pokines and Michael Bisson 2014. The Druze Marsh paleolandscape: A geoarchaeological approach to open-air paleolithic sites. *Quaternary International* 331:60–73.

Bajjali, W. and N. Abu-Jaber 2001 Climatological signals of the paleogroundwater in Jordan. *Journal of Hydrology* 243:133–147.

Bajjali, W. and K. Al-Hadidi 2005 Hydrochemical evaluation of groundwater in Azraq Basin, Jordan using environmental isotopes and GIS techniques. 25th Annual ESRI International User Conference, San Diego.

Bar-Matthews, M., A. Ayalon and A. Kaufman 1997 Late Quaternary paleoclimate in the eastern Mediterranean region from stable isotope analysis of speleothems at Soreq Cave, Israel. *Quaternary Research* 47:155–168.

Bar-Yosef, O. and A. Belfer-Cohen 1989 The origins of sedentism and farming communities in the Levant. *Journal of World Prehistory* 3(4):447–498.

Bettinger, R. L. 1987 Archaeological approaches to hunter-gatherers. *Annual Review of Anthropology* 16:121–142.

Betts, A. V. G. 1991 The late Epipaleolithic in the Black Desert, eastern Jordan. In *The Natufian Culture in the Levant*, edited by O. Bar-Yosef and F. R. Valla, pp. 217–234. International Monographs in Prehistory 1, Ann Arbor, MI.

——— 1998. *The Harra and the Hamad. Excavations and Surveys in Eastern Jordan, Volume 1*. Sheffield Academic Press, Sheffield.

Binford, L. R. 1980 Willow smoke and dogs' tails: Hunter-gatherer settlement systems and archaeological site formation. *American Antiquity* 45:4–20.

Blockley, S. P. E. and R. Pinhasi 2011 A revised chronology for the adoption of agriculture in the Southern Levant and the role of Lateglacial climatic change. *Quaternary Science Reviews* 30:98–108.

Boyer, P., N. Roberts and D. Baird 2006 Holocene environment and settlement on the carsamba alluvial fan, south-central Turkey: Integrating geoarchaeology and archaological field survey. *Geoarchaeology* 21:675–698.

Çağatay, M. N., N. Öğretmen, E. Damcı, M. Stockhecke, U. Sancar, K. K. Eriş and S. Özeren. 2014 Lake level and climate records of the last 90ka from the Northern Basin of Lake Van, eastern Turkey. *Quaternary Science Reviews* 104:97–116.

Copeland, L. and F. Hours (editors) 1989 *The Hammer on the Rock: Studies in the Early Palaeolithic of Azraq, Jordan*. B.A.R. International Series 540, Oxford.

Cordova, Carlos, April Nowell, Michael Bisson, Christopher Ames, James Pokines, Melanie Chang and Maysoon al-Nahar 2013 Glacial and interglacial desert refugia and the middle paleolithic of the Azraq Basin, Jordan. *Quaternary International* 300:194–210

Cordova, C. E., G. O. Rollefson, R. Kaichgruber, P. Wilke and L. Quintero 2008. Natural and cultural stratigraphy of 'Ayn as-Sawda, al-Azraq Wetlands reserve. 2007 Excavation report and discussion of finds. *Annual of Department of Antiquities of Jordan* 52:417–425.

Djamali, M., J. L. de Beaulieu, P. Campagne, V. Andrieu-Ponel, P. Ponel, S. A. G. Leroy and H. Akhani 2009 Modern pollen rain–vegetation relationships along a forest–steppe transect in the Golestan National Park, NE Iran. *Review of Palaeobotany and Palynology* 153:272–281.

England, A., Eastwood, W. J., Roberts, C. N., Turner, R. and Haldon, J. F. 2008 Historical landscape change in Cappadocia (central Turkey): A palaeoecological investigation of annually laminated sediments from Nar lake. *The Holocene* 18(8):1229–1245.

El-Naqa, A., M. Al-Momani, S. Kilani and N. Hammouri 2007 Groundwater deterioration of shallow groundwater aquifers due to overexploitation in Northeast Jordan. *CLEAN—Soil, Air, Water* 35:156–166.

Enzel, Y., R. Amit, U. Dayan, O. Crouvi, R. Kahana, B. Ziv and D. Sharon 2008 The climatic and physiographic controls of the eastern Mediterranean over the late Pleistocene climates in the southern Levant and its neighboring deserts. *Global and Planetary Change* 60:165–192.

Fariz, G. H. and A. Hatough-Bouran 1998 Case study: Jordan. Population dynamics in arid regions: The experience of the Azraq Oasis conservation project. In *Water and Population Dynamics: Case Studies and Policy Implications*, edited by A. de Sherbinin and V. Dompka, pp 105–35. American Association for the Advancement of Science, Washington, DC.

Field, H. 1960 *North Arabian Desert Archaeological Survey, 1925–1950. Papers of the Peabody Museum*. Peabody Museum, Harvard University, Cambridge.

Garrard, A. N. 1998 Environment and cultural adaptations in the Azraq Basin: 24,000–7,000 B.P. In *The Prehistory of Jordan*, edited by D. O. Henry. British Archaeological Reports International Series 705, Oxford, pp. 139–48.

Garrard, A. and N. Stanley-Price 1975–1977 A survey of prehistoric sites in the Azraq basin, eastern Jordan. *Paléorient* 3:109–126.

Garrard, A. N. and B. Byrd 2013 *Beyond the Fertile Crescent: Late Palaeolithic and Neolithic Communities of the Jordanian Steppe. The Azraq Basin Project Volume 1: Project Background and the Late Palaeolithic (Geological Context and Technology)*. Oxbow, Oxford.

Garrard, A. N., D. Baird, S. Colledge, L. Martin and K. Wright 1994 Prehistoric environment and settlement in the Azraq basin: An interim report on the 1987 and 1988 excavation season. *Levant* 26:73–109.

Garrard, A. N. and B. F. Byrd 1992 New dimensions to the epipalaeolithic of the Wadi el-Jilat in central Jordan. *Paléorient* 18:47–62.

Garrard, A., S. Colledge, C. Hunt and R. Montague 1988 Environment and subsistence during the Late Pleistocene and Early Holocene in the Azraq Basin. *Paléorient* 14:40–49.

Jones, J. R. 2012 Using gazelle dental cementum studies to explore seasonality and mobility patterns of the Early-Middle Epipalaeolithic Azraq Basin, Jordan. *Quaternary International* 252:195–201.

Jones, M. D. 2013 Key questions regarding the palaeoenvironment of Iran. In *The Oxford Handbook of Iranian Archaeology*, edited by D. Potts, pp. 17–28. Oxford University Press, New York.

Jones, M. and T. Richter 2011 Palaeoclimatic and archaeological implications of Pleistocene and Holocene environments in Azraq, Jordan. *Quaternary Research* 76(3):363–372

Jones, M. D., L. A. Maher, D. A. Macdonald, C. Ryan, C. Rambeau, S. Black and T. Richter 2016 The environmental setting of Epipalaeolithic aggregation site Kharaneh IV. *Quaternary International* 396:95–104.

Kelly, R. L. 2013 *The Lifeways of Hunter-Gatherers: The Foraging Spectrum*. Cambridge University Press, New York.

Lee, R. B. and I. Devore (editors) 1968 *Man the Hunter*. Chicago, Aldine.

Lourandos, H. 1997 *Continent of Hunter-Gatherers: New Perspectives in Australian Prehistory*. Cambridge University Press, Cambridge.

Maher, L. A., E. B. Banning and M. Chazan 2011 Oasis or Mirage? Assessing the role of abrupt climate in the Prehistory of the Southern Levant. *Cambridge Archaeological Journal* 21:1–29.

Maher, L. A., D. A. Macdonald, A. Allentuck, L. Martin, A. Spyrou and M. Jones. 2016 Occupying wide open spaces? Late pleistocene hunter-gatherer activities in the Eastern Levant. *Quaternary International* 396:79–94.

Maher, Lisa A., Tobias Richter, Danielle Macdonald, Matthew D. Jones, Louise Martin and Jay T. Stock. 2012. Twenty thousand-year-old huts at a hunter-gatherer settlement in eastern Jordan. *PLoS ONE* 7 (2):e31447.

Martin, L., Y. Edwards and A. Garrard 2010 Hunting practices at an eastern Jordanian Epipalaeolithic aggregation site: The case of Kharaneh IV. *Levant* 42(2):107–135.

Muheisen, M. 1988 The epipalaeolithic phases of Kharaneh IV. In *The Prehistory of Jordan: The State of Research in 1986*, edited by A. Garrard and H. Gebel, pp. 353–367. British Archaeological Reports 396, Oxford.

Munro, N. D. 2003 Small game, the Younger Dyras, and the transition to agriculture in the southern Levant. *Mitteilungen der Gesellschaft für Urgeschichte* 12:47–71.

Munro, N. 2009 Epipaleolithic subsistence intensification in the southern Levant: The faunal evidence. In *The Evolution of Hominin Diets*, edited by Jean-Jacques Hublin and Michael P. Richards, pp. 141–155. Springer, Netherlands.

Nelson, B. 1973. *Azraq: Desert oasis*. Allen Lane, London.

Ramsey, M. N., M. D. Jones, T. Richter and A. Rosen 2015 Modifying the marsh: Evaluating Early Epipaleolithic hunter-gatherer impacts in the Azraq wetland, Jordan. *The Holocene* 25:1553–1564.

Richter, T. 2009 *Marginal Landscapes? The Azraq Oasis and the Cultural Landscapes of the Final Pleistocene Levant*. University College London, Institute of Archaeology, London. http://eprints.ucl.ac.uk/18727/

Richter, T., S. Alcock, M. Jones, L. Maher, J. Stock and B. Thorne 2010a New light on Final Pleistocene settlement diversity in the Azraq Basin: Excavations at ‘Ayn Qasiyya. *Paléorient* 35(2):49–68.

Richter, T., A. Arranz, M. House and L. Yeomans 2014 The second season of excavations at Shubayqa 1. *Neo-Lithics* 1(14):10–17.

Richter, T., L. Bode, M. House, R. Iversen, A. Arranz, I. Saehle, G. Thaarup, M.-L- Tvede and L. Yeomans 2012 Excavations at the late epipalaeolithic site of Shubayqa 1: Preliminary report on the first season. *Neo-Lithics* 2(12):3–14

Richter, T., S. Colledge, S. Luddy, D. Jones, M. Jones, L. Maher and R. Kelly 2007 Preliminary report on the 2006 season at Epipaleolithic ‘Ayn Qassiya, Azraq Ash-Shishān. *Annual of the Department of Antiquities of Jordan* 51:313–328.

Richter, T., A. Garrard, S. Allcock and L. Maher 2011 Interaction before agriculture: Exchanging material and sharing knowledge in the final Pleistocene Levant. *Cambridge Archaeological Journal* 21(1):95–114.

Richter, T. and L. Maher 2013 The Late Epipalaeolithic in the Azraq Basin: A reappraisal. In *Natufian Foragers in the Levant. Terminal Pleistocene Social Changes in Western Asia*, edited by O. Bar-Yosef and F. R. Valla, pp. 429–448. International Monographs in Prehistory 19, Ann Arbor, MI.Richter, Tobias, Lisa A. Maher, Andrew N. Garrard, Kevan Edinborough, Matthew D. Jones, and Jay T. Stock

2013 Epipalaeolithic settlement dynamics in southwest Asia: New radiocarbon evidence from the Azraq Basin. *Journal of Quaternary Science* 28(5):467–479.

Richter, T., J. Stock, L. Maher and C. Hebron 2010b An early Epipalaeolithic sitting burial from the Azraq Oasis, eastern Jordan. *Antiquity* 84(324):321–334.

Roberts, N. 1983 Age, palaeoenvironments, and climatic significance of late Pleistocene Konya Lake, Turkey. *Quaternary Research* 19(2):154–171.

Rollefson, G. O. 1982 Preliminary report on the 1980 excavations at Ain El-Assad. *Annual of the Department of Antiquities of Jordan* 26:5–35.

Rollefson, G., D. Schnurrenberger, L. Quintero, R. P. Watson and R. Low 1997 Ain Soda and 'Ayn Qasiya: New late pleistocene and early Holocene sites in the Azraq Shishan area, eastern Jordan. In *The Prehistory of Jordan II. Perspectives from 1997*, edited by H. G. K. Gebel, Z. Kafafi and G. O. Rollefson, pp. 45–58. ex oriente, Berlin.

Rollefson, G. and L. Quintero and P. Wilke 1999. Bawwab al-Ghazal: Preliminary report on the testing season 1998. *Neo-Lithics* 9:2–4.

——— 2001 Azraq wetlands survey 2000. Preliminary report. *Annual of the Department of Antiquities of Jordan* 45:71–82.

Rosen, A. and N. Roberts 2006 The nature of catalhoyuk: People and their changing environemnts on the Konya Plain. In *Catalhoyuk Perspectives: Reports from the 1995–99 Seasons*, edited by I. Hodder, pp. 39–53. McDonald Institute Monographs, Cambridge and British Institute at Ankara, London.

Sorin, L., V. Anton, B. M. Miryam, P. Roi and F. Amos 2010 Late Pleistocene palaeoclimatic and palaeoenvironmental reconstruction of the Dead Sea area (Israel), based on speleothems and cave stromatolites. *Quaternary Science Reviews* 29(9):1201–1211.

Stiner, M. C., N. D. Munro and T. A. Surovell 2000 The tortoise and the hare: Small game use, the broad spectrum revolution, and paleolithic demography. *Current Anthropology* 41:39–73.

Torfstein, A., S. L. Goldstein, M. Stein and Enzel, Y. 2013 Impacts of abrupt climate changes in the Levant from Last Glacial Dead Sea levels. *Quaternary Science Reviews* 69:1–7.

Turner, R., N. Roberts, W. J. Eastwood, E. Jenkins and A. Rosen 2010 Fire, climate and the origins of agriculture: Micro-charcoal records of biomass burning during the last glacial–interglacial transition in Southwest Asia. *Journal of Quaternary Science* 25:371–386.

Waechter, J., V. Seton-Williams, D. M. Bate and L. Picard. 1938 The excavations at Wadi Dhobai 1937–1938 and the Dhobaian Industry *Journal of the Palestine Oriental Society* 18:172–186.

Whitehead, P. G., S. J. Smith, A. J. Wade, S. J. Mithen, B. L. Finlayson, B. Sellwood and P. J. Valdes 2008 Modelling of hydrology and potential population levels at Bronze Age Jawa, Northern Jordan: A Monte Carlo approach to cope with uncertainty. *Journal of Archaeological Science* 35:517–529.

Winterhalder, B. and E. A. Smith (editors) 1981 *Hunter-Gatherer Foraging Strategies: Ethnographic and Archaeological Analyses*. University of Chicago Press, Chicago.

Yeomans, L. M. and T. Richter In press Exploitation of a seasonally abundant resource: Bird hunting during the Late Natufian at Shubayqa 1. *International Journal of Osteoarchaeology*. doi: 10.1002/oa.2533

6 Living on the Edge

Pre-Columbian Habitation of the Desert Periphery of the Chicama Valley, Perú

Ari Caramanica and Michele L. Koons

Abstract

While many of the chapters in this volume provide innovative linking mechanisms in order to move beyond correlation towards statements of causation, this chapter uses local-scale and medium-term investigations to challenge a commonly posited causal relationship. We use paleobotanical and archaeological data to demonstrate that the physical constraints posed by desert environments on the North Coast of Perú were not a cause of sociopolitical division. Doing so requires adjusting the analytical scale to supplement seminal work on El Niño Southern Oscillation (ENSO) and other event-like climatic perturbations and studies of longer-term environmental change. Therefore, we propose a methodology that allows us to establish direct links between human modifications to the landscape and "medium-term" environmental change. By examining microfossil botanical remains from the surrounding landscape, we argue that the Pampa de Mocán, rather than being a desert barrier, was an agricultural landscape occupied over hundreds of years.

Introduction

On the North Coast of Perú, northern and southern cultural spheres are delineated based in large part on their geographic separation by the stretch of desert known as the Pampa de Paiján. The Paiján desert is classified by the Oficina Nacional de Evaluación de Recursos Naturales del Perú (ONERN) as a Superarid Temperate Desert (ONERN 1973). The temperatures range from 21 to 25°C and the area receives between 2–5 mm of rainfall per year, making it one of the driest deserts in the world (Wells and Noller 1999). Consequently, Paiján is characterized in the archaeological literature as a cultural, political, and linguistic barrier (Kosok 1965; Castillo and Donnan 1994; Larco 2001; Castillo and Uceda 2008).[1] In this chapter, the Pampa de Mocán, located on the southern edge of the Paiján desert and the northern edge of the irrigated Chicama River Valley, will serve a case study to demonstrate that during the pre-Columbian period some desert zones were continuously occupied, agriculturally productive landscapes, and therefore did not structure the sociopolitical divisions observed in the archaeological record (Figure 6.1). Rather, the perception that those divisions (e.g., between the Chicama-centered polities and Jequetepeque polities to the north of Paiján; see Castillo and Donnan 1994) are products of environmental conditions results from generalizations about the environments of the North Coast, which collapse under closer inspection.

Such generalizations are part of a legacy of environmental determinism on the North Coast, and together with issues of temporal and spatial scale, and the application of modern ecological categories to the past, they contribute to a confusion of correlation for causation. This is an important point of clarification for North Coast archaeology: viewing the physical constraints of the Paiján desert as causal (rather than correlative) to divisions between sociopolitical groups precludes the exploration of political, social, economic, and even environmental processes that could be at play. In this chapter, we begin by contextualizing the emergence of the commonly posited causal relationship; next, we present a methodology designed to extract data on medium-term environment change in an archaeological context; finally, we present the results of a case study in the Pampa de Mocán. The data directly link human landscape modification and environmental change, suggesting that a history of "settlement inertia" in the area led to a positive feedback loop between these variables: occupation resulted in a more amenable environment, allowing for continued occupation (see Wilkinson et al. 2004).

The Andean region has a robust legacy of environmentally determinist research. For example, the seminal work by anthropologist John Murra proposes that the distribution of natural resources structures social and political interaction at both a local and macrosocietal scale—forming what has been interpreted as a "deep structure" for Andean culture (Contreras 2010; Murra 1956; 1972; Starn 1991). Indeed, the form that this structure takes very much depends upon the bio-geographical classifications of ecological niches and life zones across the Andes. However, archaeologists working on a smaller spatial scale will find that these classifications are often too general for their purposes, and this generalization may in fact perpetuate interpretive bias.

Beginning with early bio-geographical maps, the representation of ecological diversity in the Andean region highlights a diversity of types of environmental zones but implicitly characterizes the zones themselves as ecologically homogenous. Alexander von Humboldt's work in Latin America between 1799 and 1804 resulted in one of his greatest contributions to Perú and the field of bio-geography: an innovative cross-section map that established a three-part system for the Central Andes: the coast, the highlands, and the Amazonian rainforest. This simplified three-part model has persisted into contemporary times, and as Orlove (1993:332) recognizes, Peruvian students today continue to conceive of their geography as consisting in these

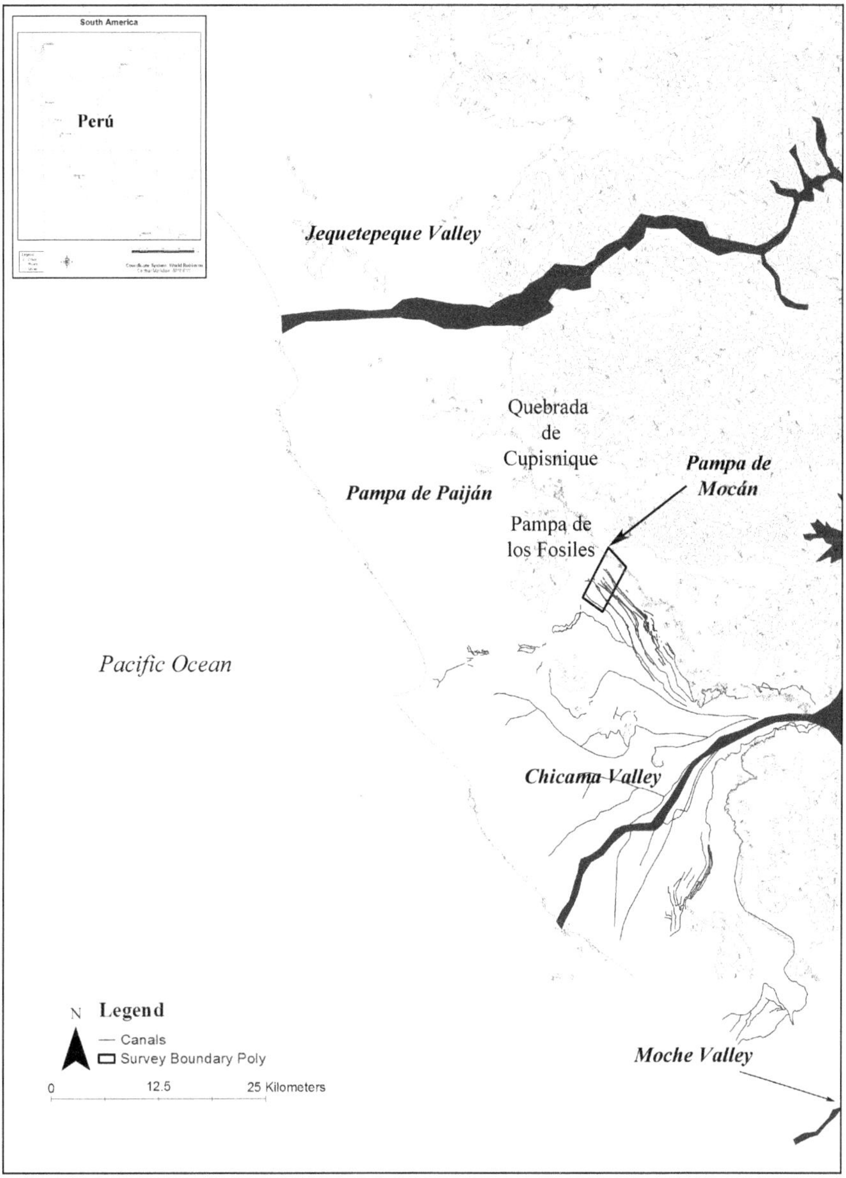

Figure 6.1 Three valleys of the North Coast, Moche Valley to the south, Chicama Valley and Jequetepeque Valley in the north. The survey area of Mocán is indicated, along with areas explored by Larco Hoyle in the 1940s. Only the irrigation system for the Chicama Valley is illustrated.

Source: Elevation data derived from NASA's Shuttle Radar Topography Mission (SRTM) (2000)

three zones. In 1938, Javier Pulgar Vidal published a new ecological map of Perú, one that explicitly reacted against the simplicity of the three-part model (Pulgar Vidal 1987). *Las ocho regiones naturales del Perú* labeled eight zones with *quechua* terms informed by elevation and human agricultural practice (Zimmerer 2011a; 2011b). However, though Pulgar Vidal nearly tripled the number of bio-geographical regions, his scheme is also necessarily a simplification. In the Pulgar Vidal system, for example, the *chala* or coast is the area below 500 masl, and the word "chala" itself refers to maize. While this area encompasses a diversity of types of landscapes (flat stretches of land, mountainous areas, deserts, and river valleys), only some of which are amenable to the production of maize, by means of this categorization, the Peruvian coast effectively becomes a monolithic category. Indeed, most archaeologists would agree that bio-geographical maps at a national scale are perhaps too simplistic for many archaeological purposes; however, even when ecological reports are at the scale of a single valley, they continue to be problematic when applied uncritically to the past.

For Pulgar Vidal, subsistence practice across the regions was an integral variable in defining the regions themselves, creating a tenuous tautology: did subsistence practice help define an ecological zone or did the zone determine practice? Moreover, conflating ecological life zones with human-use zones ties the environment to modern land-use practices, and importantly for archaeology, to legacies of past land use and environment. This has had ongoing consequences: for instance, the "land-use" variable was central to the heavily cited reports by the Oficina Nacional de Evaluación de Recursos Naturales del Perú (ONERN), which focused on smaller geographic units of typically one or two valley systems. ONERN's 1973 study of the Chicama Valley resulted in the identification of 6 life zones, but only one, "Premontane Desert" identified below 400 masl[2] (ONERN 1973). This zone was further divided into 6 areas, based largely on modern land use. ONERN depended on aerial photographs to determine the types of land use in the valley (agricultural, marginal, prehistoric habitation, etc.), which was practical because the 10,000-m elevation of the 1940s and 1960s aerial photographs[3] of the Chicama Valley resulted in a stark contrast between those areas that were and were not under cultivation. However, the accompanying report omits an important caveat: an area that is not under modern cultivation is not necessarily agriculturally useless. Additionally, large adobe mounds or *huacas* are the most visible archaeological remains in the black-and-white aerial photographs, but certainly not the only ones in the landscape. Ultimately, the Chicama Valley study depended on imagery that biased certain land-use signatures and archaeological remains. The Pampa de Paiján and the desert margins surrounding the irrigated floodplain were classified as agriculturally useless and these conclusions have profoundly contributed to the assumption that the Paiján desert was a cultural barrier in prehistory (see Castillo and Donnan 1994; Billman 2002; Toshihara 2002).

The assumption that the desert conditions of Pampa de Paiján caused a division between the cultural spheres to the north and south, derives, in part, from the difficulty of reconciling scale: modern environmental maps in Perú are drawn at either a national, department, or regional scale, while most archaeological work is concentrated on a single site or survey area. Moreover, ONERN's and other bio-geographical categories were not designed for archaeological application; while combining both environmental and human-use data into one category is not necessarily problematic for phyto- and even cultural geographers, it is a problematic conflation for archaeologists. Meanwhile, while the legacy of environmental determinism on the North Coast has produced seminal work on the sociopolitical effects of both climatic events, such as ENSO-related flooding and drought, and long-term climatic fluctuation, we propose complementary investigations into medium-term environmental change. In other words, how has human modification to landscape changed the environment over the past three millennia? Archaeological investigations require a methodology that meets both spatial and temporal scalar needs; this can be achieved by combining climate data collected from an archaeologically relevant context with settlement data. For our work in the Mocán region, a desert area at the margin of the Chicama Valley, we combined microbotanical and survey data in order to examine the past environment of Mocán empirically; this methodology allows us to avoid the pitfalls of relying on coarse-grained bio-geographical classifications as a basis for environmental reconstruction.

Mocán Case Study

Background

Throughout the early twentieth century, archaeological investigations were concentrated within the irrigated floodplain of the Chicama Valley and focused on large ceremonial centers (see Quilter and Castillo 2010). However, in 1934, Rafael Larco Hoyle ventured into the Paiján desert to the Quebrada Cupisnique where he discovered a new ceramic type at the site of La Arenita. He called the ceramic style "Cupisnique" (Larco 1938:11). He also visited the "Pampa de los Fosiles," a nearby quebrada and alluvial fan where he encountered the same ceramics. Decades later, Carlos Elera (1999) uncovered Early, Middle, and Late Cupisnique temples, domestic structures and graves at the site Puémape dating to 1000–500 B.C. Both Larco and Elera speculated about Cupisnique subsistence in the environment: Larco noted the dry "avenues of water" which drastically modified the topography (Toshihara 2002). Elera (1993) hypothesized that the Cupisnique residents withstood the semi-desertic conditions by relying on a variety of resources, most notably land snails, whose natural habitat is the *Giganton* cactus. Elera (1999; 1993) also found evidence for excellent clay resources in the dry riverbeds, anthracite quarries, and copper mines.

In terms of later cultural periods, while no published data on Moche sites in the area exists (see Leonard and Russell 1992),[4] several recent studies by Luis Jaime Castillo (2009), John Warner and Edward Swenson (Swenson 2007; Swenson and Warner 2012), and Michele Koons (2012), point to the movement of Moche fine ware across the Pampa. Herbert Eling (1977; 1987) and Richard Watson (1979) record later settlements in the Pampa de Mojucape south of the Jequetepeque Valley and in the area of Mocán in the Chicama Valley, respectively. Importantly, these studies provide evidence of agricultural communities, though both Eling and Watson hypothesize an ambitious but brief Chimu expansion followed by technological failure and later abandonment.

In his 1979 Master's thesis, Richard Watson laments the absence of fieldwork in what he called the Pampa de Mocán, an area of over 5,800 ha in the northeast margin of the Chicama Valley. Based on 1943 aerial photographs, he identifies seven large intake canals that he attributed to Chimu construction (A.D. 1100–1460). Due to the lack of evidence for settlement in the photographs, he hypothesizes that labor was imported by the Chimu state to develop a complete irrigation system and even comments that the famous Ascope aqueduct may have been constructed to feed the Mocán area. He concludes that the Chimu pushed the land beyond its physical limits and soon after abandoned the area.

Contrary to Watson's observations, after two seasons of reconnaissance and settlement survey in Mocán, our project has gathered evidence that the area was in fact occupied permanently and continuously in the ceramic periods, from the Early Horizon (900–500 B.C.) to the Late Intermediate Period (A.D. 1100–1460) and that these millennia of human settlement resulted in a significantly different environment than present.

Methodology

Recent trends in geography have emphasized the potential for instability in nature and advocated a move away from ideas of nature seeking balance or even "punctuated equilibrium" (see Moseley 1987; Zimmerer 2000). However, archaeologists working on environmental change in the coastal region of Perú have traditionally focused on catastrophic events or long-term trends. Meanwhile, the environmental information most relevant to multi-phase sites could be described as "medium-term" change. Medium-term change can be brought about in part by human input and affects the environment for several millennia. However, the potential for human action to generate significant change to the environment has yet to be fully explored on the North Coast.[5] Despite the challenges to human occupation, Mocán is an ideal area to research human-environment dynamics: the desert conditions make the Mocán region a sensitive proxy for climate change, it has been largely untouched by modern agricultural intrusions, and the dry conditions allow for exceptional preservation of buried archaeological remains.

Finally, though this area has not been archaeologically explored since the 1990s, observations made from aerial photographs reveal an extensive prehispanic canal system (see Watson 1979; Chauchat et al. 1998).

This chapter brings together the efforts of two projects. The Chicama Valley Land-Use Project, carried out in 2013, focused on the collection of sedimentary samples from cores augered in areas outside of archaeological sites, and the Proyecto Arqueologico-Ambiental del Valle Chicama (PAAVC), which conducted a full-coverage pedestrian survey over an area of 1,707 ha in 2014. Together, these projects consisted of team members from the Universidad Nacional del Trujillo, Universidad Nacional de San Marcos, the Pontificia Universidad Católica del Perú, the Universities of Harvard and Stanford, and the Denver Museum of Nature & Science. The goal of our methodology was to combine paleobotanical and settlement data from the area of Mocán.

Regional Scale Data

Regional data is often difficult to re-scale for application to local environmental reconstruction, but it provides useful information on broad trends over time. Peruvian quaternary climate change data exists almost exclusively at a regional scale. Here, we review the extant data and propose to qualify it with complementary, finer-grained evidence.

Regional data for the Andes has been gleaned from ice and lake cores, ecological analyses, and some archaeological deposits. Ice cores from Quelccaya, Huascaran, and, in Bolivia, Sajama Volcano, and several lake cores along with other climate proxies provide coarse-grained data on precipitation patterns and land cover for the broader Central Andean region (see Thompson et al. 1985; Thompson et al. 1995; Abbott et al. 1997; Contreras 2010: Figure 1, Table 1). For example, ice cores from Quelccaya indicate greater precipitation in A.D. 610–650, 760–1040, 1500–1720, and 1870–1984; similarly, periods of drought have been identified during A.D. 540–730, 1250–1310, and 1720–1860 (Thompson et al. 1985; Moore 2014). However, while such climate proxies serve as a guide to general patterns of climate change across the region, it is difficult to project the effect of such fluctuations for the Chicama Valley specifically. To reach a somewhat finer analytical scale, modern ecological data can point to persistent and long-term environmental trends. For example, botanist Augusto Weberbauer (1945) described the coast north of Cerro Campana (located at 8° latitude, north of Trujillo) as the "grand flatness." The inclination of rivers on the coast reduces significantly north of this latitude causing many of the rivers to terminate before reaching the coast. Beginning with the Chicama Valley and north to latitude 6.40, the coastal area does not receive the spring and winter *garua* or "drip-fog," with very rare exceptions. Therefore, according to Weberbauer, the famous coastal *lomas* or fog-vegetation environments cannot form in this area except on exceptionally rare and probably brief occasions. Finally, mollusk deposits from archaeological contexts can reflect the warmer currents associated with ENSO (Sandweiss et al. 2007; 2001; 1996). However, these sources of data cannot be successfully scaled down to a smaller geographic area without corroborating evidence. Instead, we propose a complementary approach, combining archaeological data with paleobotanical data in order to form a more robust reconstruction of the past local environment.

Local Scale Data

Paleobotanical Survey

When phytolith, pollen, and starch data are combined, they have the potential to provide insights into environmental change and agricultural production for the collection area (Piperno 1991; Piperno and Holst 1998). While pollen data often provides more information about plants at the species level, because pollen grains are designed for transport, their spatial resolution is often coarse. However, phytoliths and starch grains are heavier than pollen grains and tend to fall in place; therefore, they tell a more "local" story about a given collection unit, under normal depositional conditions. Several examples of sampling directly from cultural contexts or fossil fields have yielded promising results for reconstructing both agricultural practices and past environment (Pearsall and Trimble 1984; Huang and Zhang 2000; Hayashida 2006). David Beresford-Jones and colleagues' work in the Samaca Basin in the Ica Valley demonstrates the potential for methods that combine paleobotany and archaeology to reconstruct medium-term landscape transformations (Beresford-Jones et al. 2009; 2004). Deborah Pearsall, one of the first proponents of archaeological applications for microbotanical data, established a sample rate of 3–7 soil samples per field in her study of early agricultural activity in the Hawaiian Islands (Pearsall and Trimble 1984). While our project did not collect in situ samples from archaeological sites,[6] sampling locations were selected in order to capture a representative microbotanical record of both the surrounding environment and the agricultural landscape of Mocán. Cores were collected from 4 points in the landscape, one each from a northern, southern, eastern, and western point on the pampa. Two layers were recorded from each core, resulting in 8 total samples.[7] While not sufficient to characterize the area in chronological or ecological detail, this sample population provides a basis for environmental reconstruction.

Settlement Survey

The 1,707-ha survey area was selected based on observations made from 1943 aerial photographs purchased from the Servicio Aerofotografico Nacional del Perú (SAN). The purpose of this project was to test whether or not occupation was long term and if there was evidence for specialization or non-agricultural production in order to test the claim that the landscape was the result of a brief state-sponsored settlement "thrust" (Grossman 1980). Therefore we decided to carry out a full coverage survey to avoid a bias towards sites clustered around irrigation canals (see Falconer and Savage 1995; Banning 1996). For the purposes of our survey, we established 3 categories of sites: infrastructural (including isolated walls, roads), use-spaces (defined either by a concentration of artifacts related to a particular function or dwelling architecture), and complexes (a series of one or both of the first two categories that are integrally related). Originally, the survey had been planned as 22 quadrants of, on average, 75 ha each. However, site visibility was low due to erosion, dune formation, embankments, and ravines. Therefore, the quadrants were further divided by sector, based largely on topographic characteristics. The intake canals and aqueducts served as dividing lines between and within sectors. We collected only diagnostic ceramic fragments at a rate of 25 percent; 1,713 fragments were collected. A total of 64 sites were registered and their chronological and cultural association determined based on associated ceramics, construction materials and technique, and architectural features. The sites ranged in period from the Early Horizon (900–500 B.C.) to the Late Intermediate Period (A.D. 1100–1460) (Figure 6.2).

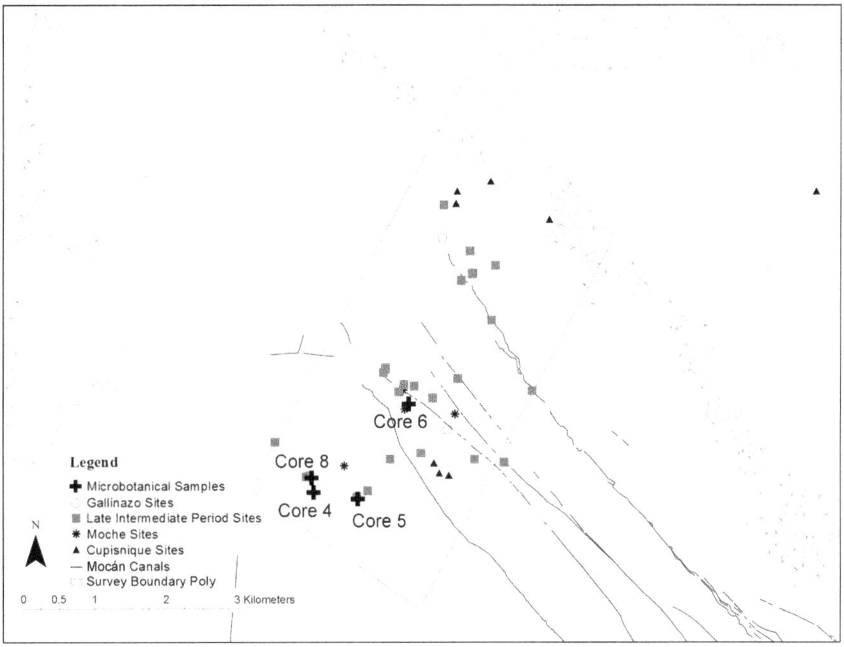

Figure 6.2 Mocán survey results. Survey area of Mocán is outlined. Late Intermediate Period sites are indicated by squares, Gallinazo sites by circle outlines, Moche sites with asterisks, and Cupisnique sites with triangles. Ancient intake canals are also outlined.

Source: Elevation data derived from NASA's Shuttle Radar Topography Mission (SRTM) (2000)

There are several important caveats to interpretation of survey results without test excavations. Importantly, we were unable to fully account for the topographic transformations that have taken place over the centuries—many sites are likely buried or destroyed by flood events, dune formations, and erosional processes, adding bias to our sample (Wilkinson et al. 2004). Moreover, processes of deflation erase vertical stratigraphy, and chronologically distinct materials can occupy the same surface. Other important pitfalls are the lack of absolute dates and microbotanical samples drawn from in situ archaeological site contexts. However, these issues do not nullify the evidence that the Pampa de Mocán was a vastly different landscape than present conditions suggest: it was occupied nearly continuously over millennia, was agriculturally productive, and was the site of specialized production and ritual.

Results

Survey Results

If occupation at Mocán was in fact the result of state-mandated settlement, we might expect a short occupational time span along with a settlement pattern that consisted of small, domestic structures with one or two administrative outposts, indicative of Chimu expansion. A ceramic signature for a state-sponsored project would involve ceramics that were overwhelmingly utilitarian and from a single time period. If the occupants were groups of corvée labor, standardized wares, such as the Mesopotamian "bevelled-rim bowls" for food rations might make up a portion of the assemblage (Millard 1988). The results of our survey demonstrate a variety of site sizes, with 3 percent of sites ranging from 1–4 m^2, 50 percent of sites ranging from 4–50 m^2, 42 percent from 51–250 m^2, and 5 percent from 251–900 m^2. Site function varied as well: 53 percent were determined to be domestic, while administrative sites, ritual architecture, burial sites, and production sites (including a ceramic workshop) made up the remaining 47 percent. Similarly, ceramics were predominantly plain-wares used for cooking and storage (73 percent), but 16 percent were elite fine wares, and 11 percent fine serving wares. The functional and temporal variety of both sites and ceramics indicate that Mocán was a place unto itself, not a "thrust" or satellite community. Inhabitants occupied property that they had some claim of ownership over, produced their own lithic tools and pottery, and lived and died there.

Evidence for Irrigation

The presence of desert varnish, field-picked stone piles, and the location of fields on top of ridge-and-swale topography confirm that the Pampa de Mocán was a desert at the time that settlers chose to develop it agriculturally.

Furthermore, water could not reach the outer limits of irrigation systems except during the summer season (December–April) when the Chicama River flow increased from a low of $264{,}130 \times 10^3$ m^3 in the winter to a high of $2{,}666{,}975 \times 10^3$ m^3 (Watson 1979: Table 7).[8] However, the construction of the seven intake canals across the 27 km-distance from the nearest point on the Chicama River would have involved a great number of laborers and a sophisticated organizational hierarchy. Ortloff et al. (1985) estimate that the construction of the Vichansao canal in the Moche Valley, which is 30 km in length, required 140,000 labor hours. Extrapolating from his calculations, we postulate that the seven intake canals alone required 882,000 labor hours. The Vichansao canal would have involved 4,500 laborers if built in one year, and 900 if built over five years (Billman 2002:383). However, given scholars' estimates for both population and levels of sociopolitical organization, especially for the earlier archaeological cultures we have recorded in Mocán, this kind of labor recruitment would be extraordinary. Still, there is a significant Cupisnique presence on the Pampa. We hypothesize that Early Horizon (900–500 B.C.) groups were using small canals to capture seasonal runoff from the nearby foothills and water table farming and later occupants continued and expanded on these systems, creating monumental irrigation networks by drawing on larger populations and collaborative labor.

Evidence for Water Table Farming

Cupisnique sites were identified both on high ground and on the alluvial plain. Two sites were located within 50 m of unusual fields. The fields were small, measuring on average 18 by 20 meters, and consisted of a structure of piled stone (1.7 m high) on the upslope side and berm-like structures consisting of small pebbles (12–40 cm high and approximately 1 m in width). Geoglyphs were often found nearby and typically formed a serpentine or snail-like figure. After comparing our mapping of this field pattern to those mapped in Llanos de Mojo, Bolivia, and Lake Titicaca, we are confident that these are in fact examples of ridged fields (Denevan 1970). Ridged fields are not unprecedented on the desert coast, and Pozorski et al. (1983) report Chimu-period rectangular ridged fields in the Casma Valley (see also West 1979; Moore 1988). Though excavation is required to confirm these hypotheses, it appears Cupisnique settlers used the water table and runoff canals to irrigate small fields and possibly marked water sources with geoglyphs (Table 6.1).

By Gallinazo and Moche periods (A.D. 50–900), canal technology was well established, and it is during this period that the first intake canals crossing Mocán were constructed. Later modifications to several of the main intake canals suggest these were earlier constructions later altered and expanded by Lambayeque and Chimu occupants between A.D. 900 and 1470. These are possibly some of the first archaeological examples of intravalley canals found in the Chicama. Based on site distribution, Lambayeque

Table 6.1 Regional chronology and irrigation

Working Chronology	*Irrigation Patterns*
Late Intermediate Period—Chimu and Lambayeque (A.D. 900–1476)	Re-use and expansion of large intra-valley canals; continued use of water table
Early Intermediate Period—Salinar, Gallinazo, and Mochica (A.D. 50–900)	Large, intra-valley canals carry water from the Chicama River
Occupation unknown (500 B.C.—A.D. 50)	Unknown
Early Horizon—Cupisnique (900–500 B.C.)	Small runoff canals and water table farming; important points in the landscape marked by geoglyphs
Occupation unknown (6300–900 B.C.)	Unknown
Paiján Complex (8800–6300 B.C.)—for further information see Chauchat (1988; 1998)	Unknown

and Chimu occupants expanded the system by constructing several new intake canals and possibly incorporating them into the Ascope aqueduct system. However, they too apparently continued to use the water table, as evidenced by several walk-in wells (Table 6.1).

Another solution for low water flow through the canals was manual. It is likely that at various times throughout the year, little water reached the fields excavated below the intake canals. At almost each intake, a small water jug or jug fragments were found. Settlers likely moved water onto their fields themselves, by hand, when low water volume failed to transport water via gravitational pull (Hatch 1976).

ENSO Evidence

Past El Niño flooding is manifested in the sharp relief of what Larco called "water avenues," or dry watercourses that cross the Mocán landscape (Larco 1938; Toshihara 2002). ENSO risk has been endemic on the North Coast throughout the latter Holocene (see Sandweiss et al. 1996), and inhabitants dealt with that reality in various ways. Canals themselves can serve as ENSO-management infrastructure: in one example, each canal berm measured approximately 1.76 m above the modern surface and 7.58 m in width, measuring between the apex of each berm—i.e., they were in some cases engineered to withstand high flows. Other strategies employed to avoid flood destruction included occupying high ground, and Mocán occupants during the Lambayeque and Chimu periods (A.D. 900–1470) constructed aqueducts to minimize the length of the canal that passed through a potential watercourse, thereby reducing the possibility of destruction to the system. Abandoned canals functioned as reservoirs and extensive fields were excavated inside the canal bed. Meanwhile, forms of social and technological infrastructure, such as water catchment constructions, opportunistic fields, and maintaining population mobility have been recorded as responses to ENSO on the North Coast (Dillehay and Kolata 2004). Ethnographic studies have recorded many farming strategies that help manage the effects of low-impact ENSO, including crop choice (see Puri 2007). Rajindra K. Puri's (2007) work with farmers in Borneo observed that despite the effects of ENSO, farmers rarely suffered crop failure or shortage. He identifies four general categories of strategies: tolerant crops, staggered planting, irrigation technology, and increased labor or dependence on kinship networks. Finally, observations made by Cesar Galvez along the Chicama Valley edges after the 1998 El Niño event demonstrate the adaptive powers of valley occupants: farmers moved from their ruined fields and flooded homes into the desert where they used the temporary water sources to continue growing their crops (Gálvez Mora and Runcio 2011). In other words, there are several potential strategies that aid in managing the effects of ENSO in its various forms (see also Brooks et al. 2005; Billman and Huckleberry 2008).

Paleobotanical Results

The two existing hypotheses about the environmental history of the Pampa de Paiján, and by implication the Pampa de Mocán, are (1) that it has been composed of xerophyte plant regimes since at least the early Holocene; or (2) that during the Late Intermediate Period, the Chimu state sponsored an agricultural program in order to meet the needs of its growing population, principally through valuable crops such as maize and cotton (Watson 1979). The paleobotanical results from our samples suggest that neither hypothesis is adequate and testify to a vastly different environment.

As discussed above, pollen can provide taxonomic precision but fails to provide spatial precision, while phytoliths and starch grain are often identified to a genus or family level but can tell a much more local story. Indeed, pollen rain is capable of traveling hundreds of miles depending on the means of transport, wind conditions, temperature, and the vegetative populations in its path (Faegri et al. 1989; Weng et al. 2004). Meanwhile, phytoliths and starch reach the sediments on the ground because a plant or plant part falls and decomposes. Water or in some cases animal (including human) or wind transport could affect their location before or after decomposition, but on a much smaller scale than pollen. In the area of Mocán, we know that water transport would be limited to irrigation and flooding events, and wind erosion has effected the preservation of the superficial soil layer. By sampling in two layers from each core, we can compare deeper microbotanical results against more recent results to test for contamination. We found no non-native (i.e., post-contact) taxa among our data, which is a testament to the lack of contamination especially considering rice, sugarcane, and various other non-native species are grown in vast quantities nearby. The lack of non-native taxa also indicates that the catchment area is limited to the upper valley and surrounding desert, rather than the lower, irrigated valley. Therefore, by considering pollen, phytolith and starch data together, we are able to draw conclusions about the vegetative population in Mocán in the past, and consequently, through comparison with modern-day conditions, about landscape change and human-environment dynamics over time.

Microbotanical results show the presence of trees, shrubs, and herbaceous plants in Mocán in the past. In the lowest strata of our cores, the presence of trees such as *Schinus molle* are abundant and later diminish, giving way to shrub and herbaceous species over time. Several taxa suggest an abundance of water at various moments in the past. Poaceae make up the majority of the microbotanicals recovered, and among these, a significant presence of Bambusoideae—which are typically forest-dwelling and require humidity. Moreover, the presence of diatoms suggests that standing water was nearby (Veintinilla 1999). Other families are more adaptive: Chloridoideae, for example, are typically weeds and/or short, drought-tolerant grasses. Meanwhile, Asteraceae are often considered invasive plants commonly found in fields. Overall, the microbotanical results represent a vegetation community made up by predominantly herbaceous and shrub plants that preferred humid environments (Figure 6.3).

The cultigens, instead of consisting of one or two taxa, make up a variety of annuals and perennials, drought-tolerant and sensitive plants, cash crops, root crops and tubers, fruit trees, medicinal plants, and dyes. Given that the pollen data largely represents plant families, we estimate that 55 percent of cultigens were perennials and 45 percent were annuals. Pollen data also records the presence of several taxa that produce fruit, including Maracuya (passion fruit) and Lucuma. Using phytolith and starch data, we can make some inferences about spatial distribution of cultigens across the landscape: for example, in one sector, crops included Quina or Cinchona, a medicinal plant famously used to treat malaria during the colonial period, legumes, maize, and cotton (Yakovleff and Herrara 1934; Ugent and Ochoa 2006; Towle 2007[1961]). In another, we find a concentration of maize, root crops (sweet potato and yuca), and *Schinus molle* (the Peruvian Pepper Tree) (see Figure 6.3).

Two of these taxa stand out for their problem-solving potential. The presence of at least two legume subfamilies (Mimosoideae, Faboideae) are pertinent in relation to the question of nitrogen deficiency. In addition to supplementary nutrients that can be added through fertilizers such as *guano*, the Fabaceae family consists of excellent nitrogen fixers, which can replenish soil N where they are grown. Meanwhile, *Schinus molle* is a woody tree that reaches up to 15 m in height and 85 cm in width (Goldstein and Coleman 2004). It has a number of uses: its fruit can be made into chicha and pepper, resin was used for ceramic and gourd fractures, and the wood and leaf products for lumber, fuel, and dyes. Crucially, in addition, the tree creates its own microenvironments: in the high Andes, its canopy protects against frost, and in the coastal deserts, it provides shade and insulation.

The crop regime found at Mocán, therefore, fits Puri's 2007 ethnographic observations of indigenous farming strategies against ENSO. A mixture of perennials and annuals were planted and staggered across the landscape.

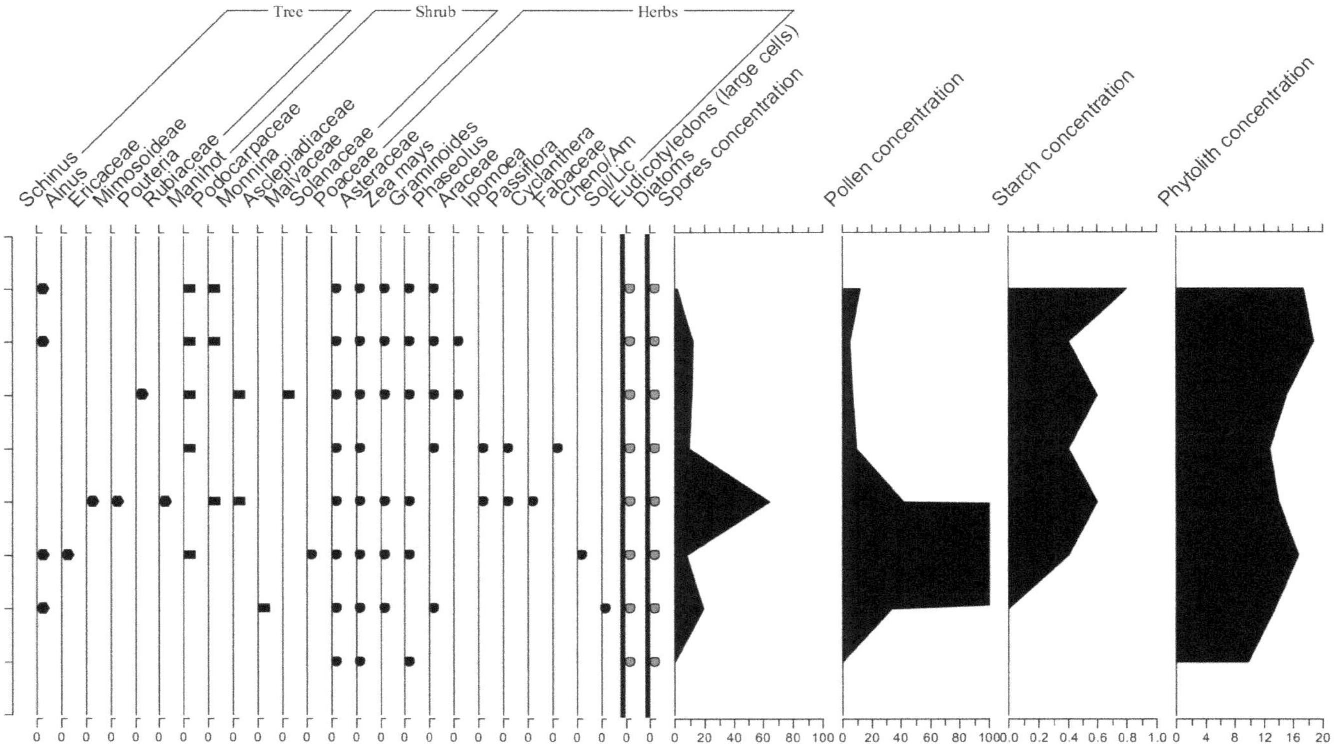

Figure 6.3 Paleobotanical results from all four sediment core samples including pollen, phytoliths, and starch remains. Symbols represent presence/absence data and shaded histograms represent sample concentrations of spores, pollen, starch, and phytolith remains.

Drought-tolerant plants were grown, as well as more water-sensitive grasses such as maize. Most important, however, the presence of *Schinus molle* likely created a microenvironment wherein a root system prevented erosion and raised the water table, and a branch system provided shading and protection against strong winds. Non-cultigen plants in the surrounding environment were dependent on significant amounts of water and humidity.

Due to human occupation, the Pampa de Mocán was apparently a humid and even riparian landscape for at least some periods during the last three millennia. Likely beginning with small fields fed by water runoff and water table farming, the environment became more favorable and motivated continued occupation, eventually changing the desert conditions significantly over the medium term. Small human modifications to the landscape aggregated over time leading to a changed environment, one that facilitated human occupation rather than constrained it.

Conclusions: Mocán Abandonment and the Importance of Analytical Scale in Coastal Perú

Today, the Pampa de Mocán is a desert in every sense, supporting very little vegetative life. The majority of the Pampa de Mocán was likely abandoned well before the Colonial Repartamiento de Aguas in 1565, given that neither the area nor its canals are considered in the reorganization of water distribution, and it is probable that abandonment led to its re-desertification.

At this time, we can only speculate as to why the area was abandoned. Perhaps sustained periods of low-water flow led to the contraction of both the water table and the population (Clement and Moseley 1991). Some scholars suggest that a combination of tectonic uplift, El Niño events, and drought can lead to agrarian collapse in coastal Perú (Moseley 1983; Moseley et al. 1983). However, Mocán settlers were well prepared to withstand ENSO effects, and Wells and Noller (1999) argue there is no evidence for tectonic uplift for the northern coast. Perhaps the highest impact—and most elusive—causes for abandonment were political. Several ancient highways cross the Paiján and traverse the Pampa de Mocán; the abandonment of older trade routes for the Camino Real during the Late Horizon could be a cause for the abandonment of what would have been the port of entry from the northern coast and highlands into the Chicama Valley (Ramírez 1995; see Ward-Perkins 1972). Susan Ramírez (1985) reviews colonial documents that suggest a pre-Columbian practice of "territorialidad salpicada" or dispersed or shared settlement, a form of Andean share-cropping where a local lord would rent land to another lord in exchange for a percentage of the yielded crop. Lords could potentially send their subjects many leagues away to work the land in another polity (see also Cock 1976; Cook 1976; Rostworowski de Diez Canesco 1977; Martinez 1981). Such practices could lead to the sudden movement of farmers. Téllez and Hayashida (2004) argue that walled fields emerge during the Late Intermediate Period (A.D. 1100–1476); a change in land tenure policies or economy could also lead to the breakdown of an agricultural community like Mocán. Finally, it is possible that polities wanted to create a "buffer zone" in order to more clearly define territorial boundaries or prevent interaction or warfare (Anderson 1996).

These hypotheses require different scalar approaches and methodologies, and some remain speculative. However, we can now say with certainty that if the Pampa de Paiján served as a cultural barrier in prehispanic times, it was not because of its physical constraints. We can also say that its presence as part of a political landscape was something significant—whether that was in the form of a borderland or frontier, as a point of cultural intersection, or simply in its potential for agricultural production. The research presented here has significant implications for the ways in which scholars view polity-formation and socioeconomic development on the North Coast; specifically, how these processes interdigitate with medium-term environmental change.

Finally, the Mocán case study employs a new set of spatial and temporal scales, and consequently challenges conventional sources of environmental data. Environmental categorization or zonation often takes into consideration modern land-use variables and agricultural practices—a problematic conflation for the study of past human-environment dynamics. Climate change records in the Andean region are abundant for both long-term and event-like temporal scales, but data that could support investigations into human-impact and environmental pushback over the medium term (an appropriate time-scale for the detailed reconstruction of landscape histories) is lacking. In reviewing the history of bio-geographical mapping in Perú, we find that ecological categories often serve as heuristic devices and are therefore necessary simplifications of the actual vegetative communities and topographic diversity of a given area. When applied uncritically to archaeological contexts, such categories can lead to an artificial sense of environmental homogeneity and stasis. By attempting a medium-term environmental reconstruction in a sub-regional, sub-valley geographical unit, the Mocán case study demonstrates that the past environment on the coast was neither homogenous nor immutable. Rather, it comprised multiple environments with rich landscape histories that are relevant for the understanding of culture history and re-draw our imaginings of pre-Columbian coastal environments.

Methodological Vignette: Landscape Paleobotany

The survey of paleobotanical records in and around relict field features allows for the reconstruction not only of agricultural practices, but also of past environments on a local geographic scale. For adequate preservation, this method requires soil contexts that are neither too basic nor too acidic, and preferably with stratigraphic integrity (Dimbleby 1957; Faegri and Iversen 1989; Piperno 2006). In designing a sampling strategy, it is important to account for potential modern contaminants, which means considering the catchment area of each location sampled for paleobotanical remains, and also the landscape transformations that have occurred over time (see Pearsall 2001). Therefore, we suggest sampling at four or more points around the area of interest and including samples from the surface around each core or test unit (Bryant and Holloway 1983). This method is particularly useful for investigating human-environment interaction because the plant fossils recovered can be linked to human impact, signaled by the presence or absence of invasive species, forest-dwelling and tree species, and water-dependent or dry-environment species (Pearsall and Trimble 1984; Huang and Zhang 2000; Hayashida 2006; Beresford-Jones et al. 2009; 2004). Crucially, pollen, phytolith, and starch data provide complementary spatial and taxonomic precision: while pollen data often provides more information about plants at the species level, their spatial resolution is often coarse; meanwhile phytoliths and starch grains tend to fall in place, therefore, they tell a more "local" story about a given collection unit (Piperno 1991; Piperno and Holst 1998). This type of methodology, however, does not address the recovery of reliable absolute dates (dating relict fields is a research problem unto itself); nor does it directly correlate plants to human behavior—this would require extensive excavation. However, this methodology does provide a means of assessing environmental changes at temporal and spatial scales directly relevant to a landscape's inhabitants.

Notes

1 Work by Claude Chauchat et al. (1998) in the area of Paiján revealed a Paleoindian occupation in the Paiján desert likely beginning 10800–8300 B.P. (see also Dillehay et al. 2003).

2 The most influential study of tropical ecosystems, which was later adopted by ONERN, is Holdridge's Life Zone system (Holdridge 1967 [1947]). The Life Zone system strove to be applicable to any area in the world, and therefore based on objective measurements. Holdridge defined the environmental conditions of an area or zone based principally on 4 factors: biotemperature, precipitation, potential evapotranspiration ratio, and elevation—subsistence practices were intentionally excluded from calculations. Critically, these life zones were considered divisible, and by adding more area-specific data, they could be qualified further to even describe land use—these are called "associations" (A.E. Lugo et al. 1999:1027).

3 Photographs taken by the Servicio Aerofotografico Nacional or SAN.

4 The Russell-Leonard Survey made reconnaissance observations about occupation in the Mocán area but published no data recording associated sites or ceramics.
5 Several studies on Peru's South Coast consider human action that is destructive to the environment, i.e., deforestation (e.g., Beresford-Jones et al. 2009).
6 The terms of our research permit did not allow for archaeological excavations, however, we positioned our cores to capture both environmental and agricultural pollen sources.
7 Microbotanical analyses were conducted at the Laboratory of Palynology and Paleobotany at the Universidad Cayetano Heredia del Perú, according to the procedures established by M. Horrocks (2005) for phytoliths and starch remains and standard procedure (see Eshet and Hoek 1996; Enciso-de la Vega 1992; Moore et al. 1991; Faegri and Iversen 1989).
8 These are the highs and lows from years from 1925 to 1950.

References Cited

Abbott, Mark B., Michael W. Binford, Mark Brenner and Kerry R. Kelts 1997 A 3500^{14}C yr High-resolution record of water-level changes in Lake Titicaca, Bolivia/Peru. *Quaternary Research* 47(2):169–180.

Anderson, David G. 1996 Chiefly cycling and large-scale abandonments. In *Political Structure and Change in the Prehistoric Southeastern United States*, edited by John F. Scarry, pp. 150–191. University Press of Florida, Gainsville.

Banning, E. B. 1996 Highlands and lowlands: Problems and survey frameworks for rural archaeology in the near east. *Bulletin of the American Schools of Oriental Research* 301:25–45.

Beresford-Jones, David G. 2004 Peru: Riparian forests in the context of environmental and cultural trajectories of the lower Ica Valley. *Horizon* 215–232.

Beresford-Jones, David G., Arce T. Susana, Oliver Q. Whaley and Alex J. Chepstow-Lusty 2009 The role of prosopis in ecological and landscape change in the samaca basin, Lowe Ica Valley, south coast Peru from the early Horizon to the late intermediate period. *Latin American Antiquity* 20(2):303–332.

Billman, Brian R. 2002 Irrigation and the origins of the southern moche state on the north coast of Peru. *Latin American antiquity* 13(4):371–400.

Billman, Brian R. and Gary Huckleberry 2008 Deciphering the politics of prehistoric El Niño events on the North Coast of Peru. In *El Niño, Catastrophism, and Culture Change in Ancient America*, edited by Daniel H. Sandweiss and Jeffrey Quilter, pp. 111–138. Dumbarton Oaks Research Library and Collection, Cambridge.

Brooks, William E., J. C. Willett, J. D. Kent, V. Vasquez and Teresa Rosales 2005 The Muralla Pircada—an ancient Andean debris flow retention dam, Santa Rita B archaeological site, Chao Valley, Northern Perú. *Landslides* 2(2):117–123.

Bryant, Vaughn M. Jr. and Richard G. Holloway 1983 The role of palynology in archaeology. *Advances in Archaeological Method and Theory* 6:191–224.

Castillo, Luis Jaime 2009 El estilo Mochica Tardío de Línea Fina de San José de Moro. In *De Cupisnique a los Incas : el arte del Valle de Jequetepeque : la donación Petrus Fernandini al MALI*, edited by Luis Jaime Castillo and Cecilia Pardo. 1. ed. Asociación Museo de Arte de Lima, Lima.

Castillo, Luis Jaime and Christopher B. Donnan 1994 Los mochica del Norte y los mochica del Sur. In *Vicús*, edited by Krzysztof Makowski, Christopher B. Donnan, and Ivan Amaro Bullon, pp. 143–181. Colección Arte y tesoros del Perú. Banco de Crédito del Perú, Lima, Perú.

Castillo, Luis Jaime and Santiago Uceda 2008 The mochicas. In *The Handbook of South American Archaeology*, edited by Helaine Silverman and William H. Isbell, pp. 707–729. Springer, New York.

Chauchat, Claude 1988 Early hunter-gathers on the Peruvian Coast. In *Peruvian Prehistory: An Overview of Pre-Inca and Inca Society*, edited by Richard W. Keatinge, pp. 41–66. 2nd ed. Cambridge University Press, Cambridge, MA.

Chauchat, Claude, César Gálvez Mora, Jesús Briceño Rosario and Santiago Uceda 1998 *Sitios Arqueológicos de la Zona de Cupisnique y Margen Derecha del Valle de Chicama*. Instituto Nacional de Cultura, Trujillo and Instituo Francés de Estudios Andinos, Lima.

Clement, Christopher Ohm and Michael E. Moseley 1991 The spring-fed irrigation system of Carrizal, Peru: A case study of the hypothesis of agrarian collapse. *Journal of Field Archaeology* 18:31–35.

Cock, Guillermo C. 1976 Los kurakas de los Collaguas: Poder politico y poder económico. *Historia y cultura* 10:95–118, Lima, Peru.

Contreras, Daniel A. 2010 Landscape and environment: Insights from the Prehispanic Central Andes. *Journal of Archaeological Research* 18(3):241–288.

Cook, Noble David 1976 La visita de los Conchucos por Cristóbal Ponce de León, 1543. *Historia y cultura* 10:23–45, Lima, Peru.

Denevan, William M. 1970 Aboriginal drained-field cultivation in the Americas. *Science, New Series* 169(3946):647–654.

Dillehay, T. D. and A. L. Kolata 2004 Long-term human response to uncertain environmental conditions in the Andes. *Proceedings of the National Academy of Sciences of the United States of America* 101(12):4325.

Dillehay, Tom D., Jack Rossen, Greg Maggard, Kary Stackelbeck and Patricia Netherly 2003 Localization and possible social aggregation in the Late Pleistocene and Early Holocene on the North Coast of Perú. *Quaternary International* 109–110:3–11.

Dimbleby, G. W. 1957 Pollen analysis of terrestrial soils. *New Phytologist* 56:12–28.

Elera Arevalo, Carlos Gustavo 1993 El Complejo cultural cupisnique: Antecedentes y desarrollo de su ideología religiosa. *Senri Ethnological Studies* 37:229–257.

——— 1999 *The Puemape Site and the Cupisnique Culture: A Case Study on the Origins and Development of Complex Society in the Central Andes, Peru*. Unpublished Ph.D. Dissertation, Department of Anthropology, University of Calgary, Canada.

Eling, Herbert Henry 1987 *The Role of Irrigation Networks in Emerging Societal Complexity During Late Prehispanic Times: Jequetepeque Valley, North Coast, Peru*. (volumes I and II). Unpublished Ph.D., University of Texas, Austin.

——— 1977 Interpretaciones Preliminares del Sistema de Riego Antiguo de Talambo en el Valle de Jequetepeque, Perú. In *El Hombre y la Cultural Andina, Actas y Trabajos*, Tomo II:401–419, Lima.

Enciso-de la Vega, Salvador 1992 Technique of sample preparation for palynological analysis: *Bol. Depto. Geol. Uni-Son* 9(2):101–107.

Eshet, Yoram and Ramses Hoek 1996 Palynological processing of organic-rich roacks, or: How many times have you called a palyniferous sample "barren"? *Review of Paleobotany and Palynology* 92:101–109.

Faegri, Knut, John Iversen, Peter Emil Kaland and Knut Krzywinski 1989 *Textbook of Pollen Analysis, IV Edition*. The Blackburn Press, New Jersey.

Falconer, Steven E. and Stephen H. Savage 1995 Heartlands and hinterlands: Alternative trajectories of early urbanization in mesopotamia and the Southern Levant. *American Antiquity* 60(1):37–58.

Gálvez Mora, César A. and María Andrea Runcio 2011 Eventos ENOS (El Niño, La Oscilación del Sur) y el cultivo de maíz en el desierto del sector medio del valle de Chicama, Perú. *Revista Archaeobios* 1(5):79–96.

Goldstein, David John and Robin Christine Coleman 2004 Schinus molle L. (Anacardiaceae) Chicha Production in the Central Andes. *Economic Botany* 58(4):523–529.

Grossman, David and Zeev Safrai 1980 Satellite settlements in western samaria. *Geographical Review* 70(4):446–461.

Hatch, John K. 1976 *The Corn Farmers of Motupe: A Study of Traditional Farming Practices in Northern Coastal Peru*. Land Tenure Center Monographs No. 1. University of Wisconsin, Madison.

Hayashida, Frances M. 2006 The pampa de chaparrí: Water, land, and politics on the north coast of Peru. *Latin American Antiquity* 17(3):243–263.

Holdridge, L. R. 1967 [1947] *Life Zone Ecology*. Tropical Science Center. San José, Costa Rica.

Horrocks, M. 2005 A combined procedure for recovering phytoliths and starch residues from soils, sedimentary deposits and similar materials. *Journal of Archaeological Science* 32(8):1169–1175.

Huang, Fei and Min Zhang 2000 Pollen and phytolith evidence for rice cultivation during the Neolithic at Longqiuzhuang eastern Jianghuai, China. *Vegetation History and Archaeobotany* 9(3):161–168.

Koons, Michele L. 2012 *Moche Geopolitical Networks and the Dynamic Role of Licapa II, Chicama Valley, Peru*. Unpublished Ph.D. Dissertation, Department of Anthropology, Harvard University, Cambridge, MA.

Kosok, Paul 1965 *Life, Land, and Water in Ancient Peru : An Account of the Discovery, Exploration, and Mapping of Ancient Pyramids, Canals, Roads, Towns, Walls, and Fortresses of Coastal Peru with Observations of Various Aspects of Peruvian Life, both Ancient and Modern*. Long Island University Press, New York.

Larco Hoyle, Rafael 1938 *Los mochicas*. Casa editora "La Crónica" y "Variedades", s. a. ltda., Lima.

——— 2001 *Los Mochicas*. Museo Arqueológico Rafael Larco Herrera, Lima.

Leonard, Banks and Glenn Russell 1992 *Informe Preliminar: Proyecto Reconocimiento Arqueológico del Chicama, Resultados de la Primera Temporada de Campo, 1989*. Instituto Nacional de Cultura, Trujillo, Peru.

Lugo, A. E., S. L. Brown, R. Dodson, T. S. Smith and H. H. Shugart 1999 The Holdridge life zones of the conterminous United States in relation to ecosystem mapping. *Journal of Biogeography* 26(5):1025–1038.

Martinez, Gabriel 1981 Espacio Lupaqa: Algunos hipótesis de trabajo. In *Ethnohistoria y antropología andina*, edited by Amalia Castelli, Marcia Koth de Paredes and Mariana Mould de Pease, pp. 263–280. Centro de Proyección Cristiana, Lima, Peru.

Millard, Alan R. 1988 The bevelled-rim bowls: Their purpose and significance. *Iraq* 50:49–57.

Moore, Jerry D. 1988 Prehistoric raised field agriculture in the Casma Valley, Peru. *Journal of Field Archaeology* 15(3):265–276.

——— 2014 *A Prehistory of South America: Ancient Cultural Diversity on the Least Known Continent*. University Press of Colorado, Boulder.

Moore, P. D., J. A. Web and M. E. Collinson 1991 *Pollen Analysis*. Blackwell Scientific Publications, Boston.
Moseley, Michael E. 1983 The good old days were better: Agrarian collapse and tectonics. *American Anthropologist* 85(4):773–779.
——— 1987 Punctuated equilibrium: Searching the ancient record for El Niño. *Quarterly Review of Archaeology* 8:7–10.
Moseley, Michael, Robert Feldman and Charles Ortloff 1983 principles of agrarian collapse in the Cordillera Negra, Peru. *Annals of the Carnegie Museum* 13:299–327.
Murra, John 1956 *The Economic Organization of the Inca State*. Ph.D. Dissertation, Anthropology Department, University of Chicago.
———1972 El "control vertical" de un máximo de pisos ecologicos en la economía de las sociedades andinas. In *Visita de la Provincia de Leon de Huánuco en 1562*, edited by John Murra, pp. 427–476. Universidad Nacional Hermilio Valdizan, Huánuco.
ONERN 1973 *Inventario, evaluación y uso racional de los recursos naturales de la costa—cuenca del río Chicama*. República del Perú, Presidencia de la República, Oficina Nacional de Evaluación de Recursos Naturales, Lima, Peru.
Orlove, Benjamin S. 1993 Putting race in its place: Order in colonial and postcolonial peruvian geography. *Social Research* 60(2):301–336.
Ortloff, Charles R., Robert A. Feldman and Michael E. Moseley 1985 Hydraulic engineering and historical aspects of the pre-Columbian intravalley canal systems of the Moche Valley, Peru. *Journal of Field Archaeology* 12(1):77–98.
Pearsall, Deborah M. 2001 *Paleoethnobotany: A Handbook of Procedures*. 2nd ed. Academic Press, Waltham, MA.
Pearsall, Deborah M. and Michael K. Trimble 1984 Identifying past agricultural activity through soil phytolith analysis: A case study from the Hawaiian Islands. *Journal of Archaeological Science* 11:119–133.
Piperno, Dolores R. 1991 The status of phytolith analysis in the American tropics. *Journal of World Prehistory* 5(2):155–191.
——— 2006 *Phytoliths: A Comprehensive Guide for Archaeologists and Paleoecologists*. AltaMira Press Press, Oxford.
Piperno, D. R. and I. Holst 1998 The presence of starch grains on prehistoric stone tools from the humid neotropics: Indications of early tuber use and agriculture in Panama. *Journal of Archaeological Science* 25(8):765–776.
Pozorski, Thomas, Shelia Pozorski, Carol J. Mackey and Alexandra M. Ilana Klymyshyn. 1983 Pre-hispanic ridged fields of the Casma Valley, Peru. *Geographical Review,* 73(4):407–416.
Pulgar Vidal, Javier 1987 *Geografía del Perú Las Ocho Regiones Naturales*. Ninth. PEISA, Lima, Peru.
Puri, Rajindra K. 2007 Responses to medium-term stability in climate. In, *Modern Crises and Traditional Strategies: Local Ecological Knowledge in Island Southeast Asia*, edited by Roy Ellen, pp. 46–83. Berghahn Books, New York.
Quilter, Jeffrey and Luis Jaime Castillo (editors) 2010 *New Perspectives on Moche Political Organization*. Dumbarton Oaks Research Library and Collection, Washington, DC.
Ramírez, Susan 1985 Social frontiers and the territorial base of Curacazgos. In *Andean Civilization and Ecology*, edited by Shozo Masuda, Izumi Shimada and Craig Morris, pp. 423–442. University of Tokyo Press, Tokyo.

——— 1995 De pescadores y agricultores: una historia local de la gente del valle de Chicama antes de 1565. *Bulletin de l'Institut Français d'Études Andines* 24(2):245–279.

Rostworowski de Diez Canesco, María 1977 *Etnía y sociedad*. Instituto de Estudios Peruanos, Lima, Peru.

Sandweiss, Daniel H., Kirk A. Maasch, Richard L. Burger, J. B. Richardson III, H. B. Rollins and A. Clement 2001 Variation in Holocene El Niño frequencies: Climate records and cultural consequences in ancient Peru. *Geology* 29:603–606.

Sandweiss, Daniel H., Kirk A. Maasch, C. Fred, T. Andrus, E. J. Reitz, M. Riedinger-Whitmore, J. B. Richardson III and H. B. Rollins 2007 Mid-Holocene climate and culture change in coastal Peru. In *Climate Change and Cultural Dynamics: A Global Perspective on Mid-Holocene Transitions*, edited by David G. Anderson, Kirk A. Maasch and Daniel H. Sandweiss. Academic Press, Burlington, MA.

Sandweiss, Daniel H., James B. Richardson III, Elizabeth J. Reitz, Harold B. Rollins and Kirk A. Maasch 1996 Geoarchaeological evidence from Peru for a 5000 years B.P. Onset of El Nino. *Science, New Series* 273(5281):1531–1533.

Starn, Orin 1991 Missing the revolution: Anthropologists and the war in Peru. *Cultural Anthropology* 6(1):63–91.

Swenson, Edward 2007 Adaptive strategies or ideological innovations? Interpretive sociopolitical developments in the Jequetepeque Valley of Peru during the late Moche period. *Journal of Anthropological Archaeology* 26:253–82.

Swenson, Edward and John P. Warner 2012 Crucibles of power: Forging copper and forging subjects at the Moche Ceremonial Center of Huaca Colorada, Peru. *Journal of Anthropological Archaeology* 31(3):314–333.

Téllez, Sandra and Frances Hayashida 2004 Campos de cultivo prehispánicos en la Pampa de Chaparrí. *Boletín de Arqueología PUCP* 8:373–390.

Thompson, L.G. et al. 1995 Late glacial stage and Holocene tropical ice core records from Huascarán, Peru. *Science* 269(5220):46–50.

——— 1985 A 1500-year record of tropical precipitation in ice cores from the Quelccaya ice cap, Peru. *Science* 229(4717):971–973.

Toshihara, Kayoko 2002 *The Cupisnique Culture in the Formative Period World of the Central Andes, Peru*. Unpublished Ph.D. Dissertation, Department of Anthropology, University of Illinois at Urbana-Champaign.

Towle, Margaret A. 2007(1961) *The Ethnobotany of Pre-Columbian Peru*. Viking Fund Publications in Anthropology. Chicago.

Ugent, Donald and Carlos M. Ochoa 2006 *La etnobotánica del Perú: desde la prehistoría al presente*. Consejo Nacional de Ciencia, Tecnología e Inovación Tecnológica, Lima, Perú.

Veintinilla, B. 1999 Análisis de opal-fitolitos en camellones del sector puntiachil, canton cayambe, provincial de Pichincha. *Memorias del Primer Congreso Ecuatoriano de Antropología* 3:149–179.

Ward-Perkins, J. B. 1972 Central authority and patterns of rural settlement. In *Man, Settlement and Urbanism*, edited by Peter J. Ucko, Ruth Tringham and G. W. Dimbleby, pp. 867–882. Shenkman Publishing Company, Cambridge, MA.

Watson, Richard Paul 1979 *Water Control and Land Use on the Arid North Coast of Peru: Prehispanic Agricultural Systems in the Chicama Valley*. Unpublished M.A. thesis, University of Texas, Austin.

Weberbauer, Agusto 1945 *El mundo vegetal de los Andes peruanos, estudio fitogeográfico*. Estación Experimental Agrícola de la Molina, Dirección de Agricultura, Ministerio de Agricultura, Lima, Peru.

Wells, Lisa Esquivel and Jay Stratton Noller 1999 Holocene coevolution of the physical landscape and human settlement in northern coastal Peru. *Geoarchaeology* 14(8):755–789.

Weng, C., Mark B. Bush and M. R. Silman 2004 An analysis of modern pollen rain on an elevational gradient in southern Peru. *Journal of Tropical Ecology* 20(01):113–124.

West, Michael 1979 Early watertable farming on the north coast of Peru. *American Antiquity* 44(1):138–144.

Wilkinson, T. J., Jason Ur and Jesse Casana 2004 From nucleation to dispersal: Trends in settlement pattern in the northern fertile crescent. In *Side-by-Side Survey: Comparative Regional Studies in the Mediterranean World*, edited by John F. Cherry and Susan E. Alcock, pp. 189–205. Oxbow Books, Oxford.

Yakovleff, Eugenio and Fernando Herrera 1934 El mundo vegetal de los antiguos personas. *Revista del Museo Nacional* 3(3):243–322.

Zimmerer, Karl S. 2000 The reworking of conservation geographies: Nonequilibrium landscapes and nature-society hybrids. *Annals of the Association of American Geographers* 90(2):356–369.

——— 2011a Mapping mountains. In *Mapping Latin America: A Cartographic Reader*, edited by Jordana Dym and Karl Offen, pp. 25–30. University of Chicago Press, Chicago.

——— 2011b Vertical environments. In *Mapping Latin America: A Cartographic Reader*, edited by, Jordana Dym and Karl Offen, pp. 263–268. University of Chicago Press, Chicago.

7 A Fine-Grained Analysis of Terra Preta Formation

Understanding Causality Through Microartifactual and Chemical Indices in the Central Amazon

Anna T. Browne Ribeiro

Abstract

Terra preta, a type of Anthropogenic Dark Earth found throughout Amazonia, has become important evidence in discussions of forms of habitation and demography in pre-Columbian Amazonia. Seen as evidence of densely settled permanent towns, terra preta is one among many indices of anthropogenic environmental engineering in what was previously seen as a pristine forest. Contemporary literature on Amazonia has demonstrated a correlation between intensification in terra preta production and settlement growth, increased investment in landscape engineering, and refinement of ceramic technologies. As a step towards understanding this transition, I conceptualize it as an ecological regime shift. I propose that every regime shift has a primary or principal causal vector, and examine the causal mechanisms of terra preta formation at Antônio Galo, a terra preta site in the Central Amazon.

I present a detailed timeline of site use and landscape transformation, constructed through a multi-proxy analysis of soils and sediments from Antônio Galo. I used chemical and microartifactual indices to disentangle cultural deposition from subsequent natural transformations and similar micro-scale data to distinguish between activity areas. This revealed temporal changes in spatial patterning. Here, I assess these in terms of social process: temporal changes in spatial organization (configuration of house and garden areas) are assessed alongside changes in the ecosystem (transition to terra preta) within a regime shift framework, which permits the isolation of variables in order to understand causality. This analysis shows that, at Antônio Galo, the gradual accumulation of miniscule or momentary encounters eventually led to ecological regime shift, which manifests as the transformation of the soil body from an Amazonian Oxisol into a terra preta.

Introduction

Terra preta, literally, "black earth," is a term used to refer to a class of archaeological sites found throughout Amazonia. The term is also often used to classify the soil or deposit found at such a site, alongside other kinds of anthropically modified or generated soils and sedimentary deposits known collectively as Amazonian Dark Earths (ADEs, see Woods and McCann 1999). Archaeologically, terra preta is known for its richness in artifacts and other kinds of occupational debris, as well as for its altered chemical and physical properties. Terras pretas are significantly darker and more nutrient-rich than the highly weathered, acidic yellow or reddish-orange native soils, and are hence sought out by contemporary Amazonian farmers for their higher yield and resilience.[1]

Known since the late 1800s, terra preta functions in the contemporary archaeological literature as one of many lines of evidence argued to indicate significant anthropic modification of Amazonia by pre-Columbian peoples (Lehmann et al. 2003; Arroyo-Kalin 2008; Erickson 2010; Browne Ribeiro 2011; Moraes and Lima 2012; Neves 2012; Rostain 2013; Schmidt et al. 2014), and also as a sign of intensive, permanent, and dense settlement (Heckenberger et al. 1999; Petersen et al. 2001; Neves 2008). Despite the widespread distribution of terra preta and other anthropic indicators, there are persistent challenges to the notion that Amazonia was densely inhabited in the pre-Columbian past (e.g., Meggers 2001; McMichael et al. 2012).

Disagreements over the overall significance of terra preta sites, which swept across Amazonia circa A.D. 1000, stem largely from difficulties in establishing the contemporaneity of habitation within or between sites. Beyond that, many of the sites used to construct regional models remain "black boxes" in terms of their particular histories. These sites, crucial to addressing questions of population density, intensity of habitation, and political integration in pre-Columbian Amazonia (Neves and Petersen 2006; Neves 2012), require closer study. Many models under dispute rely on the assumption that all terra preta sites can be treated as equivalent social and environmental phenomena. However, research has shown that terra preta sites are highly diverse and internally variable (Kampf et al. 2003). Brown-on-brown soils make distinguishing between deposits and refining chronologies challenging. Amazonian climatic conditions speed up taphonomic processes, obscuring the already subtle distinctions within the archaeological matrix and making them difficult to detect by the naked eye, a key tool in field archaeology. The physical properties of terra preta matrices and local climatic conditions thus limit the efficacy of traditional archaeological field methods.

At the majority of sites, a gap exists between characterizing a terra preta locus as a soil body—which speaks to its present conditions—and understanding the human occupation that produced the corresponding archaeological matrix, which speaks to processes that took place in the past. Geoarchaeological studies are beginning to address this gap (Kern 1988; Arroyo-Kalin 2008; Arroyo-Kalin et al. 2009; Schmidt and Heckenberger 2009; Schmidt 2010; Browne Ribeiro 2011; 2014), but terra preta still functions largely as a general sign of an evolving relationship between humans and the environment. The next step is to comprehend its diversity, variability, and significance as a product of human behavior. Understanding, for example, whether terra preta was an unintended consequence of demographic change or an intentional product is crucial for understanding its role in the evolution of social forms and technology, migration, and trade in Amazonia.

In this chapter, I build on a household-scale analysis of changing spatial organization and ecosystem change to develop a causal model for socio-ecological evolution at Antônio Galo (AM-IR-71), a terra preta site in the Central Amazon (Figure 7.1). By closely examining the archaeological matrix, and taking into account pedogenic transformations, I have elsewhere re-constructed the processes of social and landscape evolution at the ring village at Antônio Galo (Browne Ribeiro 2011; 2014). Both processes were tracked through proxy data embedded in the archaeological matrix as chemical and microartifactual traces. I used vertical patterning to discern superimposed land surfaces and internal compositional characteristics to classify them in terms of use and ecological state. Chemical indices helped distinguish between terra preta, *terra mulata* (garden or horticultural soils; see below and Sombroek 1966), and terra amarela (unmodified soils), thus characterizing the micro-ecological[2] context. Depositional practices and activities associated with each land or occupation surface were determined through microartifact patterning. By analyzing the co-variance of microartifacts (fragments of artifacts and ecofacts measuring 2–25 mm in diameter) and chemical signatures, I identified house floors and garden soils. The changing distribution of these distinct use-areas over time, along with a consistency of building and household-level practices, as indexed by these signatures, indicates the gradual growth of a village whose residents shared consistent ideas about the constitution of place.

These findings clarify relationships between soil transformation and activity areas, which entail particular types or sets of activities, moving beyond the association of terra preta formation with human activity in general. Using a regime shift framework (detailed below) to analyze events surrounding terra preta formation, I isolate the principal variables that can be implicated in a causal model, which include number and distribution of activity areas; indices of population size/density; indices of activity type; and indices of intensity of deposition. By ordering changes in these variables in a logical sequence, I demonstrate that, at Antônio Galo, spatial re-configuration and indices of population aggregation immediately precede the transformation of terra mulata *or* terra amarela into terra preta. This leads to the conclusion that population aggregation was the principal causal factor in precipitating this regime shift.

Anthropogenic Soils and Sediments in Amazonia

Amazonian Dark Earths include a range of anthropic, anthropogenic, or anthropogeomorphic soils and deposits such as mounded fields in Guyana, raised fields in the Llanos de Moxos, geoglyphs in Acre, roads and causeways along the Xingu and Beni rivers, and ringed ditches and ring villages in the Central Amazon, Central Brazilian Plateau, and Southwestern Amazonia (Petersen et al. 2001; Moraes 2006; Heckenberger et al. 2008; Erickson 2010; Schaan 2012; Prumers and Jaimes Betancourt 2014; Rostain 2013). Terra preta sites can coincide with other ADES such as earthworks and sculpted soilscapes, which include paths, ramps, ditches, mounds, plazas, terraces, and platforms created, intentionally or not, by humans (see Schmidt et al. 2014 for examples). Thus, this larger class of evidence needs to be considered in studies of terra preta, just as studies of dark earths and anthropic soils from Europe and Asia (Groenman-Van Waateringe and Robinson 1988; Frimigacci 1990; Walton and Cassels 1992; Alexandrovskaya and Alexandrovskyi 2000; Marliac 2002; Macphail 2003) can contribute comparative data and explanatory frameworks to the Amazonian context. Like other dark earths, anthrosols and anthropic deposits, terras pretas and ADEs are not only material and ecological phenomena, but are also socially constituted, meaning-laden, and often intentionally configured across a landscape (e.g., Heckenberger 2005). Amazonian Dark Earths hence constitute a crucial class of evidence for deciphering what increasingly emerges as a highly ordered and intensely inhabited landscape.

Scientific debates about the origins of terra preta began in the 1940s (e.g., Farias 1946; Gourou 1949; 1950; Sombroek 1966; Falesi 1974). By the early 1980s, an anthropic origin for these culturally significant dark soils was widely accepted (Smith 1980; Eden et al. 1984; Andrade Pérez 1986). Subsequent decades of study by soil scientists, archaeologists, and geographers, among others, have shown that terras pretas are often associated with earthworks, and most present very high densities of artifacts, mostly ceramic remains. Terra preta sites are easily recognizable on high unflooded alluvial terraces that feature ancient, brightly colored red or yellow soils, but terra preta sites have also been pedogeochemically identified on floodplains (Macedo 2009). Terras pretas are more fertile than dominant, unmodified Amazonian soils and less acidic and more stable than their precursors, which leads to better preservation of other artifact classes such as floral, faunal, and human remains (e.g., Machado 2005; Rapp Py-Daniel 2009; Tamanaha and Rapp Py-Daniel 2009; Shock et al. 2014; Silva et al. 2013). Most terra preta has a higher pH and higher amounts of organic carbon, phosphorus, calcium, magnesium, and in some cases, zinc, manganese, copper, nitrogen, potassium, and sodium than adjacent soils (Kern 1988; 1996; Kampf et al. 2003; Lehmann et al. 2003; Arroyo-Kalin 2008; Schmidt 2010; Browne Ribeiro 2011). Terras pretas also have distinct physical, mineralogical, and hydrological properties than adjacent soils, due to anthropogenic inputs (Costa et al. 2003; Lima et al. 2009; Browne Ribeiro 2011).

Terra preta is an especially dark (very dark brown through black) ADE that is especially enriched in P, Ca, and organic C (OC). *Terra mulata*, a slightly lighter, brown ADE, was originally hypothesized by Wim Sombroek (1966) as an agricultural or horticultural zone surrounding a main habitation site, the "true" terra preta. Subsequent pedoarchaeological research, revealed lower overall enrichment in terras mulatas than in adjacent terras pretas, supporting the idea that terra mulata, recognized by local farmers as a "weaker" terra preta, would have formed through a mix of slash-and-burn cultivation and soil amendment of terra amarela with terra preta deposits (Arroyo-Kalin et al. 2009).

An important agricultural resource, terra preta, though not incompatible with slash-and-burn cultivation, it does not form through farming (Petersen et al. 2001). Slash-and-burn cultivation of soils as aged as Amazonian bluff soils results in loss of soil nutrients rather than enrichment (Zech et al. 1990).[3] Additionally, terra preta presents notable enrichment in OC (Katzer 1944[1903]; Zech et al. 1979 cited in Glaser et al. 2004; Lehmann et al. 2003). The type of burning practiced in a slash-and-burn system results in a loss of most of the carbon present in plant matter, either through volatilization or transformation into ash, which quickly degrades or runs off the land surface. Rather, terra preta seems to be the result of intensive, in situ deposition of organic remains, especially pyrogenic charcoal. Arroyo-Kalin et al. (2009) emphasize that terra preta presents signs of up-building through accumulation, rather than thickening through downward particle migration, the process that characterizes terra mulata formation. Hence, terra preta differs from terra mulata in appearance, composition, and also in genesis.

Because terra preta and terra mulata are at once substrate, soil, and site, they can function, and hence be studied, in various ways. Nineteenth-century naturalists identified terra preta as "ruins" of past civilizations (e.g., Bates 1863; Smith 1879). In the 1960s and 1970s, geographers (e.g., Sombroek 1966; Smith 1980) studied terra preta as a demographic indicator for ancient Amazonia. Soil scientists study terra preta as a soil—a particulate body that developed through in situ biogeochemical and physical alteration of parent material—hence focusing on natural formation processes, rather than the formation of the original particulate body, the archaeological matrix, which likely had significant sedimentary input by humans. An agronomist studies terra preta as a growing medium and nutrient source, while a botanist might study it as a substrate for vegetation, and a microbiologist might see it as a habitat. As a dynamic biogeochemical body composed of particles that are natural and cultural, organic and inorganic, local and exotic, and that form an integral part of a socioecological system, terra preta is in every sense a hybrid phenomenon. For the sake of clarity, here "terra preta" is applied to dark, nutrient- and carbon-rich earths with Munsell color readings between 10YR 3/2 (very dark brown) and 10YR 2/1 (black). The term *terra mulata* is only used when there is reason to infer the formation processes defined above. Redeposited earths are referred to as deposits or sediments.

The Significance of Terra Preta

The term *terra preta* is used by farmers and researchers alike to emphasize its distinctness from the surrounding soil. However, the term helps to obfuscate the variability and diversity inherent in terras pretas, leading to the conflation of disparate matrices that simply *appear*—that is, can be recognized as—similar. Nevertheless, this provides insight into the fundamental characteristics of terras pretas: darker coloration than adjacent soils, which makes them recognizable; and stability, which has made them persist over hundreds or thousands of years. Both characteristics derive from the addition of organic matter[4] (Glaser et al. 2003; Teixeira et al. 2009), which contributes quickly to soil melanization (darkening); and further on, in cases of sufficient intensity of deposition, may result in a fundamental ecosystem change—a regime shift from a previous, relatively stable state to a new state. When a soil ecosystem is able to absorb the anthropogenic inputs, as occurs in slash-and-burn systems, the soil and surrounding ecosystem remain unchanged. A soil ecosystem regime shift requires deposition of specific materials through some process of sufficient *amplitude* or *persistence* to propel the ecosystem into a new *regime*, or dynamic set of relationships and processes. For terra preta, this means that people deposited a significant amount of organic matter, including pyrogenic charcoal,[5] at a rate that outpaced the soil's capacity to absorb or process these inputs. Because Amazonian soil regimes tend towards impoverishment, any process that effectively redirected the functioning of this system away from a relatively stable state must have been intensive. This reflects a major change in depositional practices or site use, which, in turn, may index a similar change in the sociocultural system. The emergence of terra preta at any given site may have been the result of technological or demographic changes, changes in social (and spatial) organization, or even of the introduction of foreign cultural elements. As a result, any study geared towards understanding the transformations entailed by the emergence of terra preta at a particular site requires a close examination of the co-evolution of two systems.

Terra preta, arguably the key type of site for understanding pre-Columbian Amazonia in the first millennium A.D. (Moraes and Neves 2012; Neves 2012), has also performed an important ideological function in Americanist archaeology, bringing renewed interest to a region once thought of as well understood. The ubiquity, size, longevity, and artifact density of terra preta sites permit claims of permanent, densely settled towns in pre-Columbian Amazonia (Heckenberger et al. 1999; Petersen et al. 2001; Neves et al. 2003), upending previous notions about Amazonia as an empty or hostile landscape (e.g., Meggers 1971). The emergence of terra preta, in its most recent form,[6] is associated with the emergence of ceramics (Lima et al. 2006; Lima 2008; Neves 2012; Neves et al. 2014). This paired development, which occurs ca. 500 B.C.E., indexes what some (Neves 2012; Neves et al. 2014) identify as a pan-Amazonian phenomenon, wherein a set of concurrent symbolic systems, technologies, and landscape management strategies, including intensive terra preta formation, begin to emerge along with sedentism on major rivers (Lima et al. 2006; Gomes 2011; Guapindaia 2011). This culture-historical reconstruction is, among other things, rooted in the idea that terra preta indexes dense, intensive, or permanent human habitation in pre-Columbian Amazonia.

But what, precisely, do the words "dense, intensive, or permanent" mean in terms of human experience? In what sense did the emergence of terra preta modify practices or the conditions of life in an Amazonian village? What possibilities did it engender, and what changes did it prompt in terms of social relations and human-environment interactions? Contemporary designations of sites as "terra preta sites" leave many questions, especially those regarding process, causality, and intentionality, unanswered. These questions, which can only be answered at temporal and spatial scales perceptible to the original inhabitants, are critical to the validity of regional analyses that use terra preta as an indicators of population density, social hierarchy, and cultural or technological affiliation.

Contemporary literature on Amazonia has confirmed a correlation between intensification in terra preta production and settlement growth, increased investment in geomorphic engineering, and refinement of ceramic technologies (Moraes 2006; 2010; Neves and Petersen 2006; Lima 2008; Moraes and Neves 2012). Some cite terra preta as evidence of social complexity, seeing these pieces of evidence as correlates for population increase and aggregation, monumental architecture, specialization, and warfare infrastructure (Petersen et al. 2001; Neves 2008). However, to state unequivocally that terra preta signifies social complexity would be irresponsible, especially given the high variability in its internal composition and its association with sites of varying size, shape, and socioecological configuration.

Instead, terra preta should be regarded as another medium in the cultural mosaic that structures Amazonia, today and in the past. Part of what makes terra preta intriguing is that it has been mapped as far into the northwest as Ecuador (Eden et al. 1984), as far south as the Upper Xingu River (Schmidt and Heckenberger 2009), and north into the Guyanas (Rostain 2013), with a long history of study in the Central and Lower Amazon (e.g., Hilbert 1968; Heckenberger et al. 1999; Petersen et al. 2001; Lima 2003; Neves et al. 2003; Neves and Petersen 2006; Lima 2008; 2013; Moraes and Neves 2012). However, if terra preta sites are known for their broad variability, it follows that the widespread occurrence of terra preta sites does not necessarily indicate uniform modes of habitation—even if a common mode of expression, through symbolic means, can be tracked across the basin. Undoubtedly, by the time terra preta had crossed the continent, people were engaged in intensive landscape management, and likely embedded in complex long-distance communication and trade networks. But understanding why and how terra preta emerged or was adopted across this heterogeneous region might provide interesting insights into how, and to what extent, a pan-Amazonian phenomenon could emerge out of such diversity.

Regime Shifts, Terra Preta, and Human Action

Tracking changes in spatial organization at terra preta sites is challenging. Brown-on-brown contexts make intra-site variability and fine-grained chronologies difficult to chart. Stone or plaster architectural features used for such purposes elsewhere are absent throughout most of the Central and Lower Amazon (see Cabral and Saldanha 2008 for exceptions). Hence, demonstrating the primary nature of any depositional context is extremely difficult using traditional methods. Additionally, most classes of artifacts are found in secondary deposits at least as often as in primary or use-related contexts, with the exclusion of mortuary contexts. Ceramics, for example, are often found in construction fill, where they are not used as actual building material (e.g., Machado 2005; Tamanaha and Rapp Py-Daniel 2009; Moraes 2013). As a result, at terra preta sites, soils and sediments—the in situ residues of dwelling—are the best means of discriminating among depositional contexts. Application of field and laboratory methods, such as granulometry, soil chemistry, micromorphology, and the analysis of microartifacts, can yield a great deal of information about occupation and ecological states (e.g., Arroyo-Kalin 2008; 2010; 2012; Schmidt and Heckenberger 2009; Schmidt 2010; Browne Ribeiro 2011; 2014; Schmidt et al. 2014).

As the most direct record of the events that transpired at a site, the terra preta matrix holds the key to understanding its own formation. The myriad studies of terra preta have shown that few properties apply across the board, and that terras pretas across Amazonia, despite looking similar, are widely variable. Kampf et al. (2003) have noted that variability within a terra preta stain can rival that between sites. Paradoxically, this internal variability may be the best clue for understanding the overall significance of terra preta.

Terra preta, as the result of human action, and as the substrate upon which human life unfolds, is also the canvas upon which the traces of these activities are inscribed. As with any other occupation space, an expanse of terra preta must contain indicators of the differentiated treatment of space over time. This spatial patterning corresponds to the social logic that organized that space when it was a living settlement—a habitation place that was also a terra preta site in the making. The staggering variability seen in dark earths is a result of the wide range of practices that, layered upon each other, left a distinctive mark on the landscape.

The similarities across terras pretas, which account for their key characteristics (darker coloration and stability), arise from refuse that provides the necessary inputs (organic matter and pyrolized charcoal). The variability speaks to how these inputs occur: process, intensity, rhythm, and spatial configuration of deposition. Hence, the study of terra preta formation is the study of a sociocultural phenomenon and must be understood as such.

Each parameter that is altered in terra preta is a signal to be deciphered. These signals are the material trace remains of an act (or set of acts) performed by a human (or set of humans) in the past. However, because depositional processes are many, and the range of distinguishable traces is small compared to potential sources, a problem of equifinality occurs. The probability of two distinct processes contributing towards the same trace signature is high. The

Amazonian climate compounds this difficulty. In such a pedogenically active zone,[7] where matter moves vertically between layers, the problem of equifinality literally takes on a new dimension: through this kind of mixing, two processes in *two distinct time periods* might combine to create a signal that resembles the signature of yet a third process.

Hence, any detailed study of terra preta needs to be carefully sited in order to minimize this kind of interference, and must also take full account of the processes that formed and modified the original depositional context. The natural processes that contribute to the current state of terras pretas include contributions of organic matter near the present-day surface, due to organic decomposition and/or contemporary and recent-historical use; downward movement of matter through percolation or physical mixing; and in situ weathering or decomposition of organic and inorganic particles. The cultural mechanisms include deposition, intentional or not, of organic and inorganic matter by humans, as well as the movement (removal and redeposition) and modification (e.g., compaction, burning, sweeping, washing) of contexts. Similar caveats should be applied to architectural remains, which are perishable or earthen, and thus subject to the same post-depositional processes.

Each terra preta site is the result of myriad habitual, structured activities and isolated acts performed by people. If spatial patterning of material traces corresponds to a social logic, then *distinct sets or configurations of signals* indicate *distinct sets of practices or spatial configurations*. These differences might co-exist within a single multi-ethnic or multicultural system, might indicate distinct cultural systems existing contemporaneously, or might even result from related or unrelated systems arising and falling in succession. Nevertheless, even though understanding the particular sources of enrichment may be quite difficult due to degradation and moving of remains or components over time, *patterning in deposition* is still accessible. A difference in patterning indicates different spatial organization. This information can be used alongside other data—such as ceramic technologies and style, house configuration, and plant selection to understand what the spatial reconfiguration might mean at the level of the household or the community.

The fundamental questions to explore the significance of changes within a site or differences between sites are:

1 To what extent are two sites or areas of terra preta similar or different?
2 How can we account for these differences, in terms of behavior?
3 To what extent, and how, are these behaviors significant in terms of social relations?

Distinct sets of signatures indicate distinct formation processes in terms of specific human *activities*. Nevertheless, the culmination of these diverse sequences of events and sets of practices led to localized ecological regime shifts—fundamental changes in ecosystem state and tendencies—that created darkened, enriched pedo-stratigraphic profiles known broadly as terra preta.

Regime Shifts

The concept of regime shifts emerged out of ecologists' recognition of the dynamic nature of ecological systems. Biggs et al. (2009:826) provide a useful general definition: "Ecological regime shifts are large, sudden changes in ecosystems that last for substantial periods of time." Theorists of non-equilibrium dynamics use the terms *regime* and *regime shift* to emphasize that natural systems, often understood as nested, hierarchical, or chain-linked sets of relationships, are dynamic. Ecologist C. S. Holling (1973) first challenged long-held principles of stable equilibria, which not only failed to describe the natural oscillations that ecological networks undergo, but also failed to provide useful information about deviations from idealized, and essentially theoretical, "stable states." Rather, Holling (1973) proposed that coupled populations interact in cycles that vary in amplitude. Depending on the particular interactions within the system, these cycles can tend towards equilibrium, in which case the system does operate at something approximating stability. More often than not, however, these cycles oscillate within a set of boundaries, which Holling denominated the "attraction domain" (Holling 1973:3). This domain, today referred to as an "attractor" or a regime, represents a relatively resilient set of interrelationships that characterizes that ecosystem. As an example of a regime, shrublands can vary slightly in species distribution, frequency, and configuration, depending on climatic regimes and multi-annual reproductive cycles. However, the replacement of shrublands with grasslands due to an increase in fire regimes is an instance of regime shift (Ludwig et al. 1997). When the amplitude of oscillations within the system rises beyond the bounds of this domain—the threshold of the attractor—what follows can range from species extinction and substitution (a reaction that demonstrates resilience at the ecosystem scale) to succession or total reconfiguration of the system. Along with recognizing ecosystems' natural oscillations, Holling's (1973) work also highlighted the ability of systems to resist complete de-structuring through alternative feedback mechanisms.

These ideas were augmented by May (1977), who contributed the notion that systems can have multiple states of relative stability, or multiple attractors. The idea that a system can spiral out of one attractor and into another rounded out the set of concepts that would eventually be encapsulated by the term *regime shift*. Regime shifts occur when a directional change in "an underlying driving variable (or set of variables)" that is internal to the system is affected by an "external shock" (Biggs et al. 2009:826). The internal variable, sometimes referred to as the *driving variable*, has a direct effect on the system; regime shift happens when the driving variable is pushed, by the external shock to the *forcing variable*, past the threshold of the current attractor. The process of a regime shift involves changes in a series of internal variables and feedbacks, pushing the system from one relatively stable state to another in a way that may or may not be reversible. Critically, the new state persists even after the kickstarting mechanism is reduced or removed (Biggs et al. 2009). Additionally, although the changes to the state of the system are major, the changes in forcing variables need not be significant, like an order of magnitude increase in harvesting, or catastrophic, like a tornado or earthquake. In fact, slight changes in one or a small class of variables can result in significant cascading ecological effects (Scheffer and Carpenter 2003:648; Biggs et al. 2009).

Terra Preta as an Example of Regime Shift

Terra preta and terra mulata exhibit distinct defining features and are formed through distinct processes. The multimodal distribution of site types and soil ecosystems on Amazonia's ancient upland alluvial terraces—i.e., the presence of terra preta, terra mulata, and the original, weathered soils, with little evidence of intermediate states—suggests that terra preta sites/soils might have formed through regime shifts. Terra preta constitutes a new state of the soil system that is stable, having lasted, in most cases, over 1,000 years. Additionally, the formation of terra preta resulted in "large, unexpected changes in ecosystem services and human livelihoods" (Brock et al. 2008).

The change in the soil ecosystem can be expressed as a transition between an initial state—occupation on an Oxisol[8]—and a final state—occupation on terra preta. Knowing that the change was precipitated by human action places the focus on the interface between humans/social systems and natural systems. The degree of difference between Amazonian Oxisols and terra preta soils demonstrates significant changes to the soil ecosystem. These changes also had cascading effects on other systems, including plant and microorganism communities, hydrological systems, and human social structure.

The probability that a human's interaction with an ecosystem will have lasting effects depends on the duration, frequency, and magnitude of the interaction. Acts that are occasional and momentary are more likely to result in major changes if they are substantial or widespread. Conversely, habitual or repeated acts can result in alteration even if they are each miniscule or imperceptible, as long as their frequency is such that their cumulative effect is significant. Hence, depending on the intensity and frequency of modification, traces of actions have a greater or lesser probability of preservation. This helps us consider what kinds of actions we will never be able to access, and how our sample may be biased towards evidence of particular types of practices or behavior. In general, the higher the frequency and the greater the intensity, which can be thought of as a "footprint" or magnitude of environmental impact, the higher the probability of persistence and detection of a particular set of traces. This idea is illustrated in Table 7.1, where variation in frequency and intensity creates signals that have higher or lower probability of persistence and detection.

Table 7.1 Probability of signal detection as a function of frequency and intensity of signal

(Intensity)		(Time) Unique →	Frequent
	Low	Very low	High
	↓		
	High	High	Very high

For the purposes of investigating change mechanisms, the concept of regime shift will be simplified to a theoretical model that can assess potential causal factors. Causal factors are either internal to the system, corresponding to potential driving variables, or external to the system but, belonging to the social system, bear a direct relationship to internal variables. The latter set of factors, which I define as the evolving traits of the social system, corresponds to the set of potential forcing variables.

Once identified, the internal and external variables relevant to causality can be combined to form a causal vector. Ecological regime shift can be seen as having a primary to causality vector, which can be described by identifying the driving variable, the direction of its initial and subsequent movement, and the cause of this change (the external shock). By studying the reconstructed sequence of events at Antônio Galo, I isolate and order changes in variables in the sociocultural system and soil ecosystem to build a causal model.

Understanding Terra Preta Formation at Antônio Galo

In the region of the confluence of the Negro and Solimões rivers, the occurrence of terra preta and its association with ceramic phases and traditions has been intensively studied (Petersen et al. 2001; Donatti 2003; Lima 2003; Neves et al. 2003; Machado 2005; Lima et al. 2006; Moraes 2006; Neves and Petersen 2006; Chirinos 2007; Lima 2008; Castro 2009; Rapp Py-Daniel 2009; Tamanaha 2012). Overall results suggest that terra preta begins to form during the Açutuba phase (300 B.C.–A.D. 400), and becomes intensified at later occupations (A.D. 600–1200) associated with Paredão ceramics (Lima 2008; Moraes and Neves 2012).

Antônio Galo, a 16-ha (Moraes 2006) terra preta site in this region, features a nearly intact 1.5-ha ring of 12 earthen mounds (Figure 7.1). Even though Antônio Galo is complex and multi-componential, the north sector has been identified as strictly Paredão (Moraes 2013). In this sector, subsurface preservation is good, and the pedo-stratigraphic sequence is straightforward. Radiocarbon dates place the occupation within the Paredão period. The pre-terra-preta occupation in this sector has produced dates that range

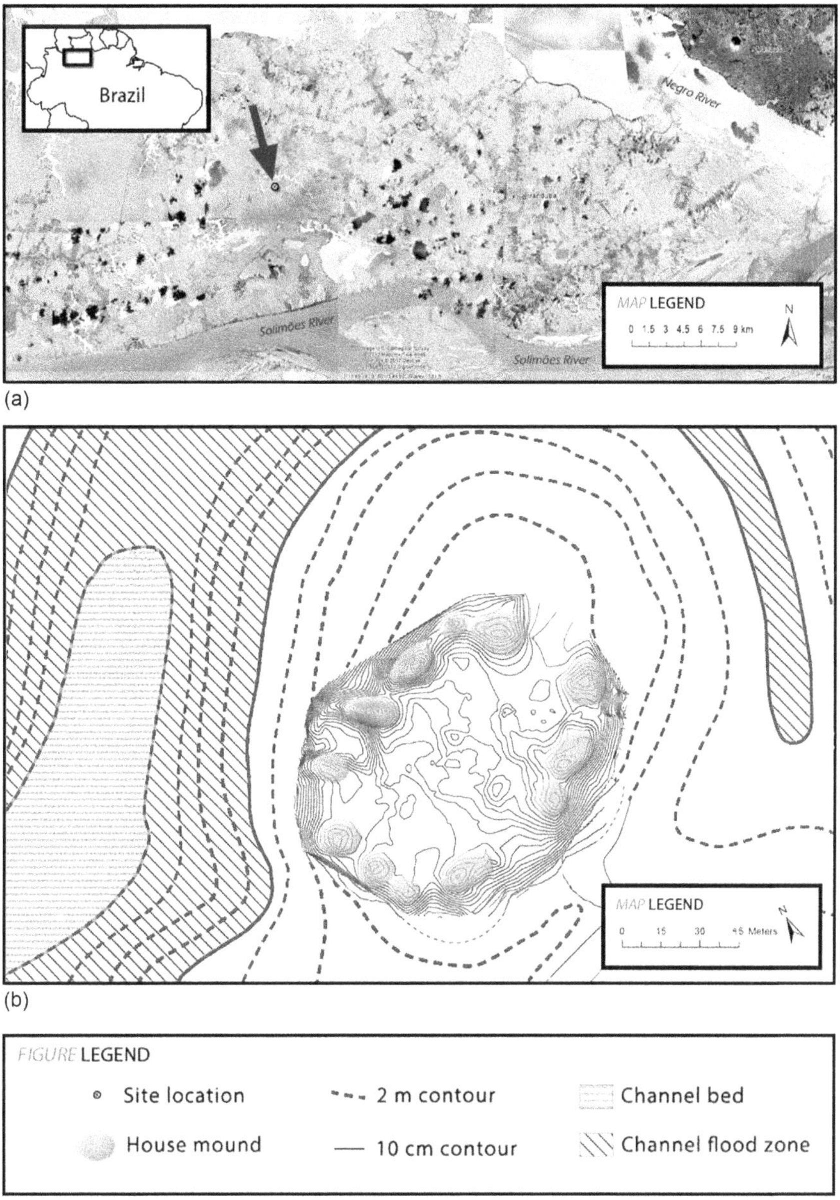

Figure 7.1 Maps showing (a) the research area (interfluvial zone between Negro and Solimões rivers, bounded to the west by the Ariaú River); (b) a map of the ring village sector showing topography, flood zone, and the remains of house platforms.

from ~A.D. 300–600 (Moraes 2013). Ring village construction, which likely begins soon after, appears to culminate ca. A.D. 850, and Moraes (2013) suggests that the village was abandoned ca. A.D. 1100.

Collaborative work at Antônio Galo (Browne Ribeiro 2011; 2014; Moraes 2006; 2013) showed that the ring village "mounds," which occupy most of a 2-ha riverine terrace (Figure 7.1), are the remains of house platforms or earthen architecture. All of the mounds excavated presented clear, visible stratigraphy, an unusual situation in this part of Amazonia due to the prevalence of highly re-configured, multicomponential occupations (see Neves et al. 2003; Machado 2005; Moraes 2006; Lima 2008; Rapp Py-Daniel 2009; Tamanaha 2012). This stratigraphic clarity permitted the correlation of mound sequences across the study area (Figure 7.2) and comparative geoarchaeological testing of deposits and soil horizons.

The aim of geoarchaeological analyses was twofold: to understand the visible distinctions in the archaeological profiles in terms of composition and structure, and to use these results to further inform interpretations of less clear distinctions or lenses perceived within larger depositional contexts. Determining whether distinct contexts were true deposits, formed through deposition, or soil Horizons, formed through pedogenesis, was crucial for testing the house platform hypothesis. A concurrent goal was to assess the degree of modification of deposits through pedogenic processes, with a focus on physical mixing, downward migration of particles, and movement of ions and chemical compounds.

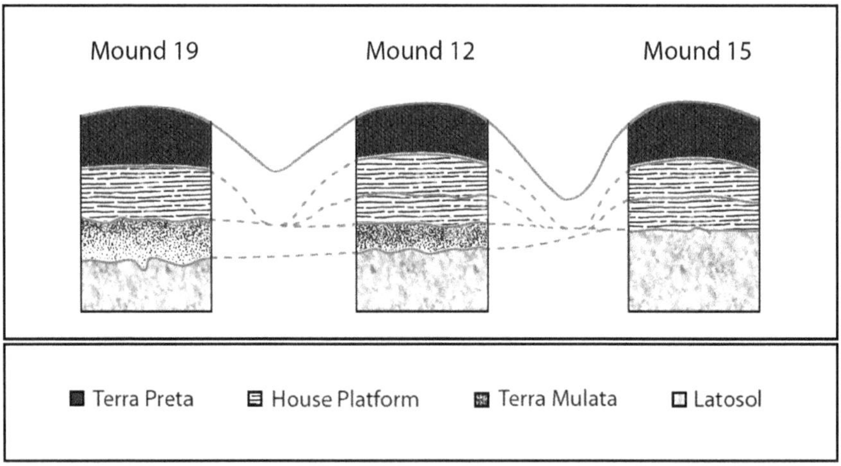

Figure 7.2 Schematic representation of the three types of pedo-stratigraphic sequences encountered. Mound 19 represents the most commonly observed sequence and Mounds 12 and 15 represent the key variations. Not to scale.

Bulk samples were collected systematically from formal 1 × 1 and 1 × 3 m excavations, for each arbitrary 10-cm level excavated (6.25 L), and also from each Horizon/deposit[9] identified in the profile. Aliquots were separated from bulk samples for soil chemical and physical testing (300 g) and curation (500 g), and the remaining sediments were floated to clean and separate light and heavy fraction. The heavy-fraction portion of floated samples was sorted according to material, size, and surface aspect (color, luster, spalling, porosity, warping, etc.), and tabulated in order to obtain quantitative data on the distribution and movement of natural particles and microartifacts.

The pedo-stratigraphic sequence for this sector attests to three major occupation phases: an early house likely built on leveled ground (Phase I); a cluster of houses at the southeast corner of the landform that correlates with terra mulata elsewhere on the landform (Phase II); and a final, platform-house ring village associated with terra preta (Phase III) that occupies the entire landform (see Figure 7.3(a), (b), (c)). The terra mulata occupation is evident in profile in two forms. Across most of the landform, it consists of a moderately melanized A Horizon with frequent inclusions of fine charcoal, which was interpreted chemically and sedimentologically as terra mulata (Figure 7.2, Mounds 19 and 12). The terra mulata A Horizon is sealed by construction fill belonging to house platforms. In the southeast portion of the landform the pedo-stratigraphic unit identified above the latosol (B Horizon of buried Oxisol) was not always a melanized A Horizon. In Mounds 13, 14, and 15 (see Figure 7.3), the stratigraphic sequence moves from (possibly truncated) latosol to a low-house platform/floor (see Figure 7.2, Mound 15) that is clearly distinct from, but similar in composition to, the later, thicker house platforms that form part of the ring village.

These earlier floors, tentatively recognized in the field, were confirmed and characterized in the laboratory through microartifactual and chemical indices in 3 of the 11 mounds excavated. In the same way that OC accumulates near natural land surfaces (Batjes 1996; Kaufmann et al. 1998), surfaces occupied by humans accumulate P (Eidt 1984), and at terra preta sites, also Ca and OC. These parameters, tested in 10-cm depth intervals, show a tendency towards enrichment of available P, Ca, and OC at the interface of all visible buried contexts (Browne Ribeiro 2014). This helped establish this triple-peak (OC-Ca-P) as a signal for buried surfaces in general. Similar kinds of enrichment below Phase III house platforms in the units lacking a clear buried A Horizon pointed to invisible buried surfaces. Vertical patterning in the distribution of microartifacts (microartifact decay signatures) at the modern-day surface as well as at depths corresponding to the chemically indexed buried surfaces further corroborated this conclusion.

Floors in both phases are composed of dense, compacted grayish-yellow sediments; earlier platform-floors are 5–10 cm thinner than later ones.

The final, ring village occupation manifests in profile as a combination of a major depositional/construction event composed of compacted, grayish-yellow sediments, identified as house platforms, and a moderately thick (20–40 cm), overlying layer of terra preta, presumably debris that accumulated over time. This final, two-layer sequence was identified in all the ring-village mounds excavated. In one instance, the buried terra mulata A Horizon is present beneath two superimposed house platforms (Figure 7.2, Mound 12), attesting to the three-phase sequence.

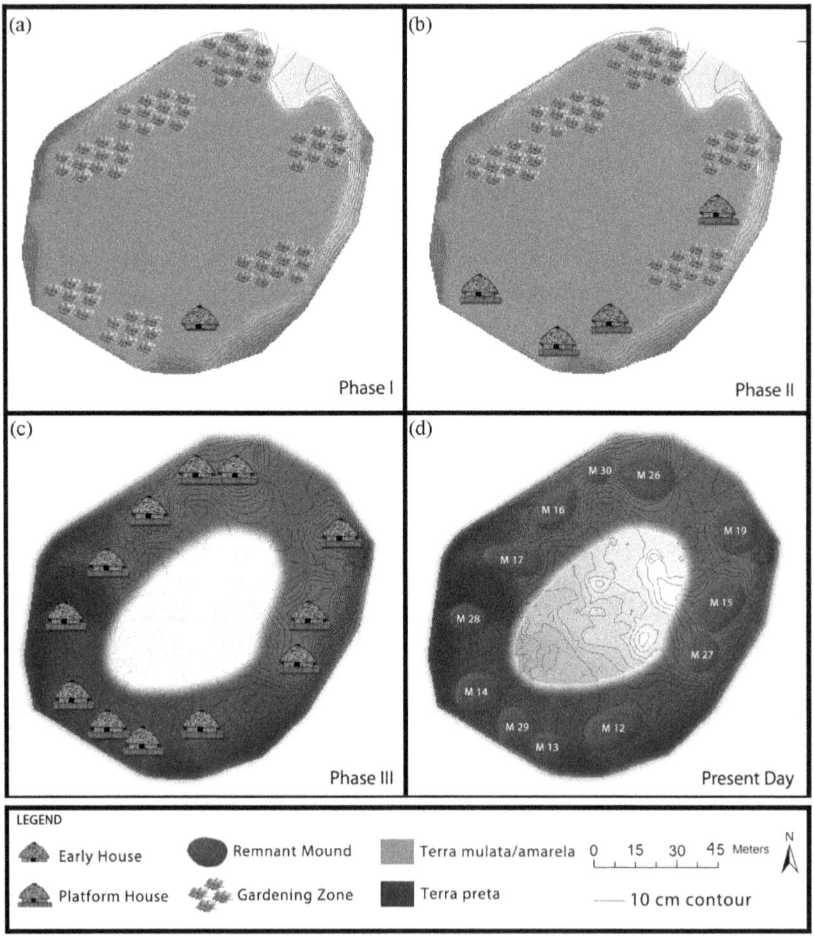

Figure 7.3 Occupation phases: (a) the early house; (b) a cluster of platform houses with terra mulata; (c) the platform-house ring village with terra preta; (d) remnant housemounds with terra preta.

In addition to chemical and microartifact decay indices, which indicated buried use-surfaces within platform deposits, these grayish-yellow sediments also presented characteristic house floors debris: comparatively higher concentrations of ceramic micro-fragments and of oxidized, low-fired clay lumps, which we associate with ovens or hearths. Additionally, these areas were higher in dense (heavy-fraction) charred plant remains, and especially in palm nut or nut endocarp, which suggests selection for economically useful plants (Browne Ribeiro 2014). The melanized buried A Horizon samples showed different patterning, with a tendency towards higher overall concentrations of finer charred plant remains with no indication of selection for particular plant parts or types. In addition, a high proportion of low-fired clay fragments presented a reduced or fire-blackened surface, suggesting that these had been subjected to high temperatures in oxygen-poor, carbon-rich environments. Elsewhere (Browne Ribeiro 2011; 2014) I have argued that these fragments were re-heated during superficial burning while embedded in carbon-rich melanized A horizons, thus acquiring a reduced or partially reduced aspect. This type of burning, consistent with clearing and maintenance practices for gardens in tropical forest regions, has been shown to transmit heating effects to mineralized particles embedded in the plough zone (Grogan et al. 2003; Buol et al. 2011). These two visibly different surface types permitted me to sort the research area into domestic and gardening activity areas. By tracking the spatiotemporal distribution of these, I was able to re-construct a sequence of occupation in which a single, very early house eventually blossomed into a ring village.

In the earliest phases, differences in predominant microartifacts and charred organic remains allowed me to outline areas likely used as gardens and house floors (Figure 7.3(a), (b)). The earliest house appears to have been built on leveled, but not necessarily raised, ground. During Phase II, house-platform technology (per Carneiro's [1957] discussion of raised, packed earth floors)[10] was being employed. Soil structure showed that, during Phase II, with the exception of the site of the earliest house, buildings were erected in areas previously occupied by gardens (Figure 7.3(b)). In the final phase, village formation entails an overall reorganization of space across the landform. A central plaza or communal space is generated through the positioning of houses in a circle. Spaces previously constituted as "house space" continue to be house sites, indicating some kind of continuity; and some of the space previously dedicated to gardening is re-signified and physically transmuted into house space through construction. Finally, in the remaining garden space, melanized soils are transformed into terras pretas through intensive deposition (Figure 7.3(c)). There is no evidence for terra preta beneath the house platforms, which shows that the soil regime shift did not occur until the entire village was in place.

My three-phase reconstruction shows the iterative nature of village expansion, which grew from a single house surrounded by a large garden expanse into a small neighborhood of three to four houses, and eventually into a densely inhabited ring village that dominates a landform previously dedicated to horticultural activities. On the experiential and ecological plane, as more space became house space, other kinds of designated space such as garden space would have been increasingly circumscribed, especially as plaza space would have been relatively open and clean; on the relational plane, as the number of households increased, levels of intimacy among some groups of inhabitants would have intensified and social networks would have increased in complexity. The significance at the community level for human subjects would have been manifest in new kinds of articulations among households and household members, and in more general spatial designations such as public and private.

The platform-building practice that characterizes the ring village occupation was already common in earlier periods. Similarities in artifact clusters and hearth features in the earliest and latest levels of the deepest excavation, which was also the site of the earliest house, also seem to point to relative stability in daily practices. Hence the process of designating, constituting, and using domestic space remained relatively stable over time. The exclusivity of Paredão ceramics (Moraes 2013) suggests a general stability in technological and stylistic systems. Taken together with evidence that village space was reorganized to accommodate more people, this suggests that changes in social structure might have been taking place at the supra-household level, but that the household remained unchanged.

Applying a Regime Shift Framework

Considering the Antônio Galo occupational sequence as a case of anthropogenic regime shift suggests one path to terra preta formation. The soil ecosystem is the primary object of study. According to the occupational sequence at Antônio Galo, the initial state was terra mulata across much of the landform.[11] The final state is terra preta. The relevant internal variables are soil characteristics or components, while potential forcing variables are aspects of human activity. Charcoal and apatite, important components in lending terra preta heightened fertility and stability (Schaefer et al. 2004), significantly alter the way native soils behave while also producing lasting signatures or traces. Hence, among the classes of pre-Columbian waste, those most relevant to terra preta formation are charred plant remains, ashes, and animal remains. Input of these kinds of organic waste is thus examined here as a potential driving variable.

Identifying the forcing variable requires thinking about how the transformation from terra amarela or terra mulata to terra preta proceeds. The two ways that native Amazonian pedogenic processes can be destabilized are mass deposition—large dumps of material that interrupt pedogenesis by sealing off the old land surface and creating a new one—and repetitive or continuous activities that introduced low-volume, high-impact inputs into the system. In the former example, the chief alteration is physical, while the latter is a primarily chemical alteration of the existing land surface. In both cases, native ecosystemic processes, including pedogenesis, suffer significant interference in water and particle movement, chemical reactions, and biotic activity, such that the soil ecosystem is transformed. At Antônio Galo, the process of excavation revealed a number of in situ or use-related contexts within the terra preta layer, showing that the former was built up iteratively over time, which matches terra preta formation processes put forth by Arroyo-Kalin et al. (2009).

The iterative accumulation of discarded material necessarily occurred according to a culturally mediated practical order. Hence, the characteristics and organization of use and discard space can be used to identify the forcing variable. I considered four possible explanatory scenarios for changes in human activities and their potential traces or signatures in the pedo-stratigraphic matrix:

1 Cultural substitution would produce
   a. completely distinctive signatures, attesting to different activities,
   b. completely different material culture,
   c. *possibly*, different configuration of signatures, indicating re-configuration of activities.

2 Technological changes would produce
   a. completely distinctive signatures, attesting to different activities,
   b. a change in the intensity of signatures, attesting to a different "mechanism,"
   c. *possibly*, different configuration of signatures, indicating re-configuration of activities.

3 Social changes would produce
   a. a re-configuration of signatures, attesting to a re-configuration of social space,
   b. *possibly*, distinctive signatures, attesting to different activities.

4 Population growth/aggregation would produce
   a. a more extensive overall signature,
   b. an increase in the number of nuclei presenting the same series of signatures.

Ceramic, architectural, and microartifactual data indicate general continuity between Phases II and III, which means cultural substitution can be readily rejected. Material culture found before and after the construction of house platforms is remarkably consistent. The architectural evidence is also compelling. During the latter period of Phase II (terra mulata) and Phase III (terra preta), houses are built on platforms, using comparable materials, and in the same places, without any pedo-stratigraphic indication of a hiatus. Hence, there is continuity not only in terms of house placement but also in terms of the process of designating, constructing, and using domestic space. This suggests relative stability in household culture, or at the very least, the absence of a major change at this social scale.

The soil ecosystem change mechanism can be explained by the expansion of the village. On this small landform, houses would have taken over garden space. This would have effectively circumscribed areas for refuse disposal and gardening, which are often one and the same in Amazonian contexts (Schmidt 2010; Schmidt et al. 2014). Without evidence for a substantial change in practices, these data suggest that the eventual transformation of the soil was a result of a change in intensity of practices, rather than in quality of practices or technological aspects.

This leaves us with the challenge of discerning whether social change or population aggregation constituted an external kick. One could just as easily argue that social change precipitated the aggregation of households on the landform as claim that increased population density engendered changes in supra-household relations. I argue that, since places were still constituted and used in the same ways, *household-level culture did not change over the course of the regime shift*. The re-arrangement in village space indicates some changes in social structure, but likely at the community level. As such, even if social changes were taking place, any such impact would have been through their potential prompting of population aggregation. Population aggregation can thus be construed as the proximal external shock.

The primary causal vector would hence look something like this:

↑ population density ⟶ ↑ deposition (inputs—e.g., charcoal and animal bone) ⟶ regime shift

An increase in population density would lead to the superposition of depositional zones, a result of circumscription. This would mean an increase in frequency of depositional events per square meter, as two or more households made use of the same space for refuse disposal. This overlap of refuse zones would thus effectively translate into a rise in intensity of deposition, or increased daily intake of relevant inputs by the soil system. At some point, the accumulated excess charcoal and apatite would "tip the scales" in the direction of a change in regimes.

This analysis has two important consequences for terra preta research specifically, and more generally, for the study of social-ecological changes. First, the process of setting up a regime shift framework helped to locate a *specific mechanism to study*: the process of transformation to terra preta.

Second, a careful analysis of exploratory data within a regime shift framework helped *produce testable hypotheses*: in this case, future excavations at Antônio Galo can target these mound contexts for:

1. Samples for radiocarbon dating that can help generate a tighter chronology
2. Micromorphological samples to clarify, for example, the number of floors associated with each house platform
3. An areal excavation of one or more domestic structures, for a better understanding of internal house dynamics
4. Bulk samples for microartifactual and chemical analyses with increased stratigraphic precision, to shed light into potential, perhaps subtle, changes in domestic practices

Final Thoughts

This household-scale case study employed a regime shift framework to understand anthropogenic transformation of a soil ecosystem as the result of changes in human habitation. In this case, the emergence of a pre-Columbian village in the Central Amazon, which re-configured and circumscribed space, resulted in anthropogenesis or the "culturing" of natural space. The value of the present analysis lies, not in the generation of a solution that can be extrapolated from Antônio Galo to all terra preta in Amazonia, but in the exemplification of a method that can generate such solutions.

The concept of regime shift provides an analytical framework that helps in parsing the complex functional relationships of an ecosystem undergoing a major transformation. Such an analysis requires a clear delimitation of the system in question and an accounting of functional relationships among variables inside and outside the system. In the case of Antônio Galo, I was able to synthesize the specific, minimal measurable parameters that transform terra amarela (or terra mulata) into terra preta in a framework that encompasses both the soil ecosystem and human activity. In the absence of such a framework, as previous descriptive and experimental work by soil scientists demonstrates, it is difficult to mobilize characterizations of terra preta to understand the internal, iterative, and miniscule processes of ecosystem transformation. Similarly, although sociocultural and demographic dimensions of terra preta are discussed by archaeologists at regional scales, the internal dynamics of the terra-amarela-occupying group that becomes a terra-preta-producing group hadn't been addressed. Understanding the social-environmental processes of terra preta formation required a spatial/temporal scale intermediate between these pedological and archaeological approaches. Relevant changes in human behavior and habitation are those perceptible at the spatial scale of the individual or household and at the temporal scale of ecological transformation.

This example demonstrates that understanding the causal mechanisms that drive changes in social and natural systems requires untangling the systems and identifying the appropriate scale of interaction vis-à-vis the scale of change observed. Establishing the appropriate scale of analysis for a model of interaction is vital. In this case, targeting the micro-scale permitted discrimination of the occupational sequence, specifically of spatial patterns associated with each phase, allowing me to track these changes opposite the transformation of non-architectural, i.e., "natural" spaces. By conducting spatially and temporally fine-grained studies of formation processes and associated human activities, I was able to look at the long-term impacts of activity areas such as domestic and garden space.

Rather than proposing a regional solution, this analysis aims at specificity. The broader question is one of coalescence: how, from a diversity of lifeways and ecosystem strategies, can an apparently homogeneous phenomenon like terra preta emerge? To what extent is terra preta, at any point, homogeneous? However, the significance of terra preta on a continental scale can only be assembled once we have unraveled its significance to particular social groups across Amazonia. In order to do this, we must generate coherent data for direct comparisons and analogous models that build from the microscopic, processual scale of ecosystem transformation, outward to regional and trans-Amazonian scales.

Methodological Vignette: Pedology for Archaeology

Soil chemistry and sedimentological methods (Shackley 1975; Hassan 1978; Folk 1980; Stein 1985; Holliday 2004) may be productively used to characterize contexts in which brown-on-brown soils make the visual identification of distinct depositional contexts and land surfaces difficult. In the case of the Central Amazonian terra preta site presented here, soil chemical indices were developed with a basis in pedological concepts (Batjes 1996; Kaufmann et al. 1998), which demonstrate that organic carbon tends to accumulate on natural land surfaces. Additional indices were developed by applying similar pedological concepts to anthropic land surfaces, which include interior and exterior surfaces that are subjected to consistent, repeated, or long-term use, along with accumulated data on the characteristics of terra preta (e.g., Lehmann et al. 2003; Teixeira et al. 2009). By extending this concept, which relies also on soil physics concepts, to the movement of particles within a soil profile,

I developed a method for identifying and characterizing use-surfaces at archaeological sites (Browne Ribeiro 2011). In order to generate indices that can pinpoint land surfaces, sampling must be systematic across a pedo-stratigraphic profile (e.g., 5 or 10 cm levels, depending on site stratigraphy); however, comparative data for pinpointing surfaces and for characterizing poorly understood contexts must come from samples that were taken judgmentally from well-understood contexts (interfaces or deposits). This method is useful for investigating human-environment interaction in contexts where traditional archaeological stratigraphic interpretation and artifact recovery through 1/4- to 1/8-inch mesh do not provide sufficient resolution. Meaningful interpretation of the data produced through these methods depends not only on a close familiarity with the contexts excavated, but also on the availability of a relevant reference collection, set of reference parameters, and local ecological and cultural knowledge base.

Notes

1 Under slash-and-burn cultivation, native soils, known locally as *terra amarela* ("yellow earth"), exhaust their plant-available nutrients in 2–3 years. These acidic soils have extremely low capacity to retain nutrients, including those from introduced fertilizers. Terra preta produces higher yields for longer periods without amendment and responds better to fertilization.
2 "Micro-ecological" indicates the area surrounding a test unit; soil type served as a proxy for local ecological conditions.
3 Research suggests that slash-and-char, an alternative to slash-and-burn, where a smoldering fire is applied, contributes pyrogenic carbon to soils (Lehmann et al. 2002), thus having a potential connection to the formation of terra mulata.
4 Organic matter comprises a range of refuse, which in turn contributed to enrichment in organic C, P, and Ca, the key elements that chemically distinguish terra preta from adjacent soils (Glaser et al. 2004; Lehmann et al. 2003).
5 Pyrogenic or anthropogenic black carbon is formed when plant remains are incompletely combusted in anoxic or low-temperature fires. Black carbon is more stable than other forms of organically derived carbon, contributing to the enhanced CEC and stability of terra preta (Liang et al. 2006; Teixeira et al. 2009).
6 Melanized soils dating to ~5,000 B.P. that have been found at archaeological sites in Rondônia are not considered related to the later spread of the terra preta phenomenon.
7 The term "pedogenically active" evokes not only accelerated chemical processes due to elevated heat and rainfall, but also higher microbial and insect activity, which intensifies vertical migration of particles.
8 Oxisols are highly weathered soils typical of warm, moist tropical regions where parent material has been stable for hundreds of thousands of years. Oxisols are characterized by a high proportion of clay minerals and iron/aluminum oxides.

Because of this, they have a low cation exchange capacity and are unsuitable for prolonged monoculture. Although Amazonia, like any other place, is a mosaic of soils, Oxisols dominate the bluff regions of the Central Amazon and the Lower Amazon River.

9 See Browne Ribeiro (2011) for a full discussion of methods. Horizons are pedogenic units created through in situ modification; deposits are generated through sedimentary processes. After Stein (1992), I consider humans as sedimentary agents who contribute to or create deposits.

10 In his study among the Kuikuru, Carneiro (1957) observes the utility of raised, packed-earth floors in preventing flooding during heavy rains and curtailing insect and rodent burrowing.

11 Little can be known about the central part of the landform; construction of the ring village houses required soil that denuded most of the central plaza of soil, removing or re-depositing a significant portion of the evidence—matrix and artifacts—that had been produced by earlier occupations.

References Cited

Alexandrovskaya, E. I. and A. I. Alexandrovskyi 2000 History of the cultural layer in Moscow and accumulation of anthropogenic substances in it. *Catena* 41(1–3):249–259.

Andrade Pérez, Angela 1986 *Estudio arqueológico de los antrosoles de Araracuara (Amazonas)*. Publicación de la Fundación de Investigaciones Arqueológicas Nacionales, Bogotá.

Arroyo-Kalin, Manuel 2008 *Towards a Historical Ecology of Pre-Columbian Central Amazonia*. Unpublished Ph.D. Thesis, Cambridge University.

——— 2010 The Amazonian formative: Crop domestication and anthropogenic soils. *Diversity* 2:473–504.

——— 2012 Slash-burn-and-churn: Landscape history and crop cultivation in pre-Columbian Amazonia. *Quaternary International* 249:4–18.

Arroyo-Kalin, M., E. G. Neves and W. I. Woods 2009 Anthropogenic dark earths of the central Amazon region: Remarks on their evolution and polygenetic composition. In *Amazonian Dark Earths: Wim Sombroek's Vision*, edited by William I. Woods, Wenceslau G. Teixeira, Johannes Lehmann, Christoph Steiner, Antoinette WinklerPrins and Lilian Rebellato, pp. 99–125. Springer, Berlin.

Bates, Henry Walter 1863 *The Naturalist on the River Amazons*. John Murray, London.

Batjes, N. H. 1996 Total carbon and nitrogen in the soils of the world. *European Journal of Soil Science* 47(15):1–163.

Biggs, Reinette, Stephen R. Carpenter and William A. Brock 2009 Turning back from the brink: Detecting an impending regime shift in time to avert it. *Proceedings of the National Academy of Sciences* 106(3):826–831.

Brock, William A., Stephen R. Carpenter and Marten Scheffer 2008 Regime shifts, environmental signals, uncertainty, and policy choice. In *Complexity Theory for a Sustainable Future*, edited by Jon Norberg and Graeme S. Cumming, pp. 180–206. Columbia University Press, New York.

Browne Ribeiro, Anna 2011 *Acts, Practices, and the Creation of Place: Geoarchaeology of a Terra Preta de Índio Site in the Central Amazon*. Unpublished Ph.D. Thesis, University of California, Berkeley.

——— 2014 Houses, hearths, and gardens: Space and temporality in a pre-Columbian village in the Central Amazon. In *Antes de Orellana. Actas del 3er Encuentro*

Internacional de Arqueología Amazónica, edited by Stéphen Rostain, pp. 183–189. IFEA/FLACSO/US Embassy.

Buol, S. W., R. J. Southard, R. C. Graham and P.A. McDaniel 2011 *Soil Genesis and Classification*. 6th ed. Wiley-Blackwell.

Cabral, Mariana Petry, and João Darcy de Moura Saldanha 2008 Paisagens arqueológicas na costa norte do Amapá. *Revista de Arqueologia* 21(1):211–214.

Carneiro, Robert L. 1957 *Subsistence and Social Structure: An Ecological Study of the Kuikuru Indians*. Unpublished Ph.D. Thesis, University of Michigan.

Castro, M. W. M. 2009 *A cronologia dos sítios Lago do Iranduba e Laguinho à luz das hipóteses da ocupação humana para a Amazônia Central*. Unpublished M.A. Thesis, Universidade de São Paulo, São Paulo.

Chirinos, R. P. 2007 *Padrões de assentamento no sítio Osvaldo, Iranduba, Amazonas*. Unpublished M.A. Thesis, Universidade de São Paulo, São Paulo.

Costa, M. L., D. C. Kern and N. Kampf 2003 Pedogeochemical and mineralogical analyses of Amazonian Dark Earths. In *Amazonian Dark Earths: Origin, Properties, Management*, edited by J. Lehman, D. C. Kern, B. Glaser and W. I. Woods, pp. 333–352. Kluwer Academic Publishers, Dordrecht.

Donatti, P. B. 2003 *A ocupação pré-colonial da área do Lago Grande, Iranduba, AM*. Unpublished M.A. Thesis,Universidade de São Paulo, São Paulo.

Eden, Michael J., Warwick Bray, Leonor Herrera and Colin McEwan 1984 Terra preta soils and their archaeological context in the caqueta basin of southeast Colombia. *American Antiquity* 49(1):125–140.

Eidt, R. C. 1984 *Advances in Abandoned Settlement Analysis: Application to Prehistoric Anthrosols in Colombia, South America*. Milwaukee: Center for Latin America University of Wisconsin-Milwaukee.

Erickson, C. 2010 The transformation of environment into landscape: The historical ecology of monumental earthwork construction in the Bolivian Amazon. *Diversity* 2(4):618–652.

Falesi, I. C. 1974 Soils of the Brazilian Amazon. In *Man in the Amazon*, edited by Charles Wagley, pp. 201–229. University Presses of Florida, Gainesville.

Farias J. B. 1946 *A Cerâmica da Tribo Uaboí dos rios Trombetas e Jamundá*. Conselho Nacional de Proteção ao Índio Rio de Janeiro, pp. 5–42.

Folk, Robert L. 1980 *Petrology of Sedimentary Rocks*. Hemphill Pub. Co, Austin.

Frimigacci, D. 1990 *Aux temps de la terre noire: ethnoarchéologie des îles Futuna et Alofi*. Peeters Press, Paris.

Glaser, B., G. Guggenberger and W. Zech 2004 Identifying the pre-Columbian anthropogenic input on present soil properties of Amazonian dark earths (terra preta). In *Amazonian Dark Earths: Explorations in Space and Time*, edited by B. Glaser and W. I. Woods, pp. 145–158. Springer-Verlag, Berlin.

Glaser, B., G. Guggenberger, W. Zech and M. L. P. Ruivo 2003 Soil organic matter stability. In *Amazonian Dark Earths: Origin, Properties, Management*, edited by J. Lehmann, D. C. Kern, B. Glaser and W. I. Woods, pp. 227–241. Kluwer, Dordrecht.

Glaser, B., W. Zech and W. I. Woods 2004 History, current knowledge and future perspectives of geoecological research concerning the origin of Amazonian anthropogenic dark earths (Terra Preta). In *Amazonian Dark Earths: Explorations in Space and Time*, edited by B. Glaser and W. I. Woods, pp. 9–18. Springer-Verlag, Berlin.

Gomes, D. M. C. 2011 Cronologia e Conexões Culturais na Amazônia: As sociedades Formativas da região de Santarém, PA. *Revista de Arqueologia* 154(1):269–314.

Gourou, P. 1949 L'Amazonie, problémes geographiques. *Les Cahiérs d'Outre-Mer* 5:1–13.

——— 1950 Observações geográficas na Amazônia. *Revista Brasileira de Geografía* 11:355–408.

Groenman-Van Waateringe, Willy and M. Robinson (editors) 1988 *Man-Made Soils*. B.A.R., Oxford, England.

Grogan, Kirsty L., Robert J. Gilkes and Bernd G. Lottermoser 2003 Maghemite formation in burnt plant litter at East Trinity, North Queensland, Australia. *Clays and Clay Minerals* 51(4):390–396.

Guapindaia, V. L. 2011 Estudos Arqueológicos na Região de Porto Trombetas, PA. *Revista de Arqueologia da Sociedade de Arqueologia Brasileira* 2450–73.

Hassan, Fekri A. 1978 Sediments in archaeology: Methods and implications for paleoenvironmental and cultural analysis. *Journal of Field Archaeology* 5(2):197–213.

Heckenberger, M. J. 2005 *The Ecology of Power: Culture, Place, and Personhood in the Southern Amazon, A.D. 1000–2000*. Routledge, New York.

Heckenberger, M. J., J. B. Peterson and E. G. Neves 1999 Village size and permanence in Amazonia: Two archaeological examples from Brazil. *Latin American Antiquity* 10(4):353–376.

Heckenberger, M. J., J. C. Russell, C. Fausto, J. R. Toney, M. J. Schmidt, E. Pereira, B. Franchetto and A. Kuikuro 2008 Pre-Columbian Urbanism, anthropogenic landscapes, and the future of the Amazon. *Science* 321(5893):1214–1217.

Hilbert, P. 1968 *Archäologische untersuchungen am mittleren Amazonas*. Marburger Studien zur Völkerkunde, Berlin.

Holliday, Vance. T 2004 *Soils in Archaeological Research*. Oxford University Press, New York.

Holling, C. S. 1973 Resilience and stability of ecological systems. *Annual Review of Ecology and Systematics* 4:1–23.

Kampf, N., W. I. Woods, Wim Sombroek, D. C. Kern and Tony J. F. Cunha 2003 Classification of Amazonian dark earths and other ancient anthropic soils. In *Amazonian Dark Earths: Origins, Properties, Management*, edited by J. Lehman, D. C. Kern, B. Glaser and W. I. Woods, pp. 77–102. Kluwer Academic Publishers, Dordrecht.

Katzer, Friedrich 1944 A terra preta [1903]. *Boletim da Secção do Fomento Agrícola no Estado do Pará* 3:35–38.

Kaufmann, S., W. G. Sombroek and S. Mantel 1998 Soils of rainforests: Characterization of major constraints of dominant forest soils in the humid tropics. In *Soils of Tropical Forest Ecosystems: Characteristics, Ecology, and Management*, edited by A. Schulte and D. Ruhiyat, pp. 9–20. Springer-Verlag, Berlin.

Kern, D. C. 1988 *Caracterização pedológica de solos com terra arqueológica na região de Oriximiná-PA*. Unpublished MSC Thesis. Soils Department, Universidade Federal do Rio Grande do Sul., Porto Alegre.

——— 1996 *Geoquímica e pedogeoquímica em Sítios Arqueológicos com Terra Preta na Floresta National de Caxiuanã (Portel-PA)*. Unpublished Ph.D. Thesis, Centro de Geociências, Universidade Federal do Pará.

Lehmann, Johannes, Dirse C. Kern, Bruno Glaser and William I. Woods 2003 *Amazonian Dark Earths: Origins, Properties, Management*. Kluwer Academic Publishers, Dordrecht.

Lehmann, Johannes, Jose Pereira da Silva Jr., Marco Rondon, M. da S. Cravo, Jacqueline Greenwood, Thomas Nehls, Christoph Steiner and Bruno Glaser 2002 Slash-and-char—a feasible alternative for soil fertility management in the central Amazon. In *Proceedings of the 17th World Congress of Soil Science*, pp. 1–12. The Congress, Bangkok.

Liang, B., J. Lehmann, D. Solomon, J. Kinyangi, J. Grossman, B. O'Neill, J. O. Skjemstad, J. Thies, F. J. Luizão, J. Petersen and E. G. Neves 2006 Black carbon increases cation exchange capacity in soils. *Soil Science Society of America Journal* 70:1719–1730.

Lima, H. N. and C. E. Schaefer, N. Kämpf, M. L. Costa and K. W. Souza 2009 Características Químicas e Mineralógicas e a Distribuição de Fontes de Fósforo em Terras Pretas de Índio da Amazônia Ocidental In *As terras pretas de índio da Amazônia: sua caracterização e uso deste conhecimento na criação de novas áreas*, edited by Wenceslau G. Teixeira, Dirse C. Kern, B. E. Madari, Hedinaldo N. Lima and William I. Woods, pp. 212–224. Embrapa Amazônia Ocidental-Livros científicos (ALICE).

Lima, Helena Pinto 2008 *História das caretas: A tradição Borda Incisa na Amazônia Central*. Unpublished Ph.D. Thesis, Universidade de São Paulo, São Paulo.

——— 2013 Variabilidade arqueológica e o estudo fronteiras culturais na região do baixo rio Urubu. Amazônico EDUA. In Fronteiras do Passado: Aportes interdisciplinares sobre a Arqueologia do baixo rio Urubu, médio Amazonas, Brasil, edited by Helena P. Lima, pp. 35–61. Museu Amazônico, Editora da Universidade Federal do Amazonas, Manaus, AM.

Lima, Helena Pinto, Eduardo Góes Neves and James B. Petersen 2006 La fase Açutuba: un novo complexo cerâmico na Amazônia central. *Arqueología Suramericana/Arqueologia Sul-Americana* 2(1):26–52.

Lima, Luiz Fernando Erig 2003 *Levantamento arqueológico das áreas de interflúvio na área de confluência dos Rios Negros e Solimões, AM*. Unpublished M.A. Thesis, Universidade de São Paulo, São Paulo.

Ludwig, Donald, Brian Walker and Crawford S. Holling 1997 Sustainability, stability, and resilience. *Conservation Ecology* 1(1):7.

Macedo, R. S. 2009 *Atributos físicos, químicos e mineralógicos de solos com horizonte antrópico (terra preta de índio) em áreas de várzea do Rio Solimões, AM*. Unpublished MSC Thesis, Universidade Federal do Amazonas, Manaus.

Machado, Juliana 2005 *Montículos artificiais na Amazônia central: um estudo de caso do sítio Hatahara*. Unpublished M.A. Thesis. Universidade de São Paulo, São Paulo.

Macphail, R. I. 2003 A future for Dark Earth? *Antiquity* 77(296):349–358.

Marliac, A. 2002 From black earth medieval urban sites in France to tropical settlement. *Anthropologie* 106(5):745–761.

May, R. M. 1977 Thresholds and breakpoints in ecosystems with a multiplicity of stable states. *Nature* 269:471–477.

McMichael, C. H., D. R. Piperno, M. B. Bush, M. R. Silman, A. R. Zimmerman, M. F. Raczka and L. C. Lobato 2012 Sparse pre-Columbian human habitation in Western Amazonia. *Science* 336(6087):1429–1431.

Meggers, B. J. 1971 *Amazonia: Man and Culture in a Counterfeit Paradise*. Aldine Atherton, Chicago.

——— 2001 The continuing quest for El-Dorado: Round two. *Latin American Antiquity* 12(3):304–325.

Moraes, B. M. and Helena Pinto Lima 2012 Antrossolos Amazônicos: Uma perspectiva histórica sobre as terras pretas de índio na Amazônia Central. *Ciência & Ambiente.* 1:87–98.

Moraes, C. P. 2006 *Arqueologia na Amazônia Central vista de uma perspectiva da região do Lago do Limão.* Unpublished M.A. Thesis, Universidade de São Paulo, São Paulo.

——— 2010 Aldeias circulares na Amazônia Central: um contraste entre fase Paredão e fase Guarita. *Arqueologia Amazônica* 2:581–604.

——— 2013 *Amazônia ano 1000: territorialidade e conflito no tempo das chefias regionais.* Unpublished Ph.D. Thesis, Universidade de São Paulo, São Paulo.

Moraes, C. P. and E. G. Neves 2012 O ano 1000: Adensamento populacional, interação e conflito na Amazônia Central. *Amazônica-Revista de Antropologia* 4(1):122–148.

Neves, E. G. 2008 Ecology, ceramic chronology and distribution, long-term history, and political change in the Amazonian floodplain. In *Handbook of South American Archaeology*, edited by Helaine Silverman and Wiliam Isbel, pp. 359–379, 380. Springer, New York.

——— 2012 *Sob os Tempos do Equinócio: Oito Mil Anos de História na Amazônia Central (6.500–1.500 DC).* Unpublished Thesis (*Tese de Livre Docência*). Museu de Arqueologia e Etnologia, Universidade de São Paulo, São Paulo.

Neves, E., H. P. Lima, L. S. Costa and J. Gomes 2014 A tradição Pocó-Açutuba e os primeiros sinais visíveis de modificações de paisagems na calha do Amazona. In *Memorias de las Conferencias Magistralesdel 3er Encuentro Internacional de Arqueología Amazónica*, edited by Stéphen Rostain, pp. 137–158. Ekseption Publicidad, Quito.

Neves, E. G. and James B. Petersen 2006 The political economy of Late pre-Columbian amerindians: Landscape transformation in Central Amazonia. In *Time and Complexity in Historical Ecology*, edited by W. Balée and C. Erikson, pp. 279–309. Columbia University Press, New York.

Neves, E. G., James B. Petersen, Robert N. Bartone and Carlos Augusto Da Silva 2003 *Levantamento Arqueológico Da Área De Confluência Dos Rios Negro E Solimões, Estado Do Amazonas: Continuidade Das Escavações, Análise Da Composição Química E Montagem De Um Sistema De Informações Geográficas. Fundação de Amparo à Pesquisa do Estado de São Paulo.* Report on file with Instituto do Patrimônio Histórico e Artistico Nacional, Manaus.

Petersen, James B., Eduardo G. Neves and Michael J. Heckenberger 2001 Gift from the past: Terra preta and prehistoric Amerindian occupation in Amazonia. In *Unknown Amazon: Culture in Nature in Ancient Brazil*, edited by Cristiana Barreto and Colin McEwan, pp. 86–105. British Museum Press, London.

Prumers, H. and C. Jaimes Betancourt 2014 100 años de investigación arqueológica en los Llanos de Mojos. *Arqueoantropológicas* 4:11–54.

Rapp Py-Daniel, Anne 2009 *Arqueologia da morte no sítio Hatahara durante a fase Paredão.* Unpublished M.A. Thesis Universidade de São Paulo, São Paulo.

Rostain, Stéphen 2013 *Islands in the Rainforest: Landscape Management in pre-Columbian Amazonia.* Left Coast Press, Walnut Creek, CA.

Schaan, Denise 2012 *Sacred Geographies of Ancient Amazonia: Historical Ecology of Social Complexity.* Left Coast Press, Walnut Creek, CA.

Schaefer, CEGR, H. N. Lima, R. J. Gilkes and J. W. V. Mello 2004 Micromorphology and electron microprobe analysis of phosphorus and potassium forms of an

Indian Black Earth (IBE) Anthrosol from Western Amazonia. *Australian Journal of Soil Research* 42(4):401–409.

Scheffer, M. and S. R. Carpenter 2003 Catastrophic regime shifts in ecosystems: Linking theory to observation. *Trends in Ecology and Evolution* 18(12):648–656.

Schmidt, Morgan J. 2010 *Reconstructing Tropical Nature: Prehistoric and Modern Anthrosols (terra preta) in the Amazon Rainforest, Upper Xingu River, Brazil.* Unpublished Ph.D. Thesis, University of Florida.

Schmidt, Morgan J. and Michael J. Heckenberger 2009 Amerindian anthrosols: Amazonian dark earth formation in the Upper Xingu. In *Amazonian Dark Earths: Wim Sombroek's Vision*, edited by William I. Woods, Wenceslau G. Teixeira, Johannes Lehmann, Christoph Steiner, Antoinette WinklerPrins and Lilian Rebellato, pp. 163–191. Springer, Berlin.

Schmidt, Morgan J., Anne Rapp Py-Daniel, Claide de Paula Moraes, Raoni B. M. Valle, Caroline F. Caromano, Wenceslau G. Teixeira, Carlos A. Barbosa, João A. Fonseca, Marcos P. Magalhães, Daniel Silva do Carmo Santos and others 2014 Dark earths and the human built landscape in Amazonia: A widespread pattern of anthrosol formation. *Journal of Archaeological Science* 42:152–165.

Shackley, Myra L. 1975 *Archaeological Sediments: A Survey of Analytical Methods.* Wiley, New York.

Shock, M. P., C. P. Moraes, J. S. Belletti, M. Lima, F. M. Silva, L. T. Lima, M. F. Cassino and A.M. A. Lima 2014 Initial contributions of charred plant remains from archaeological sites in the Amazon to reconstructions of historical ecology. In *Antes de Orellana. Actas del 3er Encuentro Internacional de Arqueología Amazónica*, edited by Stéphen Rostain, pp. 291–296. IFEA/FLACSO/US Embassy, Quito.

Silva, F. M., M. P. Shock, E. G. Neves, H. P. Lima, R. Scheel-Ybert 2013 Recuperação de macrovestígios em sítios arqueológicos na Amazônia: nova proposta metodológica para estudos arqueobotânicos. *Bol. Mus. Para. Emílio Goeldi. Cienc. Hum.*, Belém, 8(3):759–769,

Smith, Herbert Huntington 1879 *Brazil, the Amazons and the Coast.* C. Scribner's Sons, New York.

Smith, Nigel J. H. 1980 Anthrosols and human carrying-capacity in Amazonia. *Annals of the Association of American Geographers* 70(4):553–566.

Sombroek, W. G. 1966 *Amazon Soils. A Reconnaissance of the Soils of the Brazilian Amazon Region.* State Agricultural University, Wageningen, Netherlands.

Stein, J. K. 1985 Interpreting sediments in cultural settings. In *Archaeological Sediments in Context*, edited by J. K. Stein and W. R. Farrand. Center for Study of Early Man, University of Maine, Orono.

Stein, Julie K. 1992 Organic matter in archaeological contexts. In *Soils in Archaeology: Landscape Evolution and Human Occupation*, edited by Vance T. Holliday, pp. 193–216. Smithsonian Institution Press, Washington, DC.

Tamanaha, E. K. 2012 *Ocupação polícroma no baixo e médio Solimões, Estado do Amazonas.* Unpublished M.A. Thesis, Universidade de São Paulo, São Paulo.

Tamanaha, E. K. and Rapp Py-Daniel, A. 2009 Sítio Hatahara: estruturas funerárias, residenciais ou ambas? *Revista do Museu de Arqueologia e Etnologia* 8:63–73.

Teixeira, Wenceslau G., Gilvan Coimbra Martins, Rodrigo Santana Macedo, A. Ferreira Neves Junior, Adônis Moreira, V. de M. Benites and Christoph Steiner 2009 As Propriedades fisicas dos Horizontes Antropicos (Terras Pretas de Indio e Terra Mulatas) na Amazonia. In *As terras pretas de índio da Amazônia: sua caracterização e uso deste conhecimento na criação de novas áreas*, edited by Wenceslau G. Teixeira, Dirse C. Kern, B. E. Madari, Hedinaldo N. Lima and William I. Woods, pp. 242–250. Embrapa Amazônia Ocidental-Livros científicos (ALICE).

Walton, T. and R. Cassels 1992 Borrow pits and made soils near Waverley, South Taranaki. *Archaeology in New Zealand* 35(3):160–171.

Woods, William I. and Joseph M. McCann 1999 The anthropogenic origin and persistence of Amazonian dark earths. *Yearbook. Conference of Latin Americanist Geographers* 25:7–14.

Zech, W., L. Haumaier and R. Hempfling 1990 Ecological aspects of soil organic matter in tropical land use. In *Humic Substances in Soil and Crop Sciences*, edited by P. M. McCarthy, C. E. Clapp, R. L. Malcolm and P. R. Bloom, pp. 187–202. American Society of Agronomy and Soil Science Society of America, Madison, WI.

Zech, W., E. Pabst and G. Bechtold 1979 Analytische Kennzeichung der Terra preta do indo. *Mitteilungen der Deutschen Bodenkundlichen Gesellschaft* 29:709–716.

8 External Impacts on Internal Dynamics

Effects of Paleoclimatic and Demographic Variability on Acorn Exploitation Along the Central California Coast

Brian F. Codding and Terry L. Jones

Abstract

Research into human-environment interaction in California prehistory often focuses on either the internal dynamics of adaptive decisions or the external impacts of environmental change. While both processes were surely driving prehistoric variability, integrating these approaches is not altogether straightforward. Here we outline an inclusive approach examining the exploitation of acorn habitats in Central California. Acorns were critically important to many ethnographic groups in Native California, but the intensive use of acorns appears to be a Late Holocene phenomenon. Most research approaches the increased reliance on acorns as a process governed by internal dynamics linked to demographically driven resource intensification, but there are strong reasons to believe that climatic variability also structured acorn use. Here we link internal and external human-environmental dynamics through a formal behavioral ecological model. This model provides clear predictions that can be used to identify departures from expected internal dynamics linked to external factors driven by paleoenvironmental change. Results show that prehistoric settlement along the Central California coast shifts into interior oak-dominated regions with increasing population densities, consistent with model expectations of internally driven resource intensification. However, acorn use was also affected by climate: foragers were less likely to live in productive acorn habitats during periods of drought. These findings show that neither internal nor external patterns can completely account for variability in prehistoric decisions, but that integrating these through formal ecological models can provide insights into the external impacts on internal dynamics that structure broad patterns in prehistory.

Introduction

Research into California prehistory often focuses on either the internal dynamics of adaptive decisions (e.g., Basgall 1987; Hildebrandt and Jones

1992) or the external impacts of environmental change (e.g., Jones et al. 1999; Kennett et al. 2007). While both processes surely drove variability in prehistory, integrating internal and external dynamics is not a straightforward endeavor. Here we outline an inclusive approach using spatially explicit predictions from a behavioral ecological model to explain variability in the exploitation of acorn habitats along the Central California coast.

Acorns were critically important to many ethnographic groups in Native California (Gifford 1936), but the intensive use of acorns appears to be a Late Holocene phenomenon. Most research approaches the increased reliance on acorns as a process governed by internal (endogenous) dynamics linked to demographically driven resource intensification (Basgall 1987; Codding et al. 2012). Because acorns offer a low rate of nutrient return relative to the time required to process and store them (Simms 1985; Wohlgemuth 2010, also Morgan 2012), model-based predictions from behavioral ecology (e.g., MacArthur and Pianka 1966; Morgan 2015) suggest that foragers should only begin relying on such a low profitability resource when more profitable foods decline in abundance due to processes such as resource depression (e.g., Broughton 1994).

There are strong reasons to suspect that external (exogenous) factors such as climatic variability also mediated acorn use. Ecological research has shown that acorn productivity in Central California (Koenig et al. 1996) and elsewhere (Sánchez-Humanes and Espelta 2011) decreases significantly with prolonged drought. As such, reduced rainfall should not only depress the terrestrial environment overall, but could have had a particularly disruptive effect on foragers reliant on acorns as a staple food source. Indeed, periods of drought should have a differential impact on foragers depending on their subsistence strategies: those reliant on resources that are insensitive to drought, like marine foods, should be relatively resilient; while those reliant on resources that are particularly sensitive to drought, like acorns, should be dramatically impacted.

To examine these interactions, here we link internal and external human-environmental dynamics through a formal model from behavioral and population ecology. Known as the Ideal Free Distribution (IFD) model (Fretwell and Lucas 1969), recent applications have provided insights into the factors structuring land-use patterns in prehistoric California (e.g., Kennett 2005; Kennett et al. 2009; Winterhalder et al. 2010; Codding et al. 2012; Codding and Jones 2013; Jazwa et al. 2013) and elsewhere (e.g., Allen and O'Connell 2008; Kennett and Winterhalder 2008; Kennett et al. 2006; 2009). The IFD establishes clear settlement predictions based on the dynamic interactions between demography and habitat suitability. Specifically, the model predicts that individuals should choose to occupy the most suitable habitats available until a point at which demographic pressure (anthropogenic resource depression, competition with conspecifics, etc.) in those habitats causes individuals to do better by moving into less suitable and less densely occupied patches. Thus, holding climatic change constant, long-term increases in population densities should cause foragers to cascade into lower and lower suitability patches.

The greatest difficulty in applying the IFD archaeologically is that suitability must be explicitly operationalized in each case, ideally using some proxy for resource acquisition efficiency that adequately captures how an individual would be expected to evaluate the profitability of alternative habitats. To do this, we rely on logic from prey and patch choice models (MacArthur and Pianka 1966). We begin by defining habitats as aggregations of patches and patches as aggregations of resources. Resources vary in their profitability (e.g., Simms 1985) and patch profitability should vary based on the abundance of the highest profitability resources within the patch. Habitats should then vary in suitability based on the abundance of high profitability patches. Along the Central Coast of California, high profitability marine resources (e.g., fish, marine vertebrates) are only available from the littoral, and high profitability terrestrial resources should be more abundant along the coast than in the interior as a result of differences in precipitation. As such, habitats on the coastal plain with littoral patches of abundant marine resources and mixed chaparral/grassland patches with abundant terrestrial resources should be more suitable than inland habitats dominated by oak patches. While inland oak habitats are not necessarily resource poor, high profitability resources should occur in lower densities in the interior, with the most abundant food resource—acorns—offering lower net yields per time spent handling compared with resources found along the coast. As such, these inland habitats should be of lower overall suitability when compared with the coast, but provide abundant, low profitability resources that populations could exploit when the more profitable alternatives were less available.

Predictions

Linking these estimates of habitat suitability with basic IFD model dynamics, we predict that foragers should have prioritized settlement along productive coastal habitats (see also Jones 1991; Codding et al. 2012), only moving into interior oak patches once demographic pressure reduced per capita gains in high suitability coastal patches. However, this simple prediction may be confounded by climatic variability. If climatically induced reductions in environmental productivity differentially affect oak habitats relative to neighboring coastal habitats, then prolonged droughts may decrease the overall suitability of inland habitats to a point where foragers could achieve higher per capita gains even in densely occupied coastal habitats, resulting in a shift in populations out of inland habitats and into coastal ones, regardless of existing demographic pressure. Approaching the problem in this way allows for the identification of departures from expected internal dynamics linked to external factors driven by paleoenvironmental change.

After discussing the data and methods used in this chapter, we test these predictions, examining the impact of internal and external factors on individuals' decisions to settle in acorn habitats.

Methods and Data

Examining the interaction between demographic and climatic variability on habitat choice requires a combination of internal (endogenous) and external (exogenous) variables. For this study, these include proxies of human population, habitat choice, and climate.

As a proxy for population densities, we use a database of radiocarbon-dated components in Santa Cruz, Monterey, and San Luis Obispo counties (Jones et al. 2007). Occupational histories for each site are determined using all the calibrated radiocarbon dates available summed across the Holocene record in 100-year intervals (Jones et al. 2007; Codding et al. 2012). A single Paleoindian component marked by a fluted point is undated, but is included here at 12 kya. One-hundred-year bins were chosen as the minimum meaningful unit given the measurement error and calibration uncertainty inherent in radiocarbon dates; the former is a particular problem with legacy data like that incorporated into regional databases. A site was considered occupied for any given 100-year interval if that interval overlapped with the 95 percent confidence interval reading of a calibrated radiocarbon date. This broadens the number of bins considered occupied for a given radiocarbon date, but may be more archaeologically meaningful than a summed radiocarbon probability distribution as it incorporates periods of time when the site was likely occupied but for which a probability distribution around a central tendency reading would not necessarily indicate as such. This method generally provides results consistent with other methods used to estimate prehistoric populations (Williams 2012). In this way, each occupied interval is treated as an archaeological component (*sensu* Phillips and Willey 1953). To account for preservation bias that renders older components less common, we apply a taphonomic correction following Surovell et al. (2009), here set at a maximum of 100 sites per period. By empirically modeling the difference between observed volcanic events in sedimentary contexts with those in atmospheric contexts from the Greenland Ice Core record, this approach allows for a standard correction factor to be applied in order to estimate the number of actual sites from the number of observed sites. Both observed and corrected population estimates are used in quantitative models to determine the effect of demography on acorn habitat use. While a useful control on potential taphonomic bias, this approach is certainly not without its problems. This method produces a global correction assuming the equal loss of sites across all landforms, but taphonomic loss may result from specific depositional or erosional events that affect some locations and not others (e.g., sea level rise), which can only be corrected through local geomorphological research (Ballenger and Mabry 2011). Unfortunately, the work necessary to perform this analysis has not been undertaken in the study area. While taphonomy thus remains a confounding factor, and we do not account for sample size effects, our qualitative interpretations only treat high-magnitude and long-duration shifts as meaningful (see Contreras and Meadows 2014). Further research addressing taphonomy, geomorphology, and sampling, may improve the resolution of demographic patterns.

From these data, we also approximate habitat choice through time by linking the spatial location of occupied components to landscape estimates of acorn productivity (Figure 8.1). Historic potential climax vegetation from Kuchler (1977) was evaluated by Wohlgemuth (2010) to rank habitats by acorn nut productivity. While the spatial distribution of these habitats surely changed at least modestly throughout the Holocene, the relative values between patches should remain stable through the record. Because oaks of the same species should respond similarly to climatic variability, reductions in regional productivity should reduce the relative ranking of each oak habitat equally so that high-ranking areas remain high relative to low-ranking areas. Major ecotonal shifts in prehistory may invalidate this assumption, but the fine-grained paleoecological data to account for these shifts are currently unavailable. We extract vegetation data for each dated component using ArcMap (ESRI 2012). The total number of sites in highly productive acorn habitats (rank 3, see Figure 8.1) is divided by the total number of sites occupied for each 100-year interval. This results in an estimate of the proportion of sites located in highly productive nut habitats in 100-year slices across the Holocene record. As a proxy for the reliance on acorns, this is the main dependent variable used in analysis.

Two climatic proxies are used to examine the external impacts on acorn habitat settlement. First, we rely on Cook and colleague's (1999; 2004) gridded reconstructions of drought severity. Based on tree-ring reconstructions, these data provide spatially explicit estimates of drought severity normalized as the Palmer Drought Severity Index (PDSI). PDSI represents the amount of soil moisture available relative to the long-term average such that low values indicate dry conditions. The second proxy is an estimate of local sea surface temperatures from a marine core taken off the Santa Barbara coast (south of the study area; Kennett and Kennett 2000). Sea surface temperatures (SST) are approximated using oxygen isotope values from surface dwelling foraminifera (*G. bulloides*; Kennett and Kennett 2000; Kennett et al. 2007). Values are reported as the ratio of O^{16} and O^{18} relative to a standard (δO^{18} PDB‰). To make it comparable with the demographic and settlement data, SST data is averaged across 100-year intervals using the `wapply` function in the `gplots` library (Warnes et al. 2014) in R (R Development Core Team 2014). In order to examine how these two climatic proxies vary with one another over the last 2,000 years, we smooth both trends using the `loess.as` function from the `fANCOVA` package (Wang 2010) in R (R Development Core Team 2014). Smoothing terms were established using a generalized covariance function with a first-order polynomial (Figure 8.2, thick black lines). For visualization purposes, PDSI data are also shown in Figure 8.2 smoothed with a second-order degree polynomial.

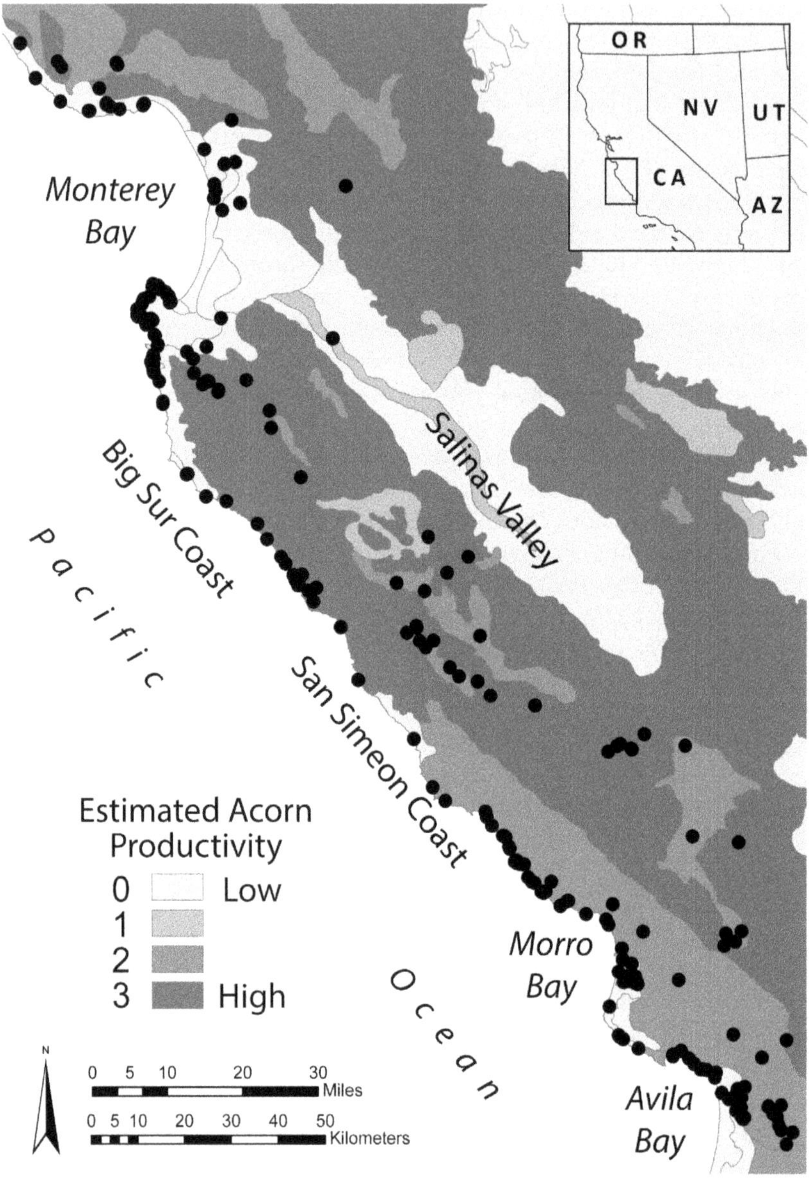

Figure 8.1 Radiocarbon-dated archaeological sites along the Central Coast of California relative to estimated historic acorn productivity.

Source: After Wohlgemuth 2010

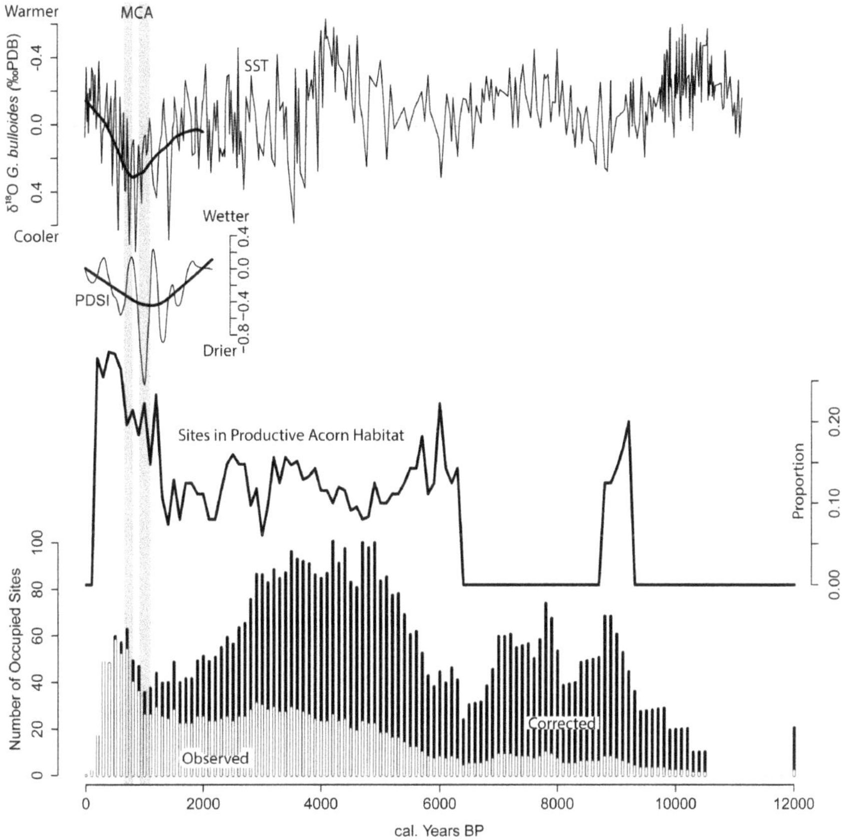

Figure 8.2 Summary of demographic, settlement, and climatic proxies for the study area across the Holocene. Population proxies include observed and taphonomically corrected occupational histories at 100-year intervals. Settlement data reports the proportion of those sites located within productive acorn habitats. Climatic data includes smoothed values of local Palmer Drought Severity Index (PDSI) from Cook et al. (1999; 2004) and inferred sea surface temperatures (SST) from Kennett and colleagues (1995; 2007; Kennett and Kennett 2000). Droughts associated with the Medieval Climatic Anomaly (MCA) are shown in grey (Stine 1994).

In order to model the effect of each independent variable (demographic and climatic proxies) on the dependent variable (the proportion of sites located in high productivity acorn habitats), we rely on generalized additive models (GAM). GAMs describe predictor-response relationships in a series of parametric fits and non-parametric smooth terms (splines) tied together at a series of "knots" (Hastie and Tibshirani 1990; Wood 2006). GAMs trade off between maximizing goodness of fit (minimizing residual deviance) and maximizing parsimony (minimizing the degrees of freedom, df).

We take a conservative approach by minimizing the number of knots used in each model in order to maximize parsimony. GAMs can examine the response of a dependent variable to an independent variable while holding the effects of additional independent variables constant. We do this below to examine the combined effects of demographic and climatic factors on the proportion of sites located within productive oak habitats. Since the primary dependent variable is a proportion that is constrained to vary between zero and one, all models assume a binomial distribution with a logit link using a quasi-likelihood estimation (unless otherwise noted). Results report the number of knots (k) used in the model, the log-likelihood r-square (R^2_L) which measures the proportion of deviance explained by the independent variable(s), the F statistic of the independent variable, and the p-value. To illustrate the model results, we present (Figure 8.3) partial residual plots that show the effect of independent variables on the response variable while holding any effect of additional independent variables constant. This approach allows us to determine how populations (the number of occupied sites) and climate (SST) influence the degree to which local populations relied on acorns (the proportion of sites located in highly productive acorn habitats). All models were implemented in R (2014).

Results

Internal Dynamics: Does Demographic Pressure Drive Acorn Exploitation?

An examination of the observed occupational sequence shows that populations increase linearly through the Holocene (Figure 8.2). When corrected for taphonomic bias, populations still fluctuate throughout the sequence, but the trend is non-linear. As shown in Figure 8.2, there are several periods of high population: one from about 10,000 to 7,000 years ago, a second from about 6,000 to 2,500 years ago, and a third beginning about 1,000 years ago.

These fluctuations in populations structured decisions to settle in highly productive acorn habitats. Throughout the Holocene, the proportion of sites located in highly productive acorn habitats increases significantly as a function of the total number of sites occupied ($k = 3$, $R^2_L = 0.43$, $F = 35.21$, $p < 0.0001$). This trend is also manifested in the taphonomically corrected estimates of population density ($k = 3$, $R^2_L = 0.16$, $F = 6.99$, $p = 0.0015$), although the proportion of the deviance explained is greatly reduced from 43 percent to 16 percent. These results confirm the prediction that individuals only move into acorn-producing habitats when population densities are high.

External Impacts: Do Drier Climates Limit Acorn Exploitation?

Over the last 2,000 years, drought severity worsened significantly when sea surface temperatures cooled along the Central Coast of California (Gaussian

GAM, $k = 3$, $R^2_L = 0.73$, $F = 2805$, $p < 0.0001$; see Figure 8.2). While sea surface temperatures in the region do not correspond perfectly with terrestrial productivity, this empirical relationship confirms previous findings that SST may be a reliable proxy for variability in local terrestrial climate (see Jones and Kennett 1999; Kennett and Kennett 2000; Kennett et al. 2007).

Using SST to examine the effect of climate on acorn habitat use across the entire Holocene record shows that the proportion of sites in productive oak habitats declines significantly with decreasing sea surface temperature ($k = 3$, $R^2_L = 0.23$, $F = 12.79$, $p < 0.0001$). Because cooler sea surface temperatures are associated with decreased terrestrial productivity, this finding suggests that drier climates limit acorn exploitation.

External Impacts on Internal Dynamics: What Are the Combined Effects of Demography and Climate?

To examine the linked effect of demographic and climatic factors, we combine these data in a single model that controls for the interaction of both predictor variables. This model shows that both demography and climate have a significant effect on decisions to move into oak habitats (Figure 8.3, $k = 3$, $R^2_L = 0.46$, $p < 0.0001$), with demography ($F = 19.15$) having a

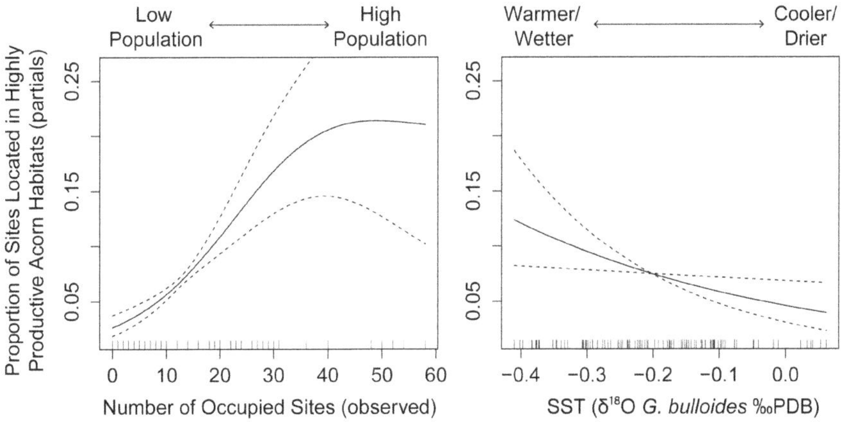

Figure 8.3 Partial residual plots of generalized additive model results showing the combined effect of observed population (left) and sea surface temperature (right) on the proportion of sites located in productive acorn habitats. These results show that as populations increase, people move into productive acorn habitats (left); but when conditions are drier, fewer people settle in productive acorn habitats (right). Data correspond to time series trends shown in Figure 8.2 sampled at 100-year intervals. These plots show the relationship between the response variable (proportion of sites) and each independent variable (number of occupied sites and sea surface temperature) holding the effect of the other independent variable constant (e.g., the effect of SST on the proportion of sites in acorn habitats while holding the effect of the number of sites constant).

greater impact than climate ($F = 6.03$). These trends hold even with the taphonomically corrected site data ($k = 3$, $R^2_L = 0.38$, $p < 0.0001$), but with climate ($F = 33.92$) having a greater effect than demography ($F = 9.15$). The results indicate that holding climate constant, demographic pressure remains a significant factor in determining settlement choice (Figure 8.3, left). Likewise, holding population constant, climate has a significant effect on settlement decisions (Figure 8.3, right). These trends and some remaining anomalies are discussed below.

Discussion

The results of this study highlight the importance of an approach that combines the effects of external and internal factors on prehistoric human decisions. Overall, the results show that both demographic and climatic factors structured settlement in productive acorn habitats through the Holocene. As predicted by the IFD, individuals chose to move into these relatively low profitability habitats as a result of demographic pressure. But, this decision was also mediated by climate: individuals were less likely to settle in oak habitats during times of decreased overall environmental productivity.

At particular points in the regional prehistory, either demography or climate appears to have been more dominant. Between about 10,000 and 6,000 years ago, individuals rarely settled in oak habitats. While climatic proxies suggest that terrestrial environment was relatively productive during this time, foragers chose not to exploit acorns due to limited demographic competition in higher suitability patches. This is supported by faunal evidence suggesting that foragers experienced relatively high returns with more profitable resources during this time (Jones et al. 2008a; 2009).

At later points, these two factors seem to have worked in concert. Beginning approximately 6,000 years ago, forager populations expanded while the climate was ameliorating. This resulted in the first large-scale population expansion into productive oak habitats. This is also the same time period when the proportion of acorn-specific processing equipment (mortars and pestles) increased above 50 percent of groundstone assemblages in the region (Codding et al. 2012; also see Stevens and McElreath 2015). Without climatic impediments, demographic pressure pushed foragers into exploiting acorns intensively for the first time.

These trends decouple beginning about 3,000 years ago, when forager populations either plateaued (observed site counts) or began to decline (corrected site counts). In this context, the proportion of sites in highly productive oak habitats remained relatively stable, despite increasing aridity. Indeed, despite radical changes in other subsistence patterns during this interval (e.g., Jones et al. 1999; 2007; Codding and Jones 2007; Codding et al. 2010), acorn habitat use remained relatively important even through the two mega-droughts associated with the Medieval Climatic Anomaly (MCA; ca. 1,100–900 and 800–650 years ago; Stine 1994). Contrary to our predictions at the outset, this may result from these extreme droughts dampening all terrestrial environments equally so that relative habitat rank remained the same. But given the response of oaks to drought, this seems unlikely. Alternatively, it could be that acorns remained important due to their storability. If the MCA increased subsistence uncertainty, foragers may have responded by maintaining fall habitation sites in highly productive oak patches where they could potentially acquire enough acorns in order to reduce subsistence variance later in the year. Finer-grained data addressing seasonality of occupation would be needed to evaluate possibilities like this one.

At the end of the MCA, the dampening effects of drought on habitat suitability were removed. Populations begin to increase during this time of climatic amelioration coincident with a dramatic increase in the proportion of sites located in oak habitats, reaching the highest point in prehistory. These late prehistoric shifts in settlement mark the onset of intensive acorn use recorded ethnographically. These patterns are likely dominated by seasonal resource extractions camps occupied in the fall (Jones et al. 2008b). Perhaps the most notable implication of this trend is that the increased use of acorn habitats exceeds what would be predicted either by demographic or climatic trends. Considering the corrected site counts, previous periods in prehistory saw greater increases in populations and the climatic records indicate higher levels of terrestrial productivity.

There are perhaps two possible explanations for this late prehistoric anomaly that center on additional constraints associated with the shift to an acorn-based economy. First, ethnographic and ecological evidence suggests that the reliable exploitation of acorns may require the frequent application of fire in order to make oak groves reliable and productive (Anderson 2005; Lightfoot et al. 2013; Anderson and Rosenthal 2015). If this is true, then the increase in oak habitat exploitation late in the record may mark the onset of anthropogenic fire regimes recorded ethnographically. The second potential explanation involves constraints on the shift from immediate to delayed return economies (Woodburn 1982). Because the intensive use of acorns requires that crops be gathered and stored, this economic shift requires paired shifts in social institutions associated with the management of private property (Bowles and Choi 2013; Bettinger 2015). It is possible that the development of social institutions designed to deal with the collective action problems associated with stored food did not emerge until late in the prehistoric record.

Conclusion

This chapter examines the external impacts of climatic variability and the internal dynamics of demography on forager settlement and subsistence. The results illustrate how incorporating both factors into analysis may be necessary to explain prehistoric decisions. By linking these factors within a formal ecological model, this approach limits the probability of spurious correlations by specifying causal predictions *a priori*. While this example illustrates the usefulness of this approach in a regional context focused on acorn use, it has the potential to be a productive framework in explaining broader and more varied patterns across other ecological, subsistence, and social settings.

Acknowledgments

Thanks to Dan Contreras for spearheading this important volume, to Eric Wohlgemuth for sharing data on acorn productivity, to Angela Barios for compiling the original radiocarbon database, and to Simon Brewer for analytical advice. Nathan Stevens, Dan Contreras, and an anonymous reviewer provided detailed comments on earlier versions of this chapter; their input is much appreciated and significantly improved the final product.

Methodological Vignette: Spatially Explicit Behavioral Ecology (SEBE)

Spatially explicit behavioral ecology links methods and models from GIS and spatial analysis to test theoretical predictions derived from behavioral ecology. The approach provides three specific advantages to help answer questions about human-environmental dynamics. First, because the BE approach reduces complex problems to their constituent parts, researchers are forced to articulate key components to be analyzed. This helps frame the question and directs research towards the specific data needed to answer the question. Second, by generating predictions from a general theory of behavior, researchers are able to nominate causal relationships a priori (O'Connell 1995; Bird and O'Connell 2006; Codding and Bird 2015), which helps to reduce spurious correlations. Finally, spatially explicit models provide an opportunity to link human behavior to specific environmental conditions. This can be analytically trying as it adds a spatial dimension to archaeological time-series analyses, but the process allows researchers to link specific aspects of ecological variability to human decisions across space and through time.

Applying this approach requires that researchers first link a research problem to an appropriate foraging model (or vice versa) depending on the scope and scale of the question of interest (for a recent review, see Codding and Bird 2015). Model variables will help researchers identify the data necessary to answer the question, which will frequently include spatially explicit time-series data that tie material proxies of human decisions (e.g., faunal remains or settlement chronologies) to paleoenvironmental data (e.g., estimates of resource abundance or environmental productivity). Key here is the compilation of data on comparable time-spaces scales that can be used to identify how humans adapt to and modify their surrounding environments.

To date, this work has been most useful in explaining hunter-gatherer foraging ecology and behavior. Examples include modeling spatial variability in men's and women's foraging returns (Zeanah 2004), hunter-gatherer foraging radii (Morgan 2008), and habitat use in response to climate change (Morgan 2009). Growing applications examine patterns of colonization across novel landscapes (e.g., Kennett 2005; Allen and O'Connell 2008; O'Connell and Allen 2012), with implications for understanding the origins of ethnic diversity (Codding and Jones 2013) and the emergence of social inequality (e.g., Winterhalder et al. 2010).

References Cited

Allen, J. and O'Connell, J. F. 2008 Getting from Sunda to Sahul. In *Islands of Inquiry: Colonization, Seafaring and the Archaeology of Maritime Landscapes*, edited by G. Clark, F. Leach and S. O'Connor, pp. 31–46. ANU E Press, Australian National University, Canberra.

Anderson, M. K. 2005 *Tending the Wild: Native American Knowledge and the Management of California's Natural Resources*. University of California Press, Berkeley.

Anderson, M. K. and J. Rosenthal 2015 An ethnobiological approach to reconstructing indigenous fire regimes in the foothill chaparral of the Western Sierra Nevada. *Journal of Ethnobiology* 35:4–36.

Ballenger, Jesse A. M. and Jonathan B. Mabry 2011 Temporal frequency distributions of alluvium in the American Southwest: Taphonomic, paleohydraulic, and demographic implications. *Journal of Archaeological Science* 38:1314–1325.

Basgall, Mark E. 1987 Resource intensification among hunter-gatherers: Acorn economies in prehistoric California. *Research in Economic Anthropology* 9:21–52.

Bettinger, R. L. 2015 *Orderly Anarchy: Sociopolitical Evolution in Aboriginal California*. University of California Press, Berkeley.

Bird, Douglas W. and James F. O'Connell 2006 Behavioral ecology and archaeology. *Journal of Archaeological Research* 14:143–188.

Bowles, S. and J.-K. Choi 2013 Coevolution of farming and private property during the early Holocene. *Proceedings of the National Academy of Sciences* 110:8830–8835.

Broughton, J. M. 1994 Late Holocene resource intensification in the Sacramento Valley, California: The vertebrate evidence. *Journal of Archaeological Science* 21:501–514.

Codding, B. F. and D. W. Bird 2015 Behavioral ecology and the future of archaeological science. *Journal of Archaeological Science* 56:9–20. doi: 10.1016/j.jas.2015.02.027

Codding, B. F. and D. W. Bird and T. L. Jones 2012 A land of work: Foraging behavior and ecology. In *Contemporary Issues in California Archaeology*, edited by T. L. Jones and J. E. Perry, pp. 115–132. Left Coast Press, Walnut Creek, CA.

Codding, B. F. and T. L. Jones 2007 History and behavioral ecology during the middle-late transition on the central California Coast: Findings from the Coon Creek site (CA-SLO-9), San Luis Obispo County. *Journal of California and Great Basin Anthropology* 27:23–49.

——— 2013 Environmental productivity predicts migration, demographic and linguistic patterns in prehistoric California. *Proceedings of the National Academy of Sciences* 110:14569–14573.

Codding, Brian F., Judith F. Porcasi and Terry L. Jones 2010 Explaining prehistoric variation in the abundance of large prey: A zooarchaeological analysis of deer and rabbit hunting along the Pecho Coast of Central California. *Journal of Anthropological Archaeology* 29:47–61.

Cook, E. R., D. M. Meko, D. W. Stahle and M. K. Cleaveland 1999 Drought reconstructions for the continental United States. *Journal of Climate* 12:1145–1162.

Cook, E. R., C. A. Woodhouse, C. M. Eakin, D. M. Meko and D. W. Stahle 2004 Long-term aridity changes in the Western United States. *Science* 306:1015–1018.

Contreras, D. A. and J. Meadows 2014 Summed radiocarbon calibrations as a population proxy: A critical evaluation using a realistic simulation approach. *Journal of Archaeological Science* 52:591–608.

ESRI 2012 Arcmap 10.0, ArcGIS Desktop 10. Environmental Systems Research Institute, Redlands, CA.

Fretwell, S. D. and H. L. Lucas 1969 On territorial behavior and other factors influencing habitat distribution in birds I. theoretical development. *Acta Biotheoretica* 19:16–36.

Gifford, E. W. 1936 Californian balanophagy. In *Essays Presented to A. L. Kroeber*, edited by R. Lowie, pp. 87–98. University of California Press, Berkeley.

Hastie, T. J. and R. J. Tibshirani 1990 *Generalized Additive Models*. Chapman and Hall Press, London.

Hildebrandt, W. R. and T. L. Jones 1992 The evolution of marine mammal hunting: A view from the California and Oregon coasts. *Journal of Anthropological Archaeology* 11:360–401.

Jazwa, Christopher S., Douglas J. Kennett and Bruce Winterhalder 2013 The Ideal Free Distribution and Settlement History at Old Ranch Canyon, Santa Rosa Island. In *California's Channel Islands: The Archaeology of Human-Environment Interactions*, edited by C. Jazwa and J. E. Perry, pp. 75–96. University of Utah Press, Salt Lake City.

Jones, Terry L. 1991 Marine-resource value and the priority of coastal settlement: A California perspective. *American Antiquity* 56:419–443.

Jones, Terry L., Gary M. Brown, L. Mark Raab, Janet L. McVicar, W. Geoffrey Spaulding, Douglas J. Kennett, Andrew York and Phil L. Walke 1999 Environmental imperatives reconsidered: Demographic Crises in western North America during the medieval climatic anomaly. *Current Anthropology* 40:137–170.

Jones, Terry L., Sebastian C. Garza, Judith F. Porcasi and Jereme W. Gaeta 2009 Another trans-Holocene sequence from Diablo Canyon: New faunal and radiocarbon findings from CA-SLO-585, San Luis Obispo County, California. *Journal of California and Great Basin Anthropology* 29:19–31.

Jones, Terry L. and Douglass J. Kennett 1999 Late Holocene sea temperatures along the Central California Coast. *Quaternary Research* 5174–82.

Jones, Terry L., Judith F. Porcasi, Jereme Gaeta and Brian F. Codding 2008a The Diablo Canyon Fauna: A coarse-grained record of trans-Holocene foraging from the Central California mainland coast. *American Antiquity* 73:289–316.

Jones, Terry L., Douglas J. Kennett, James A. Kennett and Brian F. Codding 2008b Seasonal stability in late Holocene shellfish harvesting on the Central California Coast. *Journal of Archaeological Science* 35:2286–2294.

Jones, Terry L., Nathan E. Stevens, Deborah A. Jones, Richard T. Fitzgerald and Mark G. Hylkema 2007 The central coast: A midlatitude milieu. In *California Prehistory: Colonization, Culture, and Complexity*, edited by Terry L. Jones and Kathryn Klar, pp 125–148. AltaMira Press, Walnut Creek, CA.

Kennett, D. J. 2005 *The Island Chumash: Behavioral Ecology of a Maritime Society.* University of California Press, Berkeley.

Kennett, D. J., A. J. Anderson and B. Winterhalder 2006 The ideal free distribution, food production, and the colonization of Oceania. In *Human Behavioral Ecology and the Origins of Agriculture*, edited by D. J. Kennett and B. Winterhalder, pp. 265–288. University of California Press, Berkeley.

Kennett, D. J. and J. P. Kennett 2000 Competitive and cooperative responses to climatic instability in southern California. *American Antiquity* 65:379–395.

Kennett, D. J. and B. Winterhalder 2008 Demographic expansion, despotism and the colonisation of east and south Polynesia. In *Islands of Inquiry: Colonization, Seafaring and the Archaeology of Maritime Landscapes*, p. 29 in Terra Australis. Australian National University.

Kennett, D. J., B. Winterhalder, J. Bartruff and J. M. Erlandson 2009 An ecological model for the emergence of institutionalized social hierarchies on California's northern Channel Islands. In *Pattern and Process in Cultural Evolution*, edited by S. Shennan, pp. 297–314. University of California Press, Berkeley.

Kennett, J. P., J. G. Baldauf and M. Lyle (editors) 1995. *Proceedings of the Ocean Drilling Program, Scientific Results, Volume 146 (Pt. 2). Ocean Drilling Program.* College Station, TX.

Koenig, W. D., Knops, J. M., Carmen, W. J., Stanback, M. T. and Mumme, R. L. 1996 Acorn production by oaks in central coastal California: Influence of weather at three levels. *Canadian Journal of Forest Research* 26:1677–1683.

Kuchler, A. W. 1977 *The Map of the Natural Vegetation of California. Appendix of Terrestrial Vegetation of California.* California Native Plant Society Special Publication Number 9. Sacramento, California.

Lightfoot, K. G., R. Q. Cuthrell, C. J. Striplen and M. G. Hylkema 2013 Rethinking the study of landscape management practices among hunter-gatherers in North America. *American Antiquity* 78:285–301.

MacArthur, R. and E. Pianka 1966 On optimal use of a patchy environment. *The American Naturalist* 100:603–609.

Morgan, C. 2008 Reconstructing prehistoric hunter-gatherer foraging radii: A case study from California's Southern Sierra Nevada. *Journal of Archaeological Science* 35:247–258.

——— 2009 Climate change, uncertainty and prehistoric hunter-gatherer mobility. *Journal of Anthropological Archaeology* 28:382–396.

——— 2012 Modeling modes of hunter-gatherer food storage. *American Antiquity* 77:714–736.

——— 2015 Is it intensification yet? Current archaeological perspectives on the evolution of hunter-gatherer economies. *Journal of Archaeological Research* doi: 1607 10.1007/s10814–014–9079–3, 1–51

O'Connell, J. F. 1995. Ethnoarchaeology needs a general theory of behavior. *Journal of Archaeological Research* 3:205–255.
O'Connell, J. F. and J. Allen 2012 The restaurant at the end of the universe: Modeling the colonisation of Sahul. *Australian Archaeology* 74:5–17.
Phillips, P. and G. R. Willey 1953 Method and theory in American archeology: An operational basis for culture-historical integration. *American Anthropologist* 55:615–633.
R Development Core Team 2014 *R: A Language and Environment for Statistical Computing*. R Foundation for Statistical Computing, Vienna, Austria.
Sánchez-Humanes, B. and J. M. Espelta 2011 Increased drought reduces acorn production in *Quercus ilex* coppices: Thinning mitigates this effect but only in the short term. *Forestry* 84:73–82.
Simms, S. 1985 Acquisition cost and nutritional data on Great Basin resources. *Journal of California and Great Basin Anthropology* 7:117–126.
Stevens, N. E. and R. McElreath 2015 When are two tools better than one? Mortars, millingslabs, and the California acorn economy. *Journal of Archaeological Science* 37:100–111.
Stine, S. 1994 Extreme and persistent drought in California and Patagonia during medieval time. *Nature* 369:546–549.
Surovell, T. A., J. B. Finley, G. M. Smith, P. J. Brantingham and R. Kelly 2009 Correcting temporal frequency distributions for taphonomic bias. *Journal of Archaeological Science* 36:1715–1724.
Wang, Xiao-Feng 2010 fANCOVA: Nonparametric Analysis of Covariance, v. 0.5–1. http://cran.r-project.org/web/packages/fANCOVA.
Warnes, G. R., B. Bolker, L. Bonebakker, R. Gentleman, W. Huber, A. Liaw, T. Lumley, M. Maechler, A. Magnusson, S. Moeller, M. Schwartz and B. Venables 2014 Gplots: Various R programming tools for plotting data. cran.r-project.org/web/packages/gplots/
Williams, A. N. 2012 The use of summed radiocarbon probability distributions in archaeology: A review of methods. *Journal of Archaeological Science* 39:578–589.
Winterhalder, B., D. J. Kennett, M. N. Grote and J. Bartruff. 2010 Ideal free settlement of California's Northern Channel Islands. *Journal of Anthropological Archaeology* 29:469–490.
Wohlgemuth, E. 2010 Plant resource structure and prehistory of plant use in central Alta California. *California Archaeology* 2:57–76.
Wood, S. N. 2006 *Generalized Additive Models: An Introduction with R*. Chapman and Hall Press, London.
Woodburn, J. 1982 Egalitarian societies. *Man* 17:431–451.
Zeanah, David W. 2004 Sexual division of labor and central place foraging: A model for the Carson Desert of western Nevada. *Journal of Anthropological Archaeology* 23:1–32.

9 Describing Microenvironments Used for Nomadic Pastoralist Habitation Sites

Explanatory Tools for Surfaces, Places, and Networks

Joshua Wright

Abstract

This chapter studies the detection, characterization, and modeling of microenvironments used in the present and the past. I suggest that this approach can be used to augment artifact-based descriptions and typologies of places and to approach artifact-less locales. The methods presented in this chapter are focused on strengthening inductive arguments about site functions and building tools for chronological interpretations from context. I also privilege a locale-focused, small-scale, descriptive approach to particular places. The region of Baga Gazaryn Chuluu, Dundgovi Aimag, Mongolia, provides the testing area and archaeological data for this chapter. A range of factors derived from ethnographic models, pastoral ecology, and historical archaeology are converted into simple suitability rasters that compare and highlight advantages of particular small areas and provide starting points for models and analysis of archaeological survey data.

Introduction

In his *Archaeology of Natural Places* (2000) Richard Bradley discussed the monumentalization of natural places in light of the ways in which experiences, movements, and access were directed and controlled as part of the development of culturally important locations: "Natural places have an archaeology because they acquired a significance in the minds of people in the past. That did not necessarily make any impact on their outward appearance, but one way of recognizing the importance of these locations is through the evidence of human activity that is discovered there" (Bradley 2000:34).

The examples laid out in this chapter are not focused on the numinous qualities of a place, as are Bradley's natural places, but on more quotidian values. However, these practical locations also allow people to manage experiences, movements, and access to place by people and animals.

The view in Figure 9.1 is a location in Northern Mongolia called Kholtost nuga, in the Valley of the Egiin Gol River. Elsewhere in the river valley there are other similar low-lying terraces near the river, and some of those are

Figure 9.1 The sheltered river terraces of Kholtost nuga along the Lower Egiin Gol Valley, Bulgan Aimag, Mongolia.

Source: Photo J. Wright

sheltered, but this one is special. In summer, when most areas are plagued by flies, this one has few. When it rains, the stabilized sandy soil of these terraces drains quickly. It is sheltered from storms and is close to accessible fast-moving water. This place is an excellent summer campsite and as a result has been occupied regularly for most of the Holocene and up to the present day. All the local herding families know the qualities of the place, but what is the framework of this knowledge, and how might we incorporate that into our arguments about the inhabited archaeological landscape? This chapter will use ethnographic observations to systematize and quantify choices about land use by recent pastoral nomads and use the resulting characterization of landscape as an exploratory tool and baseline for discussion of earlier pastoral nomadic land use. Specifically, I am looking at the contexts of nomadic pastoralist campsites between 1,000–3,000 years old in the desert-steppe of Mongolia. My goals here are to: (1) systematize ways of characterizing the qualities of the contexts of nomadic habitation sites, (2) build GIS models that can organize the interpretation of the elements and relationships of those qualities, (3) work towards a typological and classificatory approach to artifact-less locales and incorporate attributes of landscape into the analysis of conventionally defined sites. Ultimately such an approach can contribute to the development of tools for identifying traces of the material and landscape elements constituent of pastoralism, and more generally to building better arguments about human-environmental interaction by deconstructing the overly broad categories of "human" and "environment" and instead addressing particular human practices and particular niches/microenvironments.

Nomadic Pastoralism

The focus of this study is land use by nomadic pastoralists in the present and the past. Nomadic pastoralism is a form of subsistence in which peoples' domestic animals are systematically moved into the most advantageous locations for the animals. Human populations move along with them, picking their own favorable locations (Krader 1957; Dyson-Hudson and Dyson-Hudson 1980; Cribb 1984; Mearns 1993; Roe et al. 1998; Humphrey and Sneath 1999; Fernandez-Gimenez 2000; Gifford-Gonzalez 2005; Simukov 2007 [1934]; Fijn 2011). Over the long term a range of distinctive material culture, biotic relationships, and practices develop from these basic activities (Anthony and Brown 2007; McCorriston and Martin 2009; Shelach 2009; Outram et al. 2012; Hammer 2014; Honeychurch 2014; Makarewicz 2014). In the ethnographic present, Eurasian mobile pastoralism works well as an integrated system of herding practices, land use, storage, and crisis management that sustains populations of humans and animals. The management of risk is one of the most common ways that pastoralists' decisions are modeled. I have chosen advantage instead of risk as the central concept for the model used here. Risk management is focused on preparation for rare extreme events, the avoidance of clearly untenable situations, and choices that minimize the potential of repeated minimally dangerous chances. Examples include building a social network to recover from an out-of-season spring storm, avoiding waterless areas, or balancing animal herds to favor different species. The central aspect of modeling these behaviors is that their most effectively documented aspects are typically rare events. The pursuit of advantage includes maximizing regularly favorable choices, being positioned to benefit from rare windfalls or to mitigate extreme events. Examples include building a winter corral in a sheltered well-watered spot or camping near a raw material source. Decisions taken in pursuit of advantage are more frequent and variable than risk-management ones, and their wisdom is frequently tested. The resulting layered record of past actions provides a look at a range of variable successes, rather than focusing primarily on catastrophic failures.

Recently researchers have begun to disassemble pastoralist ethnographic analogy, questioning the assumption that the package of practices that is seen today among Eurasian nomads was completely the same in the past. Specific studies include attention to riding, carts (Jacobson-Tepfer et al. 2006; Gonchigsüren 2010) and horse domestication and use (Olsen 2006; Outram et al. 2012), dairying (Outram et al. 2009), mobile dwellings (Stronach 2004), herds, foddering and winter barns (Makarewicz 2014), and daily cooperative herding (Kazato 2005). This work has started to break apart the monolithic ethnographic whole of Eurasian nomadic pastoralism, or at least root many of its practices in the Iron Age (ca. 3000–2000 B.P.). This opening up of many variables of pastoralist practice is particularly valuable for the study of the earliest periods of nomadic pastoralism in eastern Eurasia. Archaeological traces of early herders are sparse, and using the full historic ethnographic model of pastoralism to describe these early herders is unwieldy. Addressing individual elements of pastoralist practice and adding new data developed from systematic observations of landscape will provide many more elements to use in models of early pastoralism. This study uses ethnography as a jumping-off point, and then moves on to look for key elements of relationships of pastoralists with landscape and variations from that pattern in the Bronze Age (ca. 4000–2800 B.P.).

This study emphasizes habitation sites. The ephemeral nature of most of these habitation sites has always been a challenging element of the archaeology of mobile populations. Many sites are palimpsests, some in occasional use through the entire history of pastoralism on the eastern steppe (a span of 3,500–4,000 years). As a result, most inferences are built from a few artifacts from mixed contexts. The multi-component nature of pastoralist habitation sites also includes areas used by animals. This overlap proves valuable because those animal areas can preserve traces of middens and alter soil make-up in ways that endure when other traces of habitation are gone (Wright 2016). There are occasional structural remains of dwelling footings or corral foundations of the kind that are so informative in West Asian contexts (Piggot 1944; Banning and Köhler-Rollefson 1986; Cribb 1991), but these are typically found at only a small subset of sites on Mongolian surveys.

All the examples used in this study come from the region of Baga Gazaryn Chuluu, in Dornogovi, Mongolia (see Figure 1.1 in Chapter 1). A 120-km^2 geologically distinct "island" in the desert-steppe of the North Gobi, Baga Gazaryn Chuluu has been completely surveyed (Wright et al. 2007, Amartuvshin and Honeychurch 2010) and offers an area in which it is possible to look in detail at the ways that people in different periods of the last 3,000+ years used the landscape. The locally exceptional hydrology and ecology of this area has long been attractive to humans, and there is a much denser archaeological and ethnographic record here than in the surrounding open steppe.

Pastoral Campsite Location Models from Ethnography

The starting point for this study is the ethnographic present. My colleagues and I have been able to observe and question nomadic pastoralists who live in our study areas. To complement these interviews and observations, an increasing number of publications are appearing that are specifically looking at issues of land-use decisions, campsite choices, and animal management in Mongolia (Minzhigdorj and Erdenebaatar 1993; Kazato 2005; Kakinuma et al. 2008; Upton 2010; Umekazu et al. 2010; Fijn 2011; Svoboda et al. 2011; Murphy 2012; Joly et al. 2013). From these data I have been able to build up definitions of the identifiable microenvironments used by people to sustain themselves and their herds. The selection of favorable microenvironments is the practical, small-scale reflection of the pastoralists' pursuit of advantageous locations for their herds. The definitions of advantageous places are not invariant, but are seasonal. Seasonality is one of the main factors that determine how mobile pastoralists place their habitation and manage their herds. Ethnographically summer campsites are usually the ones directly observed, but winter sites are the easiest to detect archaeologically. As a result, this study is focused on modeling the locations of winter sites.

There are many qualities in modern Mongolians' campsites that make them easy to systematically characterize. This provides a comparative baseline for discussing earlier sites. Modern campsites tend to be of roughly the same size, approximately 1 hectare or less (Umekazu et al. 2010; Svoboda et al. 2011) and systematically organized from site to site and family to family (Pedersen 2009; Wright 2016). Though there is variation, through ethnographic sketches, published images, and air-photo and satellite imagery of the study area, a range of possible configurations of campsites can be built up (Wright 2016) and, most important to this study, be examined in the context of other landscape features. Typically, a campsite includes several interior spaces (usually the white, domed, portable *gers* that are emblematic of Mongolian nomadism). Dwellings are cardinally oriented with activity areas, playgrounds, dog shades, etc., arrayed around them. Cardinal orientation and the desire not to block south-facing doorways cause dwellings to be arranged in E-W rows and arcs in most campsites.

Animal management dominates the ground around a campsite area. If possible, herders collect their animals for milking and night rest to the south of their *gers* (Umekazu et al. 2010). Those animals are quickly visible out their herders' front doors. This highlights the importance of intra-site visibility when animals are young. Surface slope is a factor in choice of areas for lambing, so flat areas are particularly valuable in spring camps (Joly et al. 2013). Animals and humans in breezy summer locations are protected from annoying flies. However, shelter from wind and weather is an important factor in most seasons and most of the time *gers* and corrals will cluster behind shelter from winter winds. Frequently this creates linear arrangements of activity areas and corals arrayed against rock faces or behind hills. Finally, studies of herder intervention with moving herds (Kazato 2005) see the most active adjustments occurring during the last approximately 1,000 meters or 30 minutes of sheep travel around the campsite. This intensive direction and sorting of animals occurs when herds are gathered close to campsites and bedded down or milked, depending on the different types of animals (female, young, etc.). All these factors combine to suggest combinations of local slope, visibility, and shelter that are important in site selection, and thus vital for describing site locations.

Terrain situations that favor efficient campsite function must be balanced against those that ameliorate risks to animals and humans. In this study I will consider four major risk factors: water, spring graze availability, windshelter, and flood protection. Many ethnographic and economic studies of nomadic pastoralists in Mongolia highlight the fact that steppe-desert grazing productivity and herd size is mainly dependent on water availability—particularly as manifested in wells, surface water, and soil moisture retention (Sternberg 2008; Zemmrich et al. 2010; Joly et al. 2013). Rain is rare in the steppe-desert where this study is based, but when it does fall, it can be very intense. This rain charges underground water reservoirs, often in valleys between bedrock outcroppings and fills pools on exposed rock surfaces. Heavy rains can also result in flash floods that flow through even short drainages and fill valley bottoms with flowing water. During flash floods in 2006 at Baga Gazaryn Chuluu the importance of subtle location choices was highlighted when all active nomads' campsites were in places unaffected by widespread flash flooding. In this region there are no flowing surface water courses, but occasional flows may fill channels or pools for a few days. Those same areas are good places to dig wells, and several modern shallow wells (ca. 3–5 meters to the water table) are found around Baga Gazaryn Chuluu today. These wells are used to water herds by filling adjacent troughs and to provide campsite water supplies by filling containers. When not requiring a motorized vehicle, this work is typically done by children or women who use hand carts or animals to move the water.

The complexity of graze resources used by Mongolian nomadic pastoralists is worthy of detailed studies itself (Simukov 2007 [1934]; Kakinuma et al. 2008; Makarewicz 2014). For this study, one particular aspect of graze usable during critical periods in the herding calendar was incorporated: access to early spring graze around winter campsites. Early growth typically occurs on sunlit slopes with water accumulated in the soil. There is also a possible feedback relationship between the slope wash of winter corral manure and the growth of early vegetation for spring graze (Makarewicz 2014) that would draw herders back to the same campsite areas.

Setting up *gers* and work areas on a flat piece of ground is typical. This is in tension with the winter and early spring camps' need for shelter from the prevailing north wind granted by topographic relief. Simply looking around the inhabited and archaeological landscape shows that wind shelter is a dominant factor in site location choices (Wright 2016). This is not simply a situation where maximum shelter is best. Wind shadow edges are also valuable spaces because wind will scour snow off the grass and provide fodder to animals. When this area is close to the shelter of their corral, the amount of exposure time for those animals that must go out to eat is reduced. In sum, the model that is used in this chapter attempts to incorporate a group of complex relationships to topography gathered around various degrees of slope, plant productivity, local topographic variation, and ground water flow.

GIS Models of Pastoral Campsite Locations

The basic structure of the model used here is a suitability raster. The approach uses simple GIS tools[1] to identify and categorize different individual characteristics of the landscape and then sum those individually classified layers to produce a combined suitability score for different pixels in the raster (Brandt et al. 1992; Dorshow 2012). In itself, this is a simple approach and does not quantify the nuances of different locales in particular detail. I use it here as a starting point to discuss model and variable choices and the relations between different elements in the landscape.

To build a suitability raster the environmental factors highlighted above were converted into GIS layers reflecting the ideal values and ranges of variation. The basic topography and pixel size used for this example study was from SRTM digital elevation data (Jarvis et al. 2008). Though not ideal, the approximately 80-m pixel size of this data is close in area to the size of many ethnographic campsites and artifact scatters. What is lost at this scale are useful breaks in slope, sharp rock edges, and narrow defiles that can be used as ad-hoc corrals and sheltered areas as well as the ability to look at variation within campsite areas. However, from this topographic data a range of characteristics can be developed, including slope, shelter, and surface water flow. Based on ethnographic models, topographic data was used to develop classified layers showing slightly sloped ground with less than 6° slope that are suitable for habitations and a layer of flatter ground with slopes between 0 and 1° that are best for herd bedding areas. These two values were linked using a sampling zone to find pixels that were very flat ground, but also have slightly sloped ground or flat ground north of them. This pair of slope characteristics makes an ethnographically ideal set of slope conditions for a herder's campsite. Topography is also an element of calculating wind shelter. Local prevailing wind data (Elliott et al. 2001:Appendix A) shows that in all but the mid-summer months, the wind in Baga Gazaryn Chuluu comes from the N–NNE. Using insolation tools adjusted to "shine" from very low on several points along the northern horizon creates a layer of shadowed areas behind hills and cliffs in which the intensity of the shadow can be used as a proxy for the quality of the wind shelter in that area (Finke et al. 2008:2789).

Surface water flow was modeled using basic GIS hydrology tools to calculate surface flow accumulation for single map cells. Mapping flood avoidance focused on areas with the lowest possible surface flow (1 pixel of area), effectively mapping local high points as well as those least likely to be inundated. Drainage flow was carried through to locate the areas of highest water accumulation and thus natural playas and potential locations for wells. Surface flow did not take into account surface absorbency, but current surface conditions map closely onto modeled flow accumulation. There are three basic types of ground at Baga Gazaryn Chuluu: granite surface with low flow accumulation, granite-derived sand with minimal soil formation, and fine-grained valley floor surfaces of clay or reworked loess. This last surface is found in areas of highest flow accumulation. GIS-derived areas of high flow accumulation were identified as potential sources of water, either as the location of possible surface water or the most effective places to dig wells. The ease of access to these potential water sources was determined using a slope-based cost distance raster around the areas divided into distance bands. This cost surface was built with a very high cost for slopes, to reflect the poor climbing ability of children pulling carts or camel-drawn wagons.

Finally, areas with locally high quantities of spring vegetation were located using the normalized difference vegetation index (NDVI), a measure of the ground surface derived from multispectral satellite imagery and reflective of types of surface vegetation cover. The layer used was developed by averaging values recorded every two weeks during April and May of 2011–2014. Pixels of this layer were reprojected and resampled to match the topographic layers and classified using their natural breaks and histograms into three different value ranges. In total 11 different raster layers were used in this model (Table 9.1).

These different raster layers were reclassified into simple binary layers (each pixel having a value of either 0 or 1) either within the basic parameters of slope or flow accumulation defined by the ethnographic model or using histograms of the values of continuous layers (distances from playa areas, for example) to break them into several binary layers that define different ranges of values. The simplest suitability raster is an equally weighted sum of all these layers (accomplished using a raster calculator tool or a weighted sum tool in a GIS), but the process of breaking continuous surfaces into separate layers for different ranges of values requires that a selection of a subset of all the layers must be made. In this case, layers were tested against the distribution of clear modern or archaeological campsites to gauge which layers best correlated with the distribution of those sites. A simple chi-square test of surface area proportions against proportions of locations was used to quickly assess whether a layer was significantly more or less favorable to

Table 9.1 Raster layers used for land-use model

Layer	*Parameters*
Highest surface flow areas	Flow accumulation higher than 24 pixels (best match for visible erosion channels and playa areas)
Low surface flow areas	Any pixel with a flow accumulation of 1 or less
Slope of 0–1°	
Slope of 1°–6°	
Maximum wind shelter	Values higher than 2 standard deviations on histogram of summed wind shelter
Moderate wind shelter	
Minimal wind shelter	Entire left side of the summer wind shelter histogram
High spring NDVI	Values higher than 2 standard deviations on histogram of NDVI surface histogram
Good spring NDVI	
Low spring NDVI	Values lower than mean of NDVI surface histogram
Well access areas	

habitation locations. A subset of the layers available was then selected based first on the most significant chi-square results and then on observations of the patterns of sites in different areas that created that significance. From this, four layers or their inverses were chosen for further use in the model.

Modern active campsite locations, drawn from a range of data sources—satellite imagery, aerial photographs, and observations by survey crews—were used to develop a layer showing current campsites in the survey area. The sample size ($n = 29$) is too small to divide them between winter and summer sites. But it is clear that the key layers for these sites were those defining areas of even moderate wind shelter ($\chi^2 = 4.19$, $p = 0.04$) along with 1°–6° slopes ($\chi^2 = 4.33$, $p = 0.037$) and staying away from areas of low spring NDVI ($\chi^2 = 4.18$, $p = 0.04$) and wind exposure ($\chi^2 = 15.2$, $p > 0.0001$). These are qualities that we might expect of winter campsite locations and not surprising considering the base data from which the model was built. The combined layer contained 15 of the known campsites with only 3 sites outside all of the defined areas ($\chi^2 = 9.1$, $p = 0.01$, df 2).

A larger sample size ($n = 154$) can be obtained from corral footings recorded by archaeological survey. The majority of these are dated by associated ceramics to the Later Medieval period (tenth–fourteenth century C.E.) and after (Wright 2016) and are most likely winter habitation sites. The same testing routine associated these much more strongly with several environmental layers. Location in areas with good springtime NDVI is the strongest factor ($\chi^2 = 77.64$, $p > 0.0001$) as well as good wind shelter ($\chi^2 = 54.33$, $p > 0.0001$) and again the middle range of slope ($\chi^2 = 7.99$, $p = 0.005$). As in the case of the active modern campsites, flat ground and unproductive spring graze were avoided. A combined model (Figure 9.2) finds 68 of 154 corral footing sites in ideal ground, where all the positive factors were present, and a strong tendency away from poor ground, where none of the positive factors were present ($\chi^2 = 73.63$, $p > 0.0001$, df 3).

It is important to specify that this is *not* an effective site location model: the areas of high suitability in both of these examples were quite large total areas (26 percent and 28 percent of the 120-km^2 survey area, respectively). What this simple test of easily identifiable sites helps to do is quantify environmental factors related to those sites and demonstrate the robustness of a systematic simplification of the landscape that represents qualities affecting site location over the last 1,000 years. One of the things that this simple systematic landscape can be used for is as a tool to study earlier arrays of sites.

One of the main foci of investigation at Baga Gazaryn Chuluu is the earliest pastoralists. The suitability model developed for historic and modern pastoralists is helpful in addressing the question of how similar or different from ethnographic models of pastoralism these first pastoralists might have been. The majority of sites associated with the pre-pastoralist Epipaleolithic do not follow the model well, being found in sheltered areas but also along the open water margins and unsheltered northern drainages. When sites that might be Early or Middle Bronze Age (4000–3300 B.P.) are isolated from this group, they conform more closely to the later model (Figure 9.3).

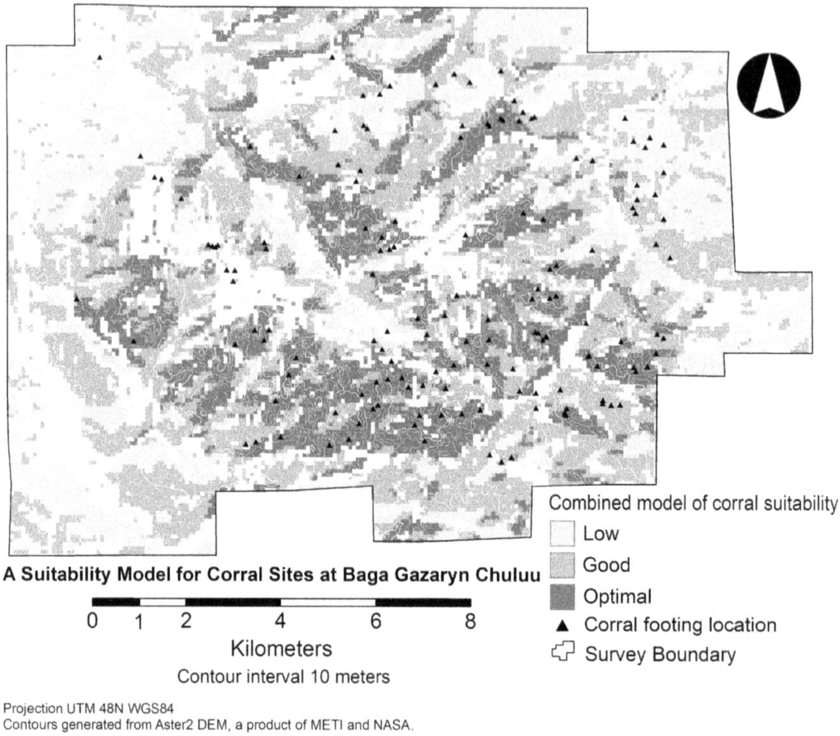

Figure 9.2 A combined suitability raster illustrating the ideal areas for Later Medieval corral sites at Baga Gazaryn Chuluu. Progressively darker grey tones show areas of progressively more favorable terrain, in this case areas sheltered from winds (i.e., wind shelter value beyond 1 standard deviation away from the mean) and close to good springtime vegetation (i.e., in the Middle Gobi, NDVI values 0.15–0.22). In this test, a significant proportion of archaeological corrals were found in optimal areas. Those outside are generally close to the edges of optimal areas, with the exception of clusters of ridge top corrals that are indicative of intensified herd management in this period.

Also in this period, small-scale metal working locales were found containing a range of traces of the smelting, casting, and recycling of bronze and copper (Park, Honeychurch, and Amartuvshin 2011). These sites lack pottery, for the most part, and the majority of other material is chipped stone. Wind is an important factor for the small furnaces that might have been used at such sites; however, the wind shelter map layers used in this study show clearly that almost every one of these sites (10 out of 12) is found in the areas sheltered from the wind ($\chi^2 = 15$, $p = 0.0006$). This suggests that control over air-flow that affected heating was important and the crafters of these small sites chose their locales with that in mind. Later periods—i.e.,

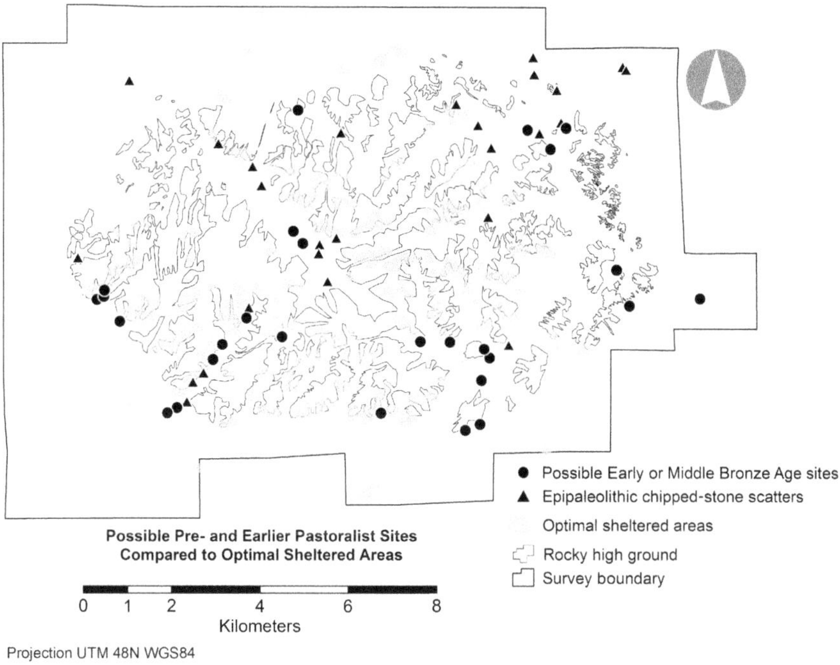

Figure 9.3 A comparison between larger Epipaleolithic chipped-stone scatters and a subset of those scatters that contain Early or Middle Bronze Age ceramics. The latter sites are found in more restricted, sheltered areas than the former. These areas are defined using the location models developed from historical pastoralist sites, suggesting that similar advantages were being sought. This phenomena is most visible in the southwest of the survey area. In contrast, large chipped-stone sites are found at the northeastern and northwestern edges of the rocky hills and in the central drainage. It is possible that seasonality is a key factor in the Bronze Age settlement pattern, but the available data does not offer any more information than site locations. It is also clear here that though most sites are in or proximate to a sheltered area, most sheltered areas do not have early sites near them. This shows a lack of precision in the model, but also raises the potentially more interesting question of why those areas do not contain archaeological material though they have the potential to shelter people and animals.

those in which pastoralism was established—show various sorts of compromises from the ideal location choices as well as illustrating only part of large pastoral movement patterns. The Iron Age Xiongnu (ca. 2300–1700 B.P.) present a pattern of mainly exposed sites along the northern edge of the rocks, suggesting distinctive seasonal use of the area, a pattern brought out further by the location of most of these sites near high-drainage flow areas (Wright 2016). During the late Medieval Khitan and Mongol Empire period (tenth–fourteenth centuries C.E.), the density of habitation traces is the highest of any time, and environmentally sub-optimal locales were being used (Figure 9.2).

This narrative is restrained by the quality of the base model. Looking at a map of ideal zones overlaid on flow accumulation areas and rock outcrops (Figure 9.3), themselves generalized from other data sources, it is clear that higher-resolution data (i.e., smaller pixels) would transform this guide to interesting areas into a map of ideal spots. Nevertheless, this simple model is still a guide to microenvironments, focusing attention onto particular areas and particular factors that seem to have been important for past land-use choices.

Discussion

A major goal of this investigation was the addition of new data to the discussion of nomadic pastoralist land use through the development of a typology of inhabited, and inhabitable, locales. This typology enables comparative measures for discussing the qualities of a place and is independent of the artifacts or architecture found in particular locales. The limited number of artifacts, the range of constructions, and the palimpsestic nature of many sites in the steppe (Houle and Broderick 2011; Wright 2012) magnify the importance of considering the places where they are found. Fortunately, places can be more than containers for cultural material: locales themselves can have attributes that can be studied and compared.

Systematically recording the elements of landscape that are part of a place's context allows those characteristics to be compared just as the artifacts recorded at that place might be. For example, a scatter of a few sherds and a scrap of slag is one thing, but two similar scatters, one found near the edge of wind shadow and one deep in a sheltered valley could have been very different types of sites in spite of their similar assemblages of material culture. Going further, a place that shares all the specific elevation, visibility, and geological characteristics of a monumental site, but does not contain any monuments, raises the question of why the place has no monuments? Systematic microenvironmental data places this question on a firm footing. Finally, a single slag fragment in a scatter whose location matches the slope and wind shelter characteristics of larger metallurgical scatters perhaps should not be discounted as intrusive, but seen as proof of the characteristic of the place predicted by its context.

Though this chapter was not directly a study of site location modeling, it would be possible to move from the description of types of locales to a search for specific places. Such models rise and fall on their effectiveness as predictive tools for finding things (Warren 1990; Finke et al. 2008; Carrer 2013; Hitchings et al. 2013, inter alia). What can we conclude when we run a model and find that the distributions of archaeological material and optimal microenvironments do not match (Figure 9.3)? On the one hand, this is a disaster for the model. On the other hand, these artifact-less places could have been inhabited and perhaps further investigation will be productive. In this case, this could be a call for a change in survey mechanics. Many of the sites I have presented here are very low impact and are defined by a handful of small artifacts or a single stone alignment. A locale that is visited less frequently may be below the threshold of detection but remain a potential site. Similarly, the model of past land use could be flawed. But, if for a moment, we set these worries aside and give primacy to the locations rather than the artifacts that they contain, these empty places remain interesting and raise the possibility of many otherwise similar locations having different habitation histories. A potentially inhabitable place without traces of habitation remains a useful data point, even if only by raising the question of why no traces of habitation were found. Formalizing this place-centric approach conceptualizes the presence of archaeological artifacts or landscape modification as a form of use-wear on place, indicative of past activities but not defining the place itself.

The model that I have just laid out in detail is one built on the characteristics of single map pixels or neighborhoods of pixels. These are relatively easy to calculate and display in a GIS raster format. However, many of the key characteristics of locations inhere not in the locations themselves, but in how those relate to other nearby locations. A flat area for a *ger* is only useful if there is a flat area for animals to its south, a winter campsite is good when there is spring graze near high ground on its windward side, and so on. In a sense, this simple GIS approach provides a starting point for reverse-engineering more complex systems of environmental categorization, niche construction, and ecological value. By using neighborhood statistics to generate raster surfaces, these proximate relationships can be attached to individual pixels by creating layers where each pixel value is assigned based on the best in its neighborhood, but the relationships themselves could be thought of as the key factors in determining the nature of a location. In this process, points (let's not call them pixels anymore) are linked to one another by their relationships. One point is sheltered by, or visible from, or above, or on the path to, another point. These relationships combined with the characteristics of the points themselves (flat, containing spring graze, etc.) can be used to create dense and complex networks of relationships (Newman 2003; Newman 2010; Robinson et al. 2013). The microenvironments that are the target of this investigation are places where particular groups of relationships are manifested. Once a node and connections structure is deployed, many actors can be incorporated into such a network (Latour 2005) along with social relationships with landscape (Basso 1996; Turnbull 2007).

Conclusion

This chapter began as a rather straightforward study using analogical reasoning (Wylie 1985) to characterize how different environmental attributes can contribute to describing land use and the past practice of pastoralism. I moved then through an interpretative or explanatory approach to micro-environment detection, what Frances Hayashida, in her closing remarks for the session, called "ethnographic microgeography." Finally, I ended with a more elaborate set of untried suggestions for other approaches to landscape modeling, making microenvironments contributing elements of typologies and networks. This is in keeping with the exploratory nature of this volume. Ultimately, I hope to provide a route towards bottom-up explanations of big cultural landscapes, an approach that systematically addresses habitation niches (Bleed and Matsui 2010; Spengler et al. 2013) in a landscape archaeological context, and describes particular experiences of macroenvironmental change in local contexts where humans' responses to larger environmental trends are played out in microenvironments (Guedes and Butler 2014). Such an approach operates in the service of strengthening inductive arguments about the chronology and function of sites and the cognitive landscapes that surround them by systematically incorporating elements of their context into their study, on an equal footing with the artifacts and architecture found there. In short, with the methods covered here, I hope to give us more data, systematically deployed for better arguments.

Methodological Vignette: Simple Suitability Rasters as Tools for Archaeological Discovery

The strengths of using simple suitability rasters as a tool for landscape classification are its exploratory nature and the way it can regularize a basic knowledge of landscape and judgments that we all use everyday in landscape archaeology. It also quantifies key factors in site location in the service of description and interpretation (Brandt et al. 1992; Dorshow 2012). This approach has two great advantages: it is simple, and it produces quickly interpretable outputs.

Its simplicity comes from the fact that it uses commonly available base data and is forgiving about the quality of that data (though more complex data could be used). These methods are also quick and not equipment-intensive; access to and familiarity with any GIS (including open-source freeware) is sufficient.

Outputs are quickly interpretable because of the thoughtful input required at most stages in the process. As one makes choices about what layers to use (or selects them quantitatively), one can watch the parameters interact, monitoring them as models that are ground-truthed.

This is a scalable tool; the same steps apply to all sizes of study area. This is also a tool that does not require spatial sampling; analyses are limited only by the extent and resolution of the base data. This is a method that works best in patchy and high-contrast landscapes. Commonly available base layers (e.g., DEMs and their derivatives, modern vegetation cover, etc.) are typical, but anything one can imagine as a raster layer can in principle be adapted.

Ethnographic data and known site locations can guide parameterization in order to avoid arbitrary or inappropriate definition and selection of layers. At the same time, users must be wary that variables can become too complex, base layers too numerous, and the interactions of factors too difficult to see for oneself without quantitative comparison.

The process of building a model of land characterization can be a powerful heuristic tool for questioning and building hypotheses in landscape archaeology. It brings into focus common interactions between humans and environment and allows us to use those in a way that moves beyond the characterization of particular environments and allows systematic investigation of the interaction of various factors at multiple scales.

Note

1 These analyses were carried out using ESRI's ArcGIS 10.2, but the tools chosen are basic and available in most other GIS packages.

References Cited

Amartuvshin, C. H. and W. Honeychurch (editors) 2010 *Dundgobi Aimagt Hiisen Arkheologiin Sudalgaa: Baga Gazaryn Chuluu [Archaeological Research in the Middle Gobi: Baga Gazaryn Chuluu]*. Mongolian Academy of Science, Institute of Archaeology., Ulaanbaatar.

Anthony, David W. and Dorcas Brown 2007 The herding-and-gathering economy at Krasnosamarskoe, Russia, and the end of the dependency model of steppe pastoralism. In *Social Orders and Social Landscapes*, edited by L. Popova, C. Hartley and A. Smith, pp. 393–415. Cambridge Scholars, Newcastle.

Banning, Edward B. and Ilse Köhler-Rollefson 1986 Ethnoarchaeological Survey in the Beda Area, Southern Jordan. *Zeitschrift des Deutschen Palästina-Vereins (1953–)* 102:152–170.

Basso, Keith H. 1996 Wisdom sits in places, notes on a western Apache landscape. In *Senses of Place*, edited by S. Feld and K. H. Basso, pp. 53–90. School of American Research, Santa Fe.

Bleed, Peter and Akira Matsui 2010 Why didn't agriculture develop in Japan? A consideration of Jomon ecological style, niche construction, and the origins of domestication. *Journal of Archaeological Method and Theory* 17:356–370.

Bradley, Richard 2000 *An Archaeology of Natural Places*. Routledge, New York.

Brandt, Roel, Bert J. Groenewoudt and Kenneth L. Kvamme 1992 An experiment in archaeological site location: Modeling in the Netherlands using GIS techniques. *World Archaeology* 24(2):268–282.

Carrer, Francesco 2013 An ethnoarchaeological inductive model for predicting archaeological site location: A case-study of pastoral settlement patterns in the Val di Fiemme and Val di Sole (Trentino, Italian Alps). *Journal of Anthropological Archaeology* 32:54–62.
Cribb, Roger 1984 Greener pastures: Mobility, migration, and the pastoral mode of subsistence. *Production Pastorale et Sociétè* (14):11–46.
——— 1991 *Nomads in Archaeology*. Cambridge University Press, Cambridge.
Dorshow, Wetherbee Bryan 2012 Modeling agricultural potential in Chaco Canyon during the Bonito phase: A predictive geospatial approach. *Journal of Archaeological Science* 39(7):2098–2115.
Dyson-Hudson, Rada and Neville Dyson-Hudson 1980 Nomadic pastoralism. *Annual Review of Anthropology* 9:15–61.
Elliott, D., M. Schwartz, G. Scott, S. Haymes, D. Heimiller and R. George 2001 *Wind Energy Atlas of Mongolia*. National Renewable Energy Laboratory, Golden, CO.
Fernandez-Gimenez, M. E. 2000 The role of Mongolian nomadic pastoralists' ecological knowledge in rangeland management. *Ecological Applications* 10:1318–1326.
Fijn, Natasha 2011 *Living with Herds, Human-Animal Coexistence in Mongolia*. Cambridge University Press, Cambridge.
Finke, P. A., E. Meylemans and J. Van de Wauw 2008 Mapping the possible occurrence of archaeological sites by Bayesian inference. *Journal of Archaeological Science* 35:2786–2796.
Gifford-Gonzalez, Diane 2005 Pastoralism and its consequences. In *African Archaeology: A Critical Introduction*, edited by A. B. Stahl, pp. 187–224. Blackwell, London.
Gonchigsüren, Nomguunsüren 2010 Preliminary study of cart wheel-bushings from Karakorum, Mongolia. In *Mongolia-German Karakorum Expedition, Excavations in the Craftsmans Quarter at the Main Road*, edited by J. Bemmann, U. Erdenebat and E. Pohl, pp. 173–211. Forschungen zur Archäologie Außereuropäischer Kulturen 8. Reichart Verlag, Weisbaden.
Guedes, Jade D'Alpoim and Ethan E. Butler 2014 Modeling constraints on the spread of agriculture to Southwest China with thermal niche models. *Quaternary International* 349:29–41.
Hammer, Emily 2014 Local landscape organization of mobile pastoralists in southeastern Turkey. *Journal of Anthropological Archaeology* 35:269–288.
Hitchings, P. M. N., K. Abu-Jayyab, P. Bikoulis and E. B. Banning 2013 A Bayesian approach to archaeological survey in north-west Jordan. *Antiquity* 87(336):project gallery.
Honeychurch, William 2014 Alternative complexities: The archeology of pastoral nomadic states. *Journal of Archaeological Research*:1–50.
Houle, Jean-Luc and Lee G. Broderick 2011 Settlement patterns and domestic economy of the Xiongnu in Khanui Valley, Mongolia. In *Xiongnu Archaeology, Multidisciplinary Persepectives of the First Steppe Empire in Inner Asia*, edited by U. Brosseder and B. K. Miller, pp. 137–152. Bonn Contributions to Asian Archaeology. vol. 5, J. Bemmann, general editor. Vor- und Frühgeschichtliche Archäologie Rheinische Friedrich-Wilhelms-Universität, Bonn.
Humphrey, Caroline and David Sneath 1999 *The End of Nomadism?: Society, State and the Environment in Inner Asia*. Duke University Press, Durham.
Jacobson-Tepfer, Esther, Vladimir Kubarev, Damdinsurenjin Tseveendorj, Gary Tepfer and James Meacham 2006 *Répertoire Des Pétroglyphs D'Asie Centrale, Mongolie Du Nord-Ouest Haut Tsagaan Gol*. Memoires de la Mission Archéologique Française en Asie Centrale. 2 vols. De Boccard, Paris.

Jarvis, A., H. I. Reuter, A. Nelson and E. Guevara 2008 *Hole-filled SRTM for the globe*. Version 4, available from the CGIAR-CSI SRTM 90m Database (http://srtm.csi.cgiar.org).

Joly, Frédéric J. C., Tulganyam Samdanjigmed, Valérie Cottereau and Claudia Feh 2013 Ecological constraints on and consequences of land use heterogeneity: A case study of the Mongolian Gobi. *Journal of Arid Environments* 95:84–91.

Kakinuma, Kaoru, Takahiro Ozaki, Seiki Takatsuki and Jonjin Chuluun 2008 How pastoralists in Mongolia perceive vegetation changes caused by grazing. *Nomadic Peoples* 12(2):67–73.

Kazato, Mari 2005 How Mongolian pastoralists manage their herds of sheep and goats: Herd formation and control in Arkhangai Province, Mongolia. *Mongolian Journal of Anthropology, Archaeology and Ethnology* 1(1):113–120.

Krader, Lawrence 1957 Culture and environment in Interior Asia. In *Studies in Human Ecology*, edited by L. Krader and A. Palerm, pp. 115–138. The Anthropological Society of Washington, Washington, DC.

Latour, Bruno 2005 *Reassembling the Social*. Oxford University Press, Oxford.

Makarewicz, Cheryl 2014 Winter pasturing practices and variable fodder provisioning detected in nitrogen (d15N) and carbon (d13C) isotopes in sheep dentinal collagen. *Journal of Archaeological Science* 41:502–510.

McCorriston, Joy and Louise Martin 2009 Southern Arabia's early pastoral population history: Some recent evidence. In *The Evolution of Human Populations in Arabia*, edited by M. D. Petraglia and J. I. Rose, pp. 237–250. Springer, Dordrecht.

Mearns, Robin 1993 Territoriality and land tenure among Mongolian Pastoralists: Variation, continuity and change. *Nomadic Peoples* (33):73–103.

Minzhigdorj, B. and B. Erdenebaatar 1993 Why Mongolians say sheep herders are lucky. *Nomadic Peoples* 33:47–49.

Murphy, Daniel J. 2012 Encountering the Franchise State: Dzud, Otor, and transformations in pastoral risk. In *Mongolia after Socialism*, edited by B. Knauft, pp. 68–80. Admon, Ulaanbaatar.

Newman, Mark 2010 *Networks: An Introduction*. Oxford University Press, Oxford.

Newman, M. E. J. 2003 The structure and function of complex networks. *SIAM Review* 45(2):167–256.

Olsen, Sandra L. 2006 Early horse domestication: Weighing the evidence. In *Horses and Humans: The Evolution of Human-Equine Relationships*, edited by S. L. Olsen, S. Grant, A. M. Choyke and L. Bartosiewicz, pp. 81–113. BAR International Series. vol. 1560, Oxford.

Outram, Alan K., Alexei Kasparov, Natalie A. Stear, Victor Varfolomeev, Emma Usmanova and Richard P. Evershed 2012 Patterns of pastoralism in later Bronze Age Kazakhstan: New evidence from faunal and lipid residue analyses. *Journal of Archaeological Science* 39:2424–2435.

Outram, Alan K., Natalie A. Stear, Robin Bendrey, Sandra Olsen, Alexei Kasparov, Victor Zaibert, Nick Thorpe and Richard P. Evershed 2009 The earliest horse harnessing and milking. *Science* 323(5919):1332–1335.

Park, Jang-Sik, William Honeychurch and Amartuvshin Chunag 2011 Ancient bronze technology and nomadic communities of the Middle Gobi Desert, Mongolia. *Journal of Archaeological Science* 38:805–817.

Pedersen, Morten Axel 2009 At home away from homes: Navigating the Taiga in Northern Mongolia. In *Boundless Worlds: An Anthropological Approach to Movement*, edited by P. W. Kirby, pp. 135–151. Berghahn Books, New York.

Piggott, Stuart 1944 Nomad house-sites in the Western Himalayas. *Man* 44:150–152.

Robinson, Ian, Jim Webber and Emil Eifrem 2013 *Graph Databases*. O'Reilly, Sebastopol, CA.

Roe, Emery, Lynn Huntsinger and Keith Labnow 1998 High reliability pastoralism. *Journal of Arid Environments* 39:39–55.

Shelach, Gideon 2009 *Prehistoric Societies on the Northern Frontiers of China*. Equinox Publishing, London.

Simukov, A. D. 2007 [1934] Труды о Монголии и для Монголии *[Works about Mongolia and for Mongolia]*. Senri Ethnological Reports. National Museum of Ethnology, Osaka.

Spengler, Robert N., Michael D. Frachetti and Gayle J. Fritz 2013 Ecotopes and herd foraging practices in the steppe/mountain ecotone of central Asia during the bronze and iron ages. *Journal of Ethnobiology* 33(1):125–147.

Sternberg, T. 2008 Environmental challenges in Mongolia's dryland pastoral landscape. *Journal of Arid Environments* 72(7):1294–1304.

Stronach, David 2004 On the antiquity of the yurt. *The Silkroad Foundation Newsletter* 2(1):9–18.

Svoboda, Jiři, Sandra Sázelová, Pavel A. Kosintsev, Vlasta Jankovská and Martin Holub 2011 Resources and spatial analysis at actual Nenets campsites: Ethnoarchaeological implications. *Journal of Anthropological Archaeology* 30:30–43.

Turnbull, David 2007 Maps narratives and trails: Performativity, hodology and distributed knowledges in complex adaptive systems—an approach to emergent mapping. *Geographical Research* 45(2):140–149.

Umekazu, Kawagishi, Sugimoto Hirofumi, Kitano Koki, Gonchigbat Ishjamts and Hirota Naoyuki 2010 Living environment of nomads residing on the outskirts of Ulaanbaatar, Mongolia Part 2: Lifestyle and living environment from the perspective of perceptions and activities. *Journal of Asian Architecture and Building Engineering* 9(1):139–146.

Upton, Caroline 2010 Wells, water rights and sustainability, New policies and practices on Mongolia's pastoral commons. In *Water, Environmental Security and Sustainable Rural Development*, edited by M. Arsel and M. Spoor, pp. 75–98. Routledge, New York.

Warren, Robert E. 1990 Predictive modelling in archaeology: A primer. In *Interpreting Space: GIS and Archaeology*, edited by K. Allen, S. Green and E. Zubrow, pp. 90–111. Taylor & Francis, London.

Wright, Joshua 2012 Temporal perspectives on the monumental constellations of inner Asia. In *"As Time goes by?" Monumentality, Landscapes and the Temporal Perspective*, edited by M. Furholt, M. Hinz and D. Mischka, pp. 73–83. Verlag Dr. Rudolf Habelt, Bonn.

——— 2016 Households without houses: Mobility and moorings on the Eurasian Steppe. *Journal of Anthropological Research* 72(2): 133–157.

Wright, Joshua, William Honeychurch and Chunag Amartuvshin 2007 Initial findings of the Baga Gazaryn Chuluu archaeological survey (2003–2006). *Antiquity* 81(313):Project Gallery.

Wylie, Alison 1985 The reaction against analogy. In *Advances in Archaeological Method and Theory*, edited by Schiffer, pp. 63–111. vol. 8. Academic Press, New York.

Zemmrich, A., M. Manthey, S. Zerbe and D. Oyunchimeg 2010 Driving environmental factors and the role of grazing in grassland communities: A comparative study along an altitudinal gradient in Western Mongolia. *Journal of Arid Environments* 74:1271–1280.

10 Soil Geochemistry and the Role of Nutrient Values in Understanding Archaic State Formation

A Case Study from Kaupō, Maui, Hawaiian Islands

Alexander Baer

Abstract

Anthropologists have long looked to the Pacific Islands as model systems for the interaction between people and their environments. With a settlement history less than a thousand years, the Hawaiian Islands in particular offer an unprecedented opportunity to examine how cultural elaboration and increasing social complexity are influenced by environmental conditions. Emphasizing soil geochemical analyses from transects across the district of Kaupō in arid, southeastern Maui, this chapter demonstrates how this region features a uniquely nutrient-rich mosaic of soils. Merging these data with GIS modeling, field survey, and excavation reveals how early Hawaiian farmers recognized that despite its limited rainfall, Kaupō was ideally suited to growing dryland crops such as sweet potato and taro. With an increasingly centralized political system, Maui elites moved to control the region, rapidly surrounding it within a network of major temples and intensifying the fertile interior with a broad zone of agricultural walls and embankments. By employing the soil geochemical analyses described here, we are afforded insights into why Kaupō, in a seemingly undesirable location, became home to the only dryland field system on the island, and why the district was ultimately made the royal seat of the entire Maui polity.

Introduction

As archaeological tools for understanding past environments grow more robust, we are afforded new ways to explore the link between ecology and dynamic cultural practices. Soil geochemical analyses are one method through which quantifiable environmental metrics can inform our interpretations of past activities. Using the district of Kaupō, Maui, I demonstrate that by examining soil nutrient values we can understand why certain regions throughout the Hawaiian archipelago became the most densely populated and agriculturally intensified. In the case of Kaupō, exceptionally high nutrient values can be linked to observed settlement patterns, largely defined by an intensified agricultural field system and core of commoner households set within a network of massive temples bounding the edges of the productive center. Correlating these nutrient values with the construction of ritual, residential, and agricultural features demonstrates that as intensification of the field system began, local elites rapidly constructed this new series of monumental structures, effectively proscribing and controlling the highly valuable region. Here, soil geochemistry affords a direct, quantifiable link between favorable environmental conditions and the increasingly complex sociopolitical practices of an emerging archaic state.

Kaupō, a semi-arid, marginal hinterland turned agricultural powerhouse and sociopolitical nucleus, serves as an ideal region in which to examine the role of ecology in a nascent archaic state. With relatively late settlement (fourteenth–fifteenth century) and the subsequent adoption of the district as royal seat in the 1700s, we are afforded a small chronological window through which to assess changes in agricultural practices and settlement patterns. Previous archeological studies have demonstrated the presence of an intensified dryland field system by the 1600s (Kirch et al. 2009) and the construction of numerous ritual and residential sites from the same time period (Kolb 1994; Kirch and Sharp 2005; Baer 2015; Kirch et al. 2015). These coincide with a larger shift throughout the archipelago towards centralized political control and the transition from complex chiefdom to a number of competing archaic states (Hommon 1976, 2013; Kirch 2010, 2012). By the arrival of Europeans in 1778, Kaupō had become the center of the Maui polity, moving from political hinterland to a powerful core in a very short space of time. Through the use of geochemical analyses we can see direct evidence for the region's productive potential, effectively linking Kaupō's rapid ascent and infrastructure investments to the recognition by Hawaiian elites of the district's value.

For Kaupō, soil geochemical analyses provide direct evidence for the district's high nutrient values and productive capacity. Beyond broad models of substrate age, rainfall, and other factors, sampling throughout the district demonstrates that the intensive dryland field system was built on a particularly rich portion of Maui. This offers a connection between environmental factors and cultural change, as the early settlement of the region was rapidly followed by intensive, monumental infrastructure development in a way unlike anywhere else on the island. As with soil geochemical studies elsewhere, Kaupō benefits from a context that has remained largely undisturbed since its abandonment in the mid-1800s. Even elsewhere on Maui, more modern practices such as sugarcane, banana, and pineapple cropping have not only tilled the soil to an untestable level, but also added and removed suites of nutrients to create a unclear amalgamation of values. Fortunately, as the district has remained grazing land for cattle, little has disturbed the subsurface nutrient pool (cow waste, which does not get deep into the soil, is virtually all a recycled version of the plants upon which the cows feed).

This chapter uses the formerly agrarian district of Kaupō, Maui, to show how geochemistry has shaped our interpretations of a uniquely settled and intensified section of the Hawaiian Islands. By integrating soil geochemical data with oral traditions, GIS modeling, and excavated materials, I argue that despite its relatively late settlement, Kaupō's productive potential was quickly recognized. Hawaiian elites rapidly constructed a network of monumental temples enclosing a core of commoner residences and intensive agricultural features, effectively creating a burgeoning sociopolitical center tied directly to a discrete area of ecologically prime soils.

Political Ecology of Agriculture in the Hawaiian Islands

Like most of the other leeward, arid parts of the islands, Kaupō features a slightly shorter settlement history than the windward valleys, creating a dichotomy between older, wetter areas and the younger, drier parts. The first settlers, arriving around A.D. 1000 (Athens et al. 2014) scattered throughout the archipelago's wet, windward valleys. They transformed landscapes and biota, recreating familiar conditions suitable for traditional Polynesian lifeways (Kirch 1985; 2000; 2012; Cordy 2000). Small, diffuse communities practiced subsistence agriculture, tending irrigated terraces of taro and yam. Over time, however, the population swelled into the tens of thousands, forcing people beyond the valleys and into the dry stretches of the islands (Kirch 2007). By the fifteenth to sixteenth centuries, people were living throughout the islands, and formerly disparate groups began organizing in greater numbers. Ultimately, a small number of chiefs managed to consolidate entire islands, assuming divine kingship and shifting political organization towards archaic statehood (Kirch 2010).

By the time Captain James Cook made contact in 1778, two main polities remained—one based on Hawai'i's Big Island and another on the island of Maui. Because of their geological youth, these islands lacked extensive valleys or permanent water flows suited to irrigated agriculture, features commonly found on the older islands of Kaua'i and O'ahu. To feed the demands of a burgeoning non-productive class (including chiefs, priests, soldiers, and artisans), the people of Hawai'i and Maui had intensified dryland field systems, emphasizing broad fields of sweet potato over irrigated taro and yam (Kirch 1994; 2007; Vitousek et al. 2004). Despite requiring more labor than irrigated systems, the sheer volume of dryland sweet potato production was critical in supporting the rise of Hawaiian archaic states (Kirch 2012; 2015; Hommon 2013).

While the cultivation of sweet potatoes allowed for production outside of well-watered valleys, modeling by Ladefoged et al. (2009) has demonstrated that only select zones across the islands would have been suitable for dryland cropping. Using data on rainfall, elevation, and geological substrate age, they identified broad areas on Hawai'i and Maui, as well as smaller areas on Moloka'i, Lana'i, and O'ahu, in which dryland agriculture was environmentally feasible. Fieldwork by numerous teams has revealed that many of these locations did feature intensified production systems, with dryland fields present in Kohala, Kona, and Ka'ū districts, and at Waimea, on Hawai'i

(Clark and Kirch 1983; Schilt 1984; Burtchard and Tomonari-Tuggle 2004; Lincoln and Ladefoged 2014), in Kalaupapa, Moloka'i (McCoy 2005), and Kaupō, Maui (Kirch et al. 2009; Baer 2015). Of these, the most extensively studied is the Leeward Kohala Field System (LKFS) on Hawai'i's Kohala Peninsula (Ladefoged et al. 1996; 2003; Ladefoged and Graves 2008; Kirch 2010). Researchers participating in the Hawai'i Biocomplexity Project have demonstrated that productivity within this 60-km^2 (Ladefoged and Graves 2010) system of reticulate walls is highly constrained by a combination of substrate age and rainfall.

Through in situ weathering of basalt parent materials and the aeolian deposition of fine-grained sediments, areas such as the LKFS and Kaupō feature nutrient-rich soils ideal for sweet potato, dryland taro, and other dryland crops. Within these zones, however, variations in precipitation and substrate age create a mosaic of microenvironments, creating small regions of differential productive capacity. While younger geologic substrates tend to have higher nutrient concentrations, very young substrates (including the jagged 'a'a lava flows under 5 kyr) have not had enough time to either decompose sufficiently or accumulate airborne sediments. Conversely, very old substrates (usually greater than 120 kyr) tend towards lower nutrient values due to leaching and weathering (Jenny 1980; Vitousek 2004). As rainfall is a key factor in nutrient loss, precipitation must be correlated with substrate age to identify zones featuring conditions appropriate for the accumulation of rich soils (Chadwick and Chorover 2001; Chadwick et al. 2003).

The Natural and Cultural History of Kaupō, Maui

In the leeward southeast of Maui, Kaupō was one of 12 semi-autonomous political districts. While each district (or *moku*) featured its own internal sociopolitical organization, by the time Captain James Cook arrived in 1778, paramount rulers had come to control entire islands, installing their own supporters as the heads of various districts. Occupying a unique environmental niche, Kaupō was a highly productive agricultural region, making the *moku* politically and economically valuable to the competing polities of Maui and Hawai'i. By the mid-eighteenth century, the region had become the political and productive center of late pre-contact Maui—a uniquely important locale from which King Kekaulike and other Maui rulers orchestrated their wars against Hawai'i. The rise of this district reflects an early recognition by Hawaiian elites of the enormous capacity for localized production, tying the development of the district to environmental factors. Merging modern geology and hydrology with the soil geochemistry described here provides a causal link between local ecology and the investment in monumental structures and landscape modifications designed to bound and control the productive core.

Geography and Environment

On the southeastern coast of Maui, the *moku* of Kaupō straddles the boundary between the lush, wet districts of Kīpahulu and Hāna to the east, and the extremely dry region of Kahikinui to the west. Bounded geographically by the gulches Kālepa and Wai'ōpai, Kaupō stretches approximately 13 km at its widest extent, and climbs 5 km inland up the slope of the volcano Haleakalā. The district's highest peak, at Pōhaku Pālaha, reaches 2,470 m. While these high points tower above the region, Kaupō's upper climes are notable less for their peaks than for a broadly incised, erosional valley breaching the southern face of Haleakalā Crater (Figure 10.1). Known as the "Kaupō Gap," this rift in the crater wall is the result of erosion during the Pleistocene. During a rejuvenation phase of volcanism ~120 kya (Stearns and MacDonald 1942), the gap allowed both lava and mud to flow out of the volcano down to the sea, creating a vast accretionary fan of nutrient-rich lavas and sediments. Unlike the mosaic of predominantly leached sediments in Kahikinui on Kaupō's western border (Dixon et al. 1999; Coil and Kirch 2005) or the overly wet, incised valleys of Kīpahulu to the east, Kaupō's situation on the Hāna Volcanic Series (Stearns and MacDonald 1942; Sherrod et al. 2007) placed it within a set of conditions predicted by Ladefoged et al. (2009) as prime for dryland agriculture.

While the predictive model created by Ladefoged et al. (2009, Figure 1) identified a broad swath of land extending from Kaupō through the districts

Figure 10.1 Oblique view of southeastern Maui with prominent locations. White lines across the landscape represent remains of a formerly intensified dryland agricultural field system.

of Kahikinui, Honua'ula, and Kula, the variable nature and ages of discrete mud and lava flows actually make this zone far less continuous than the Ladefoged model alone would suggest. Remote sensing and survey reveal that, of the four districts identified by the Ladefoged model, only Kaupō contained a field system in the vein of Kohala, Kona, and Kalaupapa's networks of broad, reticulate walls (Holm 2006; Kirch et al. 2009). Though agriculture was practiced throughout the dry length of Kahikinui and further west, the mosaic of sediments and low rainfall resulted in discontinuous production emphasizing intensive cultivation in discrete swales (Kirch 1997; 2010; Coil 2004; Kirch et al. 2004; Coil and Kirch 2005; Hartshorn et al. 2006; Holm 2006).

Ecologically, Kaupō and its accretionary fan of lavas and sediments offered the single largest planting area in the region, but even this was subject to internal substrate variability (see Figure 10.2). The fan dates to the post-shield building phase of Haleakalā Volcano (Stearns and MacDonald 1942), with the deposition of Hana Volcanic Series flows overlying older Kula Series lava (Sherrod et al. 2007). The older Kula surfaces (generally > 140 kyr) can be seen on both the eastern and western margins of Kaupō, with the interior covered in flows from the past 140 kyr. Within this fan, Sherrod et al. (2007) have identified discrete flows, mapping their boundaries and individual ages. The western portion features younger 'a'a flows, including the Pu'u Maile basanite (3–5 kyr) and the Pu'u Nole basanite (0.75–1 kyr). In the east, older flows of Loaloa (5–15 kyr) and Kaupō basanites (13–30 kyr) have had more time to weather, both releasing nutrients into the developing soils as well as accumulating aeolian sediments. The center of the fan contains two substrates, Mamalu Bay basanite and Kamanawa Bay basanite, with ages between 50–140 kyr.

Along with substrate age, rainfall is critical in determining productive potential, both in watering crops and weathering surfaces to allow for nutrient release. Kaupō sits at a transition point for southern Maui, as the rain shadow created by Haleakalā Volcano pushes extensive precipitation east into Kīpahulu, while greatly limiting rainfall to the western districts of Kahikinui and Honua'ula (Giambelluca and Schroeder 1998:56; Giambelluca et al. 1986, Figure A.79). Between them, rainfall decreases significantly, with Kaupō's eastern uplands receiving > 2,000 mm/year compared to only 700 mm at Nu'u Bay on the western coast (Giambelluca et al. 2013). This precipitation gradient again affects the region during both annual planting cycles as well as long-term weathering and leaching patterns. Virtually the entire fan falls within the range of annual rainfall required for sweet potato cultivation (determined to be 760–1,270 mm/year by Purseglove et al. [1968:82]), indicating that outside of droughts affecting the western portion of Kaupō, annual precipitation was likely not a limit factor.

Though critical for individual cropping cycles, annual precipitation alone does not account for Kaupō's level of agricultural intensification. Equally important is the combination of rainfall and substrate age, allowing for

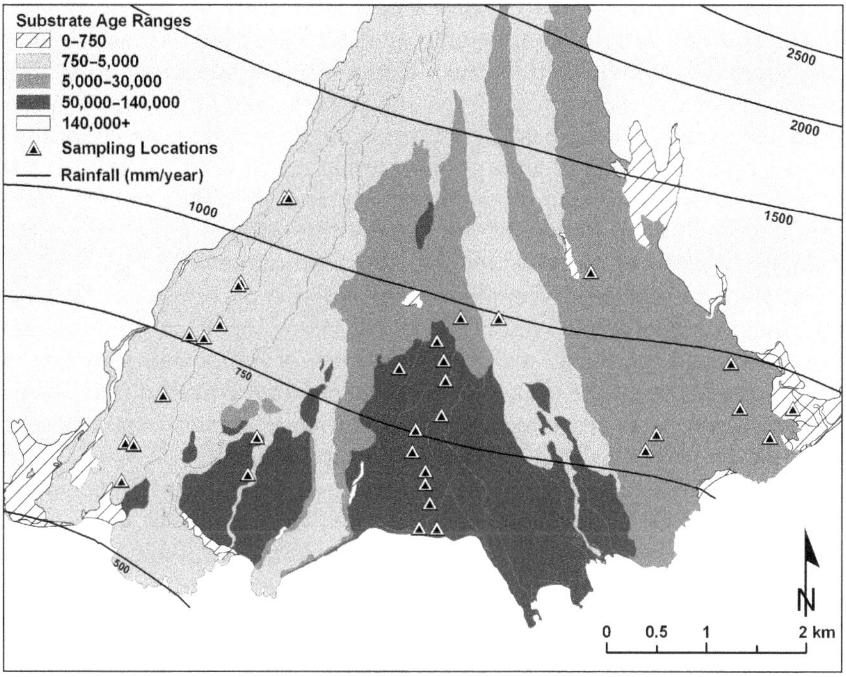

Figure 10.2 The locations of 30 soil samples and their distribution across Kaupō's geological substrates indexed by age.

Source: Substrate information from Sherrod et al. (2007) and rainfall from Giambelluca et al. (2013)

enough weathering of the young Hana Volcanics (releasing nutrients into the developing topsoil), but not enough time to begin leaching. Described further below, the range of substrates found throughout the accretionary fan do create some level of nutrient variability, but in comparison with similar age-rainfall locations throughout the archipelago, such as the Lower Kohala Field System, Kaupō provides an ideal setting for sweet potato and other dryland crops.

Regional Ethnohistory

Historical research in Hawai'i benefits from both a relatively short settlement history, as well as a strong, early commitment by Hawaiians and Europeans to record oral traditions and practices. The district of Kaupō holds an interesting place in these historical traditions, as it goes virtually unmentioned through most of the early Hawaiian histories, genealogies, and stories, before suddenly becoming central to the struggle for control of the entire archipelago around the turn of the seventeenth to eighteenth centuries. It became home to Maui's kings, and the site of numerous intra- and inter-island battles, set among broad fields famed for their sweet potatoes. Even into the twentieth century, despite massive population loss, both locally and archipelago-wide, Kaupō was remembered as a *moku* of great productivity; ethnologist E.S.C. Handy noted, "Kaupō has been famous for its sweet potatoes, both in ancient times and in recent years" (1940:161). It is this capacity for dryland production that presumably attracted the attention of Maui's King Kekaulike, ultimately transforming the district from unknown hinterland to the center of Maui's political power.

Through the nineteenth-century records of Native Hawaiian and Western scholars, including Samuel Kamakau, David Malo, John Papa ʻĪʻī, Abraham Fornander, and others, we are afforded a view of Kaupō as it emerged in a context of war, production, and chiefly expansion. Powerful island-wide polities, ruled by divine kings and classes of chiefs, priests, and artisans, came to dominate in the late sixteenth century (Kirch 2010). Kaupō itself was largely invisible during this primary phase of increasing social complexity, but within two generations of Maui's first unification, the district was adopted as the royal seat of the Maui paramount, King Kekaulike (Kamakau 1992; Fornander 1996; Cachola-Abad 2000).

The grandson of Kiha-a-Piʻilani, Maui's unifying *aliʻi nui* (king), Kekaulike rose to kingship following two battles (Ka-eulu and Ka-hale-mamalakoa) at Mokulau in Kaupō. In his ambition to expand beyond Maui, Kekaulike then shifted the center of the island's political power from Wailuku, in the West, to Kaupō, where he began to plot an invasion of Hawaiʻi Island. From Mokulau he launched his attack, raiding from Kona all the way north to Kohala before returning to Kaupō, where his army waited for a counter-attack by Hawaiʻi's King Alapaʻi-nui. Before the fighting began on Maui, however, Kekaulike fled Kaupō for Wailuku where he "was seized with a violent illness" (Kamakau 1992:69) and died. While this eased tensions with Alapaʻi-nui, it created a power struggle within Maui itself. Once again, Mokulau served as the battlefield on which Kamehameha-nui, one of Kekaulike's junior sons, emerged victorious to claim the Maui kingship. Following his ascension in ~1730 (Cachola-Abad 2000; Kirch 2010), the islands of Maui and Hawaiʻi enjoyed a brief peace, though the subsequent generation of kings would again make Kaupō the center of a new war.

In the 1770s, Hawaiʻi's ruler Kalaniʻōpuʻu brought his armies against Maui and its new king, Kahekili. Kalaniʻōpuʻu had previously fought with Kamehameha-nui over the westernmost district of Hāna, but this time, his armies raided Kaupō. Moving just inland from Mokulau, the Hawaiian forces were met by those of Maui at Puʻumaneoneo and Ka-puka-ʻauhuhu in the battle Ka-lae-hohoa ("forehead beaten with clubs"). The Hawaiian army was pushed back to Ka-lae-o-ka-ʻilio in western Kaupō before fleeing, wholly defeated, to their home island. Accounts of this battle, notable partly for the first appearance of the young Kamehameha (the first Hawaiian king to unify the islands), offer an early description of the district's extensive "potato hills" and the furrows between them, noting the density of crops which caused warriors to become "entangled in the vines" (Kamakau 1992:84).

While Kaupō's centrality to the ongoing wars between Maui and Hawai'i is demonstrated through numerous battles described in oral traditions, increasing references to the district's developing agronomic and sociopolitical control systems are also evident (Maunupau 1998). Kekaulike is credited in the oral tradition with the construction of a number of major structures, including the war temples (*luakini*) of Pu'u-maka'a and Lo'alo'a, and a royal center at Pōpōiwi (Kamakau 1992)—the latter two measuring among the largest structures in the archipelago. He is similarly associated with other temples, indicating that much of Kaupō's ritual network (discussed further below) was completed during Kekaulike's reign. What remains less clear, however, is the extent to which the field system had been intensified prior to his arrival. The sheer presence of Kekaulike implies that the district was already productive enough to support the king and his royal court by the early 1700s, while the traditions recorded by Kamakau (1992) and others note numerous hills, furrows, and vines associated with sweet potato production in the period surrounding his rule.

Early ethnographic work from Kaupō similarly speaks to the region's high productive capacity. In addition to heavy sweet potato cultivation up to an elevation of approximately 2,000 feet (Yen 1974), ethnographer E.S.C. Handy noted "formerly great quantities of dry taro were planted in the lower forest belt from one end of the district to the other" (1940:113; Handy and Handy 1972:507–508). As a secondary crop in the uplands, dryland taro would have produced lower yields, but could grow in conditions too wet and too nutrient poor for sweet potato. Combined, these crops spread from Kaupō through Kahikinui, Honua'ula, and Kula, creating what Handy called "the greatest continuous dry planting area in the Hawaiian Islands" (Handy 1940:161). While research in Southern Maui (described above) has demonstrated that the agricultural practices to the west of Kaupō were more diffuse and swale-oriented than the formalized field system found within the accretionary fan, this broad area from Kaupō to Kula was no doubt rich with dryland output.

Archaeological Research

In 2003, archaeologists Patrick Kirch and John Holson began conducting research in a western section of Kaupō called Nu'u, undertaking intensive survey to chronicle settlement practices throughout a single *ahupua'a* (a narrow land division traditionally running from the coast inland and upland). I joined Kirch and Holson in 2007, conducting remote sensing analyses and further field surveys to identify dryland field walls and remaining archaeological sites across the accretionary fan (Kirch et al. 2009). Using high-resolution aerial photographs, we remotely identified linear features representing the remains of an intensified dryland field system similar to that of Kohala. Kaupō's field system features a series of long, dark lines running up- and down-slope (roughly north-south), cross cut by generally shorter, more narrowly spaced linear alignments (east-west), creating numerous small square and rectangular plots. Ground survey revealed the longer north-south lines to be stone walls, and/or vegetation growing on linear collections of mounded stone and dirt. The more closely spaced lines, perpendicular to the longer north-south walls, were less commonly formal walls, but largely subtle elevation changes of aggregated stone and earth. These embankments, or terraces, were visible remotely as dense vegetation grows selectively along the field-plot boundaries.

Our analyses described Maui's first field system, covering between 12.5–15 km^2 across the lower stretches of the Kaupō fan (Kirch et al. 2009). Examining the areas most highly intensified, however, revealed discrepancies largely defined by differential substrate ages. As described above, the fan itself comprises a number of different flows, dating between .75–140 kyr. On the very youngest substrate (Pu'u Nole basanite, .75–1 kyr) virtually no agricultural features are discernable remotely, with ground survey revealing the presence of some residential sites but no field walls or other evidence of production. In contrast, the highest density of features falls in the central portion of the fan where the Mamalu Bay and Kamanawa Bay basanites date between 50–140 kyr. The other substrates, of ages from 3–50 kyr, contain walls associated with intensified agriculture, but not to the same extent as the central, older section.

Having identified the extent of the field system, along with some of its environmentally defining characteristics, we then embarked on a comprehensive study combining extensive field survey and excavations, GIS analyses, and soil nutrient testing to better understand the settlement patterns and history of the district (Baer 2015). We began by conducting an intensive survey across 5 km^2 of Kaupō, identifying over 1,000 discrete sites, including ritual and residential structures, along with agricultural features and the broad network of walls and embankments making up the field system. Residential sites demonstrate a clear association with agricultural features, both of which are most dense in the central, coastal area known as Paukū. This locus of production and dense habitation is also where we find the majority of small agricultural temples, which present a stark contrast to the monumental ritual structures built along the outer borders of the Kaupō fan. As I discuss in the following section, the pattern of major *heiau* ("temples") bounding Kaupō's nutrient-rich, productive core presents a compelling link between agricultural production and sociopolitical efforts to control valuable territory.

To add a diachronic perspective on local development (particularly how settlement patterns and consumptive practices may have changed during the time of increasing sociopolitical complexity), we then conducted a series of targeted excavations at a variety of ritual and residential sites throughout the district. Twenty-seven different sites were sampled, including 7 of the large border temples, 3 of the interior temples, and 17 residential structures, ranging from large, multi-room complexes to smaller shelters and cookhouses. In conjunction with Michael Kolb's excavations at two additional major *heiau* on Kaupō's eastern boundary (Kolb 1991; 1994), we now have data from 29 sites, including nearly 70 radiocarbon samples (Baer 2015). These ^{14}C dates bracket the temporal span over which the local settlement pattern developed, along with, presumably, the dryland field system.

Our third phase of research involved the systematic analyses of soil nutrients from throughout the district, which in conjunction with the radiocarbon dating of ritual structures, enabled us to directly connect geochemistry and potential productivity to the expansion of elite power across the southeastern Maui landscape. The results of this soil geochemical testing are the primary focus of this chapter.

Soil Geochemical Analyses

In 2013, we undertook soil sampling across Kaupō to explore how the district's nutrient profile compared with other Hawaiian regions demonstrating similar pre-contact agricultural intensification. Employing a shovel-testing collection methodology developed by Vitousek et al. (2004), we collected 36 samples representing a range of the region's environmental zones. All samples incorporated materials from both A and B horizons, integrating soils and sediments from 0–30 cm below the surface. Rock percent by volume was estimated independently by two team members and cross-checked for comparability, while moist soil color was assessed with a Munsell soil chart.

To understand nutrient variation across the accretionary fan, we conducted three transects, two along the eastern and western edges, and one more through the center, with a small number of other individual locations of interest sampled separately. All sampling sites were recorded using a Trimble Juno 3B GPS unit using the WGS-84 datum as a base. While the resulting transects were not strictly linear or perfectly spaced (the result of terrain, vegetation, and property access), they represent a strong coverage of the fan, emphasizing differently aged substrates and rainfall regimes from east to west.

Figure 10.2 shows all the sampling locations mapped onto substrate age classes based on Sherrod et al. (2007), with rainfall derived from Giambelluca et al. (2013). On the eastern edge of the fan, Transect 1 includes five samples from approximately sea level up to an elevation of 300 m. All of these samples fall within the 5–13 kyr Loaloa basanite, representing relatively young soils subject to annual rainfall significantly higher than in the west (particularly in the upper elevations). Transect 2 runs through the center of Kaupō, directly through the area known traditionally as Paukū. Aerial analyses demonstrate this to have been the core of the intensified field system. Through this zone of dense walls and embankments we took

14 samples, again from the coast up to 300 m in elevation. Nearly all of these samples are from the older Mamalu Bay and Kamanawa Bay basanites, dating between 50–140 kyr. Transect 3 follows the western edge of the Kaupō fan in an area known as Nuʻu. From the coast, at Nuʻu Bay, up to an elevation of 400 m, we took 13 samples. Of these, 11 are from the young substrate of Puʻu Maile basanite (3–5 kyr), and 2 are just off the fan, located within the older Kula Volcanics dating to more than 140 kyr. Transects 1, 2, and 3 represent both east-west and north-south coverage, while 4 additional samples create something of a fourth transect running east-west across an elevation of approximately 175 m.

Laboratory Analyses

To generate a set of data comparable with other soil geochemical analyses from agricultural contexts, we focused on a suite of nutrient values tied closely to substrate ages and rainfall (Vitousek 2004; Vitousek et al. 2004; 2014). Prior to testing, soils were air dried and sieved through 2-mm mesh to separate out the fine-earth fraction (< 2 mm). This fine portion was then subjected to standard treatments described in the Department of Agriculture's Kellogg Soil Survey Laboratory Methods Manual (2014).

We focused our chemical analyses on measures that had been previously shown to be sensitive to differences in lava flow age and rainfall and indicative of agricultural potential (Vitousek and Chadwick 2013; Vitousek et al. 2014). The fine-earth fractions (< 2 mm) were analyzed using procedures described in Soil Survey Staff (2014): (1) exchangeable bases (calcium (Ca), magnesium (Mg), potassium (K), and sodium (Na)) were extracted using 1 molar NH_4OAC buffered at pH 7, measured on an optical inductively coupled plasmaspectrometer, and presented as meq•100 g^{-1} soil; (2) cation exchange capacity (CEC) was measured by displacing NH_4^+sorbed onto soil exchange with 1 molar KCl, measured by ion chromatography, and presented as meq•100 g^{-1} soil; (3) pH was measured in 1:1 soil:water ratio; and (4) the total amount of Ca in the fine-earth fraction was measured on a borate fusion using X-ray fluorescence and presented in its oxide form (CaO). Percent base saturation was calculated as the ratio of the sum of exchangeable bases to CEC. The ratio of exchangeable Ca to total Ca was calculated after converting CaO to meq•100 g^{-1}.

While analyses of soil nutrient values have proven critical to archaeological understandings of Polynesian agriculture (not only in Hawaiʻi, but similarly on Easter Island [Vitousek et al. 2014]), the application of the methodology described above is by no means universal, and must be tailored to specific research questions. Projects designed to understand historical or modern patterns of agriculture, or even basic plant dispersal (in the case of hunting and gathering ranges), may benefit greatly from geochemical nutrient analyses but should explore other tests not employed here. Particulate size and values describing the percentage of organic inclusions are two measures of further potential interest, but with all of these analyses it is critical to design any testing with a deep knowledge of local conditions, as interpretation is largely dependent on understanding the release of specific elements and molecules, and their mobility or stability in relation to local environmental factors (Ågren and Bosatta 1998; Chadwick et al. 1999; Vitousek 2004).

Within the context of Kaupō, we were afforded a broad spectrum of over 30 discrete variables (see Baer et al. 2015 online Supplementary Information 1, for full list). To understand the impact of soil nutrients on the region's agricultural and sociopolitical development, I focus here on the variables known to most directly affect cropping practices, namely, calcium (Ca), exchangeable bases (calcium, magnesium, potassium, and sodium), base saturation (a combined measure of exchangeable bases and their cation exchange capacity), soil pH, and total rock percentage (Vitousek and Chadwick 2013; Vitousek et al. 2014). While phosphorus (particularly as resin-p) often serves as a good metric for assessing agricultural potential (Vitousek et al. 2010), the presence of extensive cow manure across this heavily grazed ranchland resulted in unreliable P measurements. Outside of this specific location, however, the relative immobility of phosphorus (as compared to calcium and others) may be of interest to researchers in both Hawai'i and elsewhere (Harrington et al. 2001).

Calcium

Work elsewhere in Hawai'i has demonstrated that calcium serves as the single best nutrient in predicting the presence or absence of dryland agricultural practices (Vitousek et al. 2014). Exchangeable calcium (as $Ca{++}$) is released through the weathering of parent materials, with concentrations determined largely by rainfall and substrate age (Vitousek 2004). Initially a part of CaO molecules, trapped calcium is unusable by plants until weathering releases the exchangeable calcium into developing soils. Overall values are tied closely to precipitation, with areas receiving > 2,000 mm of annual rainfall drastically losing nutrients to leaching, and areas of extreme aridity maintaining high levels of mineral-bound calcium in proportion to the exchangeable variety. Within both of these rainfall regimes, however, substrate age will affect the amount of available calcium, as young surfaces (even in areas of heavy rain) will not have had enough time to release nutrients, and older surfaces reflect enough accumulated weathering to have few nutrients of any variety remaining.

In their studies of the Leeward Kohala Field System, Vitousek et al. (2014) determined that intensified dryland production was only conducted in areas with exchangeable calcium values above 10.2 meq•100 g^{-1}. Table 10.1 displays values of $Ca{++}$ from each sampling location throughout Kaupō. Across the fan, exchangeable calcium concentrations are uniformly high, averaging 23.4 ± 10.4 meq•100g^{-1}. All values are above the threshold established by

Table 10.1 Soil sampling nutrient values and environmental data

FIELD ID	*Geologic substrate*	*Rainfall (mm/ year)*	*Ca++ (meq•100g⁻¹)*	*CaO (%)*	*% Base saturation*	*pH*	*Rock (%)*
KSS-2	Pu'u Maile basanite	1092	13.98	1.06	30.4	5.8	65
KSS-3	Pu'u Maile basanite	1115	22.86	2.78	41.2	5.9	
KSS-4	Pu'u Maile basanite	872	23.42	2.19	42.1	5.6	55
KSS-5	Pu'u Maile basanite	872	19.27	1.72	41.4	6.1	
KSS-7	Older alluvium	945	25.19	2.58	47.1	5.5	
KSS-8	Pu'u Maile basanite	809	49.68	4.52	82.8	7.2	75
KSS-9	Pu'u Maile basanite	748	31.90	2.80	57.4	6.4	35
KSS-10	Pu'u Maile basanite	792	30.02	2.42	54.1	6.4	50
KSS-11	Pu'u Maile basanite	644	21.11	2.12	47.4	5.5	60
KSS-12	Pu'u Maile basanite	602	30.51	2.61	59.3	6.1	40
KSS-13	Pu'u Maile basanite	602	24.78	2.69	58.4	5.8	65
KSS-14	Pu'u Maile basanite	548	13.37	2.13	49.4	6.3	75
KSS-15	Kaupo Mud flow	666	15.03	0.77	42.0	6.6	40
KSS-16	Kaupo Mud flow	606	12.71	0.60	39.1	6.3	20
KSS-17	Mamalu Bay basanite	719	26.17	1.46	43.7	6.0	35
KSS-18	Mamalu Bay basanite	683	19.93	1.08	47.7	6.5	30
KSS-19	Mamalu Bay basanite	653	20.38	1.20	81.9	6.4	60
KSS-20	Mamalu Bay basanite	626	14.50	2.73	80.8	6.4	50
KSS-21	Kaupo Mud flow	626	14.98	2.11	73.1	6.6	20
KSS-22	Oili Pu'u basanite	989	67.92	4.36	55.8		75
KSS-23	Kaupo Fan basanite	960	28.86	4.61	45.2	7.1	65
KSS-24	Kamanawa Bay basanite	894	20.52	1.02	39.0		20
KSS-25	Kaupo Mud flow	833	17.35	0.55	57.1	6.6	15
KSS-26	Kamanawa Bay basanite	908	20.35	1.10	44.8	6.4	10

FIELD ID	*Geologic substrate*	*Rainfall (mm/ year)*	*Ca++ (meq•100g^{-1})*	*CaO (%)*	*% Base saturation*	*pH*	*Rock (%)*
KSS-27	Kamanawa Bay basanite	859	15.43	0.98	36.2	6.6	30
KSS-28	Mamalu Bay basanite	812	27.00	1.40	55.6	6.5	20
KSS-29	Mamalu Bay basanite	758	19.31	0.97	57.7	6.4	20
KSS-30	Mamalu Bay basanite	709	17.05	0.80	44.2	6.4	30
KSS-31	Loaloa basanite	877	15.95	2.64	45.7	5.9	90
KSS-32	Loaloa basanite	905	23.05	5.59	46.5	6.1	70
KSS-33	Loaloa basanite	947	19.82	4.04	37.9	5.9	70
KSS-34	Loaloa basanite	1183	25.72	4.21	43.0	5.9	80
KSS-35	Kaupo Fan basanite	818	31.85	4.37	54.3	6.3	85
KSS-36	Kaupo Fan basanite	775	18.73	3.24	58.8	6.6	40

Substrate ages from Sherrod et al. (2007), and rainfall from Giambelluca et al. (2013)

Vitousek et al. (2014), with many of these locations featuring Ca++ at levels that could have supported intensified agriculture well into the future.

While all the values are relatively high, relative abundances can be related to both substrate age and rainfall. The center of the fan, featuring the greatest concentration of intensified field system, is dominated by older Mamalu Bay and Kamanawa Bay basanites, dating from 50–140 kyr. It also receives approximately 900–1,100 mm of annual rain, directly between the highly watered eastern edge (~1,500 mm) and the dryer west (~700 mm). On these older, medium-rainfall surfaces, exchangeable calcium averages 19.3 ± 4.8 meq•100g^{-1}. This figure is close to the 21.9 ± 4.1 meq•100g^{-1} average of the samples taken from the more eastern Loaloa and Kaupō basanites (5–30 kyr), but as these younger soils are subjected to greater rainfall and weathering, the similar amount of Ca++ is unsurprising. On the younger substrates of Pu‘u Nole and Pu‘u Maile basanites, however, we see an increase in overall released calcium, with an average of 29.1 ± 15.6 meq•100g^{-1}. This figure is somewhat inflated by the high concentration in sample KSS-22, but this in itself is interesting as KSS-22 was taken from a relatively young flow (compared to most of the other samples from Transect 2 on older materials) that branched east into an area of higher rainfall. The rest of the Transect 3 samples from young parent materials still demonstrate a relatively high average of exchangeable calcium (25.5 ± 10.1 meq•100g^{-1}).

The availability of exchangeable calcium serves as a central factor for examining the distribution of intensified dryland cropping, but values for the total amount of calcium (as the oxide CaO) and the ratio of the two also provide information surrounding nutrient exchange and availability. In addition to exchangeable calcium, Table 10.1 shows total soil calcium (CaO percent of all nutrients) from which the ratio of available to trapped calcium can be calculated. Regarding percentage of CaO, we again see values highly dependent on substrate age. Samples taken from substrates > 50 kyr average 1.12 percent, compared to 3.06 percent from any younger parent materials. Finally, in converting CaO into meq•100 g^{-1} we are afforded a ratio of how much calcium has been released compared to how much remains trapped. In this instance, we find that while the older substrates feature lower volumes of CaO, they are proportionately more dense with available calcium than younger regions. In practice, these combined calcium results would have meant that the center of the accretionary fan contained the most accessible calcium for crops, but the rest of the region would have been suitably rich to support an intensified field system.

Base Saturation

Outside of calcium, one of the most important metrics for understanding agricultural potential is the combined measurement of base saturation. Base saturation looks at surface sorbed and plant available calcium (Ca), magnesium (Mg), potassium (K), and sodium (Na) against cation exchange capacity (CEC), producing percentages comparable across cropping locations throughout the world (Palmer et al. 2009). In Hawai'i Island's highly productive LKFS, Vitousek et al. (2014) found that while calcium was one of the field system's bounding factors, base saturation also limited the extent of cropping capabilities. As demonstrated elsewhere in Hawai'i, there is a strongly non-linear relationship between base saturation and rainfall (Chadwick and Chorover 2001; Chadwick et al. 2003; Vitousek and Chadwick 2013). Regardless of substrate age, when rainfall reaches a certain threshold, base saturation drops precipitously. In the LKFS, this occurs around 1,500 mm of annual rain (Vitousek et al. 2010, Figure 2.3). As levels of rainfall across Kohala are largely correspondent with elevation (as opposed to Kaupō, with a gradient from northeast to southwest), the LKFS features an upper cropping boundary where rainfall exceeds 1,500 mm, resulting in a base saturation below 29.5 percent (Vitousek et al. 2014). This 29.5 percent represents the minimum value with which sweet potato and dryland taro could be successfully grown, providing a figure (similar to the minimum Ca++ threshold) against which Kaupō's soils can be tested. While all of our sampling locations fall below the rainfall threshold, we do see a slight increase in base saturation as rainfall decreases, with all four of the highest values located in zones receiving less than 800 mm per year.

Table 10.1 provides base saturation values for each of the sampling locations. Most striking about these data are that every one of these numbers is above the 29.5-percent threshold defined for the LKFS, meaning everywhere throughout the Kaupō fan would have been more than suitable for production. Most values are over 40 percent, with a mean value of 51.2 ± 12.9 percent, while some reach as high as 81 percent. Relating these numbers to substrate ages, we find the highest values concentrated on old flows in the center of the fan. This area, Paukū, not only features an extremely high level of base saturation, but the most dense agricultural walls and residential sites in the entire district. To the west, the figures from Transect 3 represent samples from younger flows, yet show base saturation levels comparable to the rest of the fan. This may be a function of terrain and sampling, as this younger landscape features more arable swales and small valleys than the broad slopes to the east, but it nonetheless demonstrates that the entirety of the Kaupō fan features a high degree of nutrient availability. Similarly, samples from Transect 1 show more than adequate values, though, perhaps due to increased rainfall, they average slightly lower base saturation than the samples from the two westerly transects.

Soil pH

Behind calcium and base saturation, pH is the third most reliable correlate of the presence or absence of intensified agriculture in Kohala, with values below 5.7 too acidic to support much cropping (Vitousek et al. 2014, Table 1). In Kaupō, we find a range of pH values largely commensurate with those of the LKFS, averaging 6.2 ± 0.4 across the fan. A small number of samples, however, do fall just below the 5.7 figure, though it seems that high levels of other nutrients mitigated this acidity, allowing for cropping with a slightly lower pH floor. Overall, pH levels do not fluctuate widely, maintaining a relative consistency across substrates and rainfall levels.

Rock Percent by Volume

In collecting each of our samples, units were dug to a depth of 30–40 cm. During these excavations, the percentage of rock (pebbles, cobbles, and boulders) within these removed materials was visually estimated. While specific sampling locations or local variables may skew individual samples, rock percent by volume serves as a proxy for the degree of lava flow decomposition. With very low levels of colluvial transportation and deposition, lower rock percentage indicates more time for both parent materials to break down in situ and for the accumulation of fine aeolian particles. Higher percentages of rock therefore evince younger source material. These age-rock percent correlations are borne out across Kaupō. With the 14 samples taken from older substrates, the average rock percentage is 29 percent, compared to 71 percent for medium-aged substrates (7 samples), and 60 percent for samples less than 5 kyr (from 10 samples, and all excluding 3 locations in which rock percent was not estimated).

Controlling Kaupō

In the early twentieth century, the archaeologist and ethnographer Winslow Walker traveled throughout Maui recording the names, histories, and architectural features of prominent sites (Walker 1930; Sterling 1998). In conjunction with the early ethnohistories of Kamakau (1992) and Fornander (1996), these studies revealed that by the time of European contact in 1778, Kaupō had become central to the political rule of Maui: home to the island's king, a large population, numerous ritual sites, and a constant locus of the war between polities on Maui and Hawai'i. Archaeologically, we see the evidence for massive sociopolitical infrastructure in the form of an intensified field system, monumental architecture, and extensive residential sites, but by combining these data with ethnohistoric knowledge and regional soil geochemical analyses we are afforded a view of Kaupō's history that directly ties ecology to the development of Hawaiian social complexity.

Throughout his exploration of Maui, Walker chronicled a variety of structures, including some of Kaupō's ancient residential features and nearly all of the district's large *heiau* (Sterling 1998). Relying primarily on local informants rather than any systematic survey, Walker's work provides an interesting, if incomplete, view of the sites with enough importance to be remembered well into the twentieth century. Of particular interest are the *heiau* whose specific names and uses were still built into social memory more than a century after their supposed desanctification by royal decree in 1821 (Kamakau 1992). In addition to survey throughout Kaupō's various environmental zones, I relocated 21 of the 24 temples recorded by Walker, adding the ethnohistoric information he gathered on these sites to my own spatial studies of Kaupō.

Integrating these sites with survey data from throughout the district reveals a striking pattern of dense agricultural production and habitation within the core of Kaupō. The intensive field system becomes more diffuse towards the edges of the fan (see Figure 10.3), while residential sites are similarly most dense in the central Paukū region (Baer 2015). More compelling for understanding sociopolitical control, however, are the presence of two lines of monumental *heiau* running along the geologic edges of the fan, effectively bounding the productive core (Figure 10.3). Of the 13 temples clearly defining the nucleus of Kaupō, 11 retained their traditional names into the twentieth century (Walker 1930). Conversely, only two other *heiau* throughout the entirety of Kaupō demonstrated enough cultural significance to maintain their names as long. In > 5 km^2 of intensive survey through the district's interior, other ritual structures have been identified, but almost none are on the scale of the perimeter *heiau*, appearing instead to be smaller, likely family-scale agricultural temples (Baer 2015).

While prior studies have demonstrated the role of *heiau* as markers delineating land ownership (Mulrooney and Ladefoged 2005; McCoy et al.

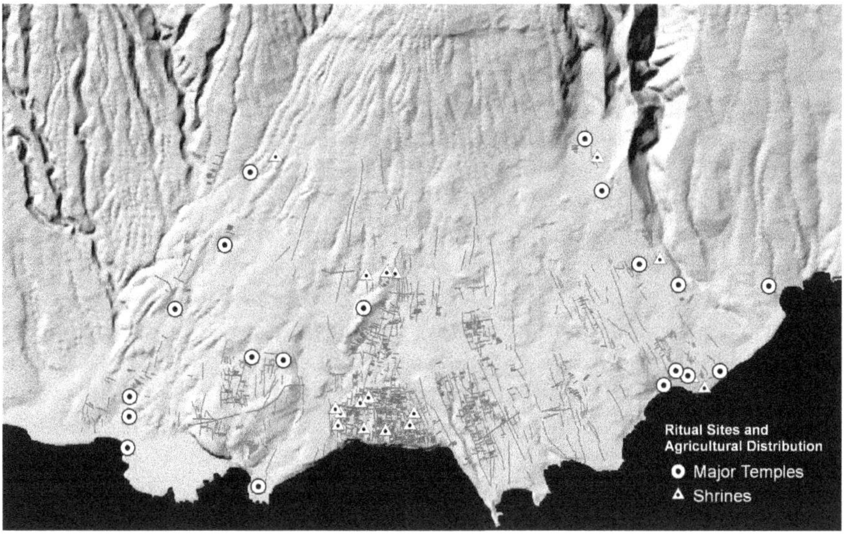

Figure 10.3 Digital elevation model of the district with agricultural walls and ritual structures. Major temples run along both the eastern and western edges of the accretionary fan, while numerous small shrines dominate the heavily intensified interior (see Baer [2015] for discussion of ritual site classification).

2011), the lines of structures along Kaupō's geologic edges represent the cultural construction of space on a scale previously unseen in Hawai'i. These temples define the core of the district, bounding the entirety of the formal field system and effectively isolating the interior. This not only supports Hommon's (2008; 2009) ideas of the "salubrious core" (broad, inter-district zones of low population density punctuated by centralized population aggregations), but speaks to the recognition by early Hawaiians of Kaupō's productive value. Through the early and rapid construction of these monumental *heiau*, we have evidence that environmentally rich locations were being consciously proscribed, with massive structures defining not only agricultural lands, but the community involved in their management.

Additionally, while oral traditions link some of the large temples with Kekaulike in the early 1700s, radiocarbon and uranium-thorium dating of these ritual structures (along with residential sites) indicate that the construction of the temple network began well prior to his arrival, in the mid- to late 1500s, and continued in an intensive flurry of building for the next 80–100 years (Kolb 1991; 1994; Baer 2015; Kirch et al. 2015). This coincides with the sixteenth- to seventeenth-century phase of temple construction proposed by Kirch and Sharp (2005), which they argue was a reflection of emergent divine kingship and the push to establish formalized sociopolitical controls.

Even those temples associated with the eighteenth-century king appear to have existed well before his arrival, albeit in smaller forms (Kolb 1994). Through the excavation, soil geochemical testing, and dating of these numerous sites, we now know that while Kaupō was not settled as early as many of the other locations throughout the archipelago (predominately the wetter, windward valleys capable of supporting irrigated agriculture), the development of its temple system was both early and rapid, mirroring the rise of monumental structures elsewhere in the Hawaiian Islands (Kirch and Sharp 2005; Kirch 2010), but at a uniquely rapid rate. Furthermore, evidence for the construction of large temples early within the settlement sequence of Kaupō differs from patterns seen in Kohala, where McCoy et al. (2011) have demonstrated that small shrines dominated the early phases of construction before the subsequent development of larger *heiau*.

With a chronological understanding of Kaupō's temple construction beginning some 100–150 years prior to the arrival of King Kekaulike in the early 1700s, we are presented with an interesting and potentially unique district-wide development. Prior to its adoption as the royal seat, Kaupō is virtually unmentioned in the ethnohistoric record, with almost all references to paramount chiefs and the consolidation of power taking place in the wetter, windward portions of the archipelago (Kamakau 1992; Fornander 1996). At the same time, in arid southeast Maui, contemporaneous with increasing social stratification and the rise of divine kings elsewhere, we find a formalized set of monumental structures exactly bounding the only dryland field system on the island. The development of Kaupō is then a clear demonstration that this region's agricultural value was recognized, and that despite being a relative hinterland, large-scale labor was being organized to maximize production. Whether this organization was run by some independent local authority or under the aegis of a leader elsewhere on the island remains unclear, but in either case, massive amounts of labor were being mobilized in the development of the region.

In Kaupō, the rapid rise of the temple network and intensified field system can be directly tied to increasing sociopolitical centralization. To understand why Kaupō alone was subject to such a formal proscription and labor investment, however, we must examine the local factors that would have allowed for such a development. Soil geochemical analyses conducted elsewhere in Hawai'i reveal that a combination of environmental conditions may dictate the presence, absence, and fecundity of cropping practices (Chadwick et al. 2003; Vitousek 2004; Vitousek et al. 2004; 2010; 2014; Palmer et al. 2009; Vitousek and Chadwick 2013). To test Kaupō's agricultural potential, we undertook extensive soil analyses from throughout the district to better understand if local environmental conditions influenced the unique development of Kaupō's field system. We also hoped to more thoroughly examine the microenvironments associated with differential substrate ages and rainfall patterns.

The results of these nutrient analyses demonstrate that Kaupō's soils are of exceptional quality (see Vitousek et al. [2004; 2014], Palmer et al. [2009] for comparative values elsewhere in Hawai'i). Across the district, nutrient values, particularly for calcium and base saturation, well exceed the minimum thresholds identified from Kohala (Baer et al. 2015; Vitousek et al. 2014). In some cases, the nutrient availability matched that described in areas with irrigated wetland agriculture (Palmer et al. 2009), long believed to be the more rich and efficient method of cropping. The universally high nutrient levels from throughout Kaupō speak to the district's agricultural potential and explain the presence of the intensified field system. This system of walls and embankments not only structured the region by creating defined social and political boundaries, but also served to maximize yields by creating increased water capture, wind breaks, and embankments upon which crops such as sugarcane could be planted (Handy 1940; Kirch et al. 2009; McCoy and Graves 2010). This agricultural intensification thus shows an early recognition of Kaupō's unique potential, and the landesque capital investments made to ensure its sociopolitical control and productivity (Brookfield 1972; Baer 2015).

While overall nutrient levels show how valuable the district would have been as a production center, the range of samples taken throughout the interior may indicate why Kaupō itself was the only portion of southern Maui to be agriculturally intensified and tightly bounded by ritual structures. Of the three age groups dominating the accretionary fan, the central features the oldest soils (50–140 kyr) along with the most available calcium and highest average base saturation. Where it is low, however, is in categories such as trapped calcium, meaning that many of its nutrients have already been released through various weathering and leaching processes. This is significant, as the land surfaces to both the east and west of Kaupō are of older Kula substrate that experienced longer periods of leaching and have nutrient levels insufficient to support extensive agriculture. While the model of Ladefoged et al. (2009) predicts all of southern Maui to have been cropped, we know agricultural production from Kahikinui and Honouli to have been patchy and discontinuous (Holm 2006; Kirch 2010).

By conducting these soil geochemical analyses we can see how important this particular section of Maui truly was. Much like the importance of Kohala in supporting Hawai'i's kings (most notably Kamehameha, who was born there), Kaupō was a critical asset to the Maui polity. When the region's productive potential was realized sometime in the sixteenth century (perhaps partly due to demographic pressure pushing people from the wet valleys into the leeward sections of the archipelago), elites hastened to control it by building a network of corporate temples associated primarily with the major gods Lono and Kane. In contrast, the center of the accretionary fan features the highest densities of agricultural intensification and commoner habitation, but also an emphasis on small family shrines. By understanding how local environmental conditions offered Hawaiian planters a productive advantage, we can directly connect various lines of evidence to show that soil nutrient values were intimately tied to the rise of Kaupō as a sociopolitical center. This provides more than ad hoc explanations for why this region came to prominence, instead defining a causal link between local sociopolitical development, settlement patterns, and the intensification of an area recognized for its productive potential.

Conclusion

Early settlement throughout the Hawaiian archipelago emphasized small groups practicing irrigated agriculture in windward valleys. By the 1500s, however, demographic pressures and an increasingly powerful class of elites pushed people towards the drier parts of the islands and into the more labor-intensive practice of dryland cropping. The lava and mudflows of Kaupō were quickly recognized as highly conducive to sweet potato and dryland taro production, and the region was consolidated through the construction of numerous monumental temples bounding this agriculturally rich section of southeastern Maui. Through soil testing, we have demonstrated that Kaupō's nutrient availability exceeds that of Kohala (Baer et al. 2015), the largest and best-known dryland field system in the islands, and that it potentially featured a geochemical profile as rich as an irrigated system (Palmer et al. 2009).

By integrating these nutrient values with the ethnohistoric and archaeological data, we can see a direct connection between Kaupō's rise from relative obscurity and the environmental conditions that made this possible. Where the districts on either side were unsuitable for this type of production, ideal rainfall and the relatively young mud and lava outflows from the Kaupō Gap created an ecological setting ripe for intensified agricultural production at a broad scale. Despite a few areas identified as potentially able to support this type of rain-supported cropping (Ladefoged et al. 2009), no other dryland systems have been identified on Maui. While it is possible other Maui field systems have been lost to modern land uses, the lack of both historical and archaeological references to any such planting regimen strongly points towards Kaupō as the single most intensively farmed dryland location on the island.

The geochemical analyses described here offer clear evidence that Kaupō's soils were rich enough to support an intensified field system similar to Kohala. On Hawai'i Island, however, numerous other locations, including Kona, Ka'ū, and Waimea, were similarly focused on dryland crop production. As perhaps the only large area suitable for this type of agricultural intensification, Kaupō's value to the increasingly powerful Maui polity would have been immense.

The first radiocarbon dates for the district place early settlement at the beginning of the fifteenth century (Baer 2015), but within 50 years, many of the large temples bounding the productive core had been founded. Additionally, the total number of ^{14}C dates for the district increase dramatically, indicating a surge in human presence and general cultural activity. In the dryland systems of Hawai'i we see a similar, if slightly earlier, increase in radiocarbon dates from dryland contexts (Ladefoged and Graves 2008; Kirch 2010; Field et al. 2011; McCoy et al. 2011), but do not have a system of major temples proscribing any large planting region. Smaller temples do serve as the markers for territorial divisions (Mulrooney and Ladefoged 2005; McCoy et al. 2011), but these exist within the field systems themselves, rather than around the entire productive zone. Kaupō, then, was uniquely marked through the modification of the agricultural landscape along with a bounding network of monumental structures.

In understanding the nutrient values from throughout the district, we can see how Kaupō's productive potential was equal or greater to that of Kohala. As a smaller region, however, and perhaps the only large dryland planting location on Maui, it was controlled and intensified in a way unique throughout the archipelago. This emphasis on Kaupō at a time in which new, independent states were emerging on each of Hawai'i's main islands (Kirch 2010) reflects how critical environmental factors were in the consolidation of early state power. While Kaupō would not be made the royal center until Kekaulike's reign in the early 1700s, the district was nonetheless subject to an unprecedented investment in infrastructure and oversight from early on. By employing explicitly environmental approaches and methodologies, such as soil geochemical analyses demonstrated here, it becomes clear that Kaupō's productive potential made it a location of singular importance to the Maui polity.

Methodological Vignette: Soil Geochemical Analyses in Archaeology

Soils and sediments have long been of interest to archaeologists, but advances in geochemical analyses and nutrient modeling have now made their study more accessible and quantifiable (Walker and Syers 1976). With simple field sampling protocols (see the "Soil Geochemical Analyses" section in this chapter) and subsequent analyses common in most soil laboratories (see Soil Survey Staff 2014), it is possible for archaeologists to collect and compare quantitative data on, for instance, Ca, soil pH, and base saturation.

Increasing understandings of nutrient release, input, cycling, and loss influence how we think about past agricultural practices, land use patterns, and even broad social collapses (Chadwick and Chorover 2001; Gomez-Pompa et al. 2003; Vitousek 2004). In the example presented here, testing soil nutrient values across multiple large transects offers insights into not only surrounding regional geochemistry but also its impacts on agriculture and sociopolitical development. Taking samples from throughout the study region guaranteed full coverage of the entire area at a range of rainfall and elevation gradients and across substrates, ultimately showing a uniformly high suite of nutrient values. This broad method of sampling is by no means the only way to employ soil geochemical analyses, however, as nutrient values across both large and small areas can provide archaeologists with valuable information. These values may speak to differentially productive zones or to the types of crops potentially grown, while unexpected values may indicate nutrient drawdown (associated with either specific crops or the length of time under cultivation), the presence of nitrogen- and other element-fixing plants, enrichment through mulching, and more (Hillel 1991). Of course, interpreting the similarities or discrepancies between various samples (or against expected values) requires an understanding of environments, plants, and possible human practices specific to the study region, ensuring that all samples are taken with a critical knowledge of their context (Bell and Boardman 1992). Overall, soil geochemical analyses offer access to a variety of nutrient values and other metrics, providing a quantifiable way to describe subsurface contexts from within a single site or across an entire landscape.

References Cited

Ågren, Goran I. and Ernesto Bosatta 1998 *Theoretical Ecosystem Ecology: Understanding Element Cycles*. Cambridge University Press, Cambridge.

Athens, J. Stephen, Timothy Reith and Thomas Dye 2014 A paleoenvironmental and archaeological model-based age estimate for the colonization of Hawai‘i. *American Antiquity* 91(1):144–155.

Baer, Alexander 2015 *On the Cloak of Kings: Power, Agriculture, and Community in Kaupo, Maui*. Ph.D. Dissertation, University of California, Berkeley.

Baer, Alexander, Oliver Chadwick and Patrick V. Kirch (2015). Soil nutrients and intensive dryland agricultural production in Kaupō, Maui, Hawaiian Islands. *Journal of Archaeological Science Reports* 3:429–436.

Bell, M. and J. Boardman (editors) 1992 *Past and Present Soil Erosion: Archaeological and Geographical Perspectives*. Oxbow monograph 22. Oxbow Books, Oxford.

Brookfield, H. C. 1972 Intensification and disintensification in Pacific agriculture: A theoretical approach. *Pacific Viewpoint* 13:30–48.

Burtchard, G. C. and M. J. Tomonari-Tuggle 2004 Agriculture on leeward Hawaii Island: The Waimea agricultural system reconsidered. *Hawaiian Archaeology* 9:50–73.

Cachola-Abad, Carolyn Kēhaunani 2000 *The Evolution of Hawaiian Socio-Political Complexity: An Analysis of Hawaiian Oral Traditions*. University of Hawaii at Manoa, Honolulu.

Chadwick, Oliver A. and Jon Chorover 2001 The chemistry of pedogenic thresholds. *Geoderma* 100(3–4):321–353.

Chadwick, Oliver A., L. A. Derry, P. M. Vitousek, B. J. Huebert and L. O. Hedin 1999 Changing sources of nutrients during four million years of ecosystem development. *Nature* 397(6719):491–497.

Chadwick, Oliver A., Robert T. Gavenda, Eugene F. Kelly, Karen Ziegler, Carolyn G. Olson, W. Crawford Elliott and David M. Hendricks 2003 The impact of climate on the biogeochemical functioning of volcanic soils. *Chemical Geology* 202(3):195–223.

Clark, Jeffrey and Patrick V. Kirch (editors) 1983 *Archaeological Investigations in the Mudlane-Waimea-Kawaihae Road Corridor, Island of Hawaii: An Interdisciplinary Study of an Environmental Transect*. Bernice P. Bishop Museum, Honolulu.

Coil, J. H. 2004 *"The beauty that was": Archaeological Investigations of Ancient Hawaiian Agriculture and Environmental Change in Kahikinui, Maui*. Ph.D. Dissertation, University of California, Berkeley.

Coil, J. H. and P. V. Kirch 2005 An Ipomoean landscape: Archaeology and the sweet potato in Kahikinui, Maui, Hawaiian Islands. In *The Sweet Potato in the Pacific: A Reappraisal*, edited by C. Ballard, P. Brown, R. M. Bourke and T. Harwood, pp. 71–84. Oceania Publications, Sydney.

Cordy, Ross 2000 *Exalted Sits the Chief*. Mutual Publishing, Honolulu.

Dixon, B., P. J. Conte, V. Nagahara and W. K. Hodgins 1999 Risk minimization and the traditional ahupua'a in Kahikinui, Island of Maui, Hawai'i. *Asian Perspectives* 38:229–255.

Field, J. S., T. N. Ladefoged, W. D. Sharp and P. V. Kirch 2011 Residential chronology, household subsistence, and the emergence of socioeconomic territories in Leeward Kohala, Hawai'i Island. *Radiocarbon* 53:605–627.

Fornander, Abraham 1996 *Ancient History of the Hawaiian People to the Times of Kamehameha I*. Mutual Publishing, Honolulu.

Giambelluca, T. W., Q. Chen, A. G. Frazier, J. P. Price, Y. L. Chen, P. S. Chu, . . . D. M. Delparte 2013. Online rainfall atlas of Hawaii. *Bull. Amer. Mereor. Soc.* 94:314–316. doi: 10.1175/BAMS-D-11–00228.1

Giambelluca, T. W., M. A. Nullet and T. A. Schroeder 1986 *Rainfall Atlas of Hawai'i*. Department of Land and Natural Resources, Division of Water and Land Development, Honolulu.

Giambelluca, T. W. and T. A. Schroeder 1998 Climate. In *Atlas of Hawai'i*, edited by S. P. Juvik and J. O. Juvik, 3rd ed., pp. 49–59. University of Hawaii Press, Honolulu.

Gomez-Pompa, A., M. F. Allen, S. L. Fedick and J. J. Jimenez-Osornio (editors) 2003 *The Lowland Maya Area: Three Millennia at the Human-Wildlife Interface*. Food Products Press, Binghamton, NY.

Handy, E. S. C. 1940 *The Hawaiian Planter-Volume I, his Plants, Methods and Areas of Cultivation* (Vol. 161). Bernice P. Bishop Museum, Honolulu.
Handy, E. S. Craighill and Elizabeth Green Handy 1972 *Native Planters in Old Hawaii: Their Life, Lore, and Environment*. Bishop Museum Press, Honolulu.
Harrington, R. A., J. H. Fownes and P. M. Vitousek 2001 Production and resource-use efficiencies in N- and P-limited tropical forest ecosystems. *Ecosystems* 4:646–657.
Hartshorn, A. S., O. A. Chadwick, P. M. Vitousek and P. V. Kirch 2006 Prehistoric agricultural depletion of soil nutrients in Hawai'i. *Proceedings of the National Academy of Sciences* 103(29):11092–11097.
Hillel, D. 1991 *Out of the Earth: Civilization and Life of the Soil*. University of California Press, Berkeley.
Holm, Lisa A. 2006 *The Archaeology and the Aina of Mahamenui and Manawainui, Kahikinui, Maui Island*. Ph.D. Dissertation, University of California, Berkeley.
Hommon, Robert J. 1976 *The Formation of Primitive States in Pre-Contact Hawai'i*. University of Arizona, Tucson.
——— 2008 Watershed: Testing the limited land hypothesis. In *Research Designs for Hawaiian Archaeology: Agriculture, Architecture, Methodology*, edited by T. S. Dye, pp. 1–92. Society for Hawaiian Archaeology, Honolulu.
——— 2009 *The Emergence of Primary States in Ancient Hawaii*. Paper presented at the Theoretical Archaeology Group Meeting, Palo Alto, California.
——— 2013 *The Ancient Hawaiian State: Origins of a Political Society*. Oxford University Press, Oxford.
Jenny, H. 1980 *Factors of Soil Formation: A System of Quantitative Pedology*. McGraw-Hill, New York.
Kamakau, Samuel M. 1992 *Ruling Chiefs of Hawaii* (Rev. ed.). Kamehameha Schools Press, Honolulu.
Kirch, Patrick V. 1985 *Feathered Gods and Fishhooks: An Introduction to Hawaiian Archaeology and Prehistory*. University of Hawaii Press, Honolulu.
——— 1994 *The Wet and the Dry: Irrigation and Agricultural Intensification in Polynesia*. University of Chicago Press, Chicago.
——— (editor) 1997 *Na-mea kahiko o Kahikinui: Studies in the archaeology of Kahikinui, Maui*. Oceanic Archaeology Laboratory Special Publication No. 1 Archaeological Research Facility, University of California, Berkeley.
——— 2000 *On the Road of the Winds: An Archaeological History of the Pacific Islands before European Contact*. University of California Press, Berkeley.
——— 2007 Hawaii as a model system for human ecodynamics. *American Anthropologist* 109(1):8–26.
——— 2010 *How Chiefs Became Kings: Divine Kingship and the Rise of Archaic States in Ancient Hawai'i*. University of California Press, Berkeley.
——— 2010 *Roots of Conflict : Soils, Agriculture, and Sociopolitical Complexity in Ancient Hawai'i*. School for Advanced Research Press, Santa Fe, NM.
——— 2012 *A Shark Going Inland Is My Chief: The Island Civilization of Ancient Hawai'i*. University of California Press, Berkeley.
——— 2015 *Unearthing the Polynesian Past*. University of Hawai'i Press, Honolulu.
Kirch, P. V., A. S. Hartshorn, O. A. Chadwick, P. M. Vitousek, D. R. Sherrod, J. Coil, and W. D. Sharp 2004 Environment, agriculture, and settlement patterns in a marginal Polynesian landscape. *Proceedings of the National Academy of Sciences of the United States of America* 101(26):9936–9941.
Kirch, P. V., John Holson and Alexander Baer 2009 Intensive dryland agriculture in Kaupō, Maui, Hawaiian Islands. *Asian Perspectives* 48(2):265–290.

Kirch, Patrick V., Regina Mertz-Kraus and Warren D. Sharp 2015 Precise chronology of Polynesian temple construction and use for southeastern Maui, Hawaiian Islands determined by 230 Th dating of corals. *Journal of Archaeological Science* 53:166–177.

Kirch, P. V. and W. D. Sharp 2005 Coral ^{230}Th dating of the imposition of a ritual control hierarchy in precontact Hawaii. *Science* 307(5706):102–104.

Kolb, M. J. 1991 Social Power, Chiefly Authority, and Ceremonial Architecture, in an Island Polity, Maui, Hawaii. Ph.D. Dissertation, University of California, Los Angeles.

Kolb, Michael J. 1994 Monumentality and the rise of religious authority in precontact Hawai'i. *Current Anthropology* 35(5):521–547.

Ladefoged, T. N. and M. W. Graves 2008 Variable development of dryland agriculture in Hawai'i: A fine-grained chronology from the Kohala field system. *Current Anthropology* 49:771–802.

Ladefoged, T. N. and M. W. Graves 2010 The leeward Kohala field system. In *Roots of Conflict: Soils, Agriculture, and Sociopolitical Complexity in Ancient Hawai'i*, edited by P. V. Kirch, pp. 89–110. School for Advanced Research Press, Santa Fe.

Ladefoged, T. N., M. W. Graves and M. D. McCoy 2003 Archaeological evidence for agricultural development in Kohala, Island of Hawai'i. *Journal of Archaeological Science* 30(7):923–940.

Ladefoged, Thegn N., Michael W. Graves and R. Jennings 1996 Dryland agricultural expansion and intensification in Kohala, Hawai'i Island. *Antiquity* 70(270):861–880.

Ladefoged, T. N., P. V. Kirch, S. M. Gon, O. A. Chadwick, A. S. Hartshorn and P. M. Vitousek 2009 Opportunities and constraints for prehistoric intensive agriculture in the Hawaiian archipelago. *Journal of Archaeolical Science* 36:2374–2383.

Lincoln, Noa and Thegn Ladefoged 2014 Agroecology of pre-contact Hawaiian dryland farming: The spatial extent, yield and social impact of Hawaiian breadfruit groves in Kona, Hawai'i. *Journal of Archaeological Science* 49(0):192–202. doi: http://dx.doi.org/10.1016/j.jas.2014.05.008

Maunupau, T. K. 1998 *Huakai Makaikai a Kaupo, Maui* (N. Losch & R. G. Rose Eds.). Bishop Museum Press, Honolulu.

McCoy, Mark D. 2005 The development of the Kalaupapa field system, Moloka'i Island, Hawai'i. *The Journal of the Polynesian Society* 114(4):339–358.

McCoy, Mark D. and Michael W. Graves 2010 The role of agricultural innovation on Pacific Islands: A case study from Hawai'i Island. *World Archaeology* 42(1):90–107. doi: 10.1080/00438240903430340

McCoy, Mark D., Thegn N. Ladefo ged, Michael W. Graves and Jesse W. Stephen 2011 Strategies for constructing religious authority in ancient Hawai'i. *Antiquity* 85(329):927–941. doi:10.1017/S0003598X0006840X

Mulrooney, Mara A. and Thegn N. Ladefoged 2005 Hawaiian heiau and agricultural production in the Kohala dryland field system. *The Journal of the Polynesian Society* 114(1):45–67.

Palmer, Molly A., Michael Graves, Thegn N. Ladefoged, Oliver A. Chadwick, T. Ka'eo Duarte, Stephen Porder and Peter M. Vitousek 2009 Sources of nutrients to windward agricultural systems in pre-contact Hawai'i. *Ecological Applications* 19(6):1444–1453. doi: 10.1890/08–0983.1

Purseglove, J. W. 1968 *Tropical Crops: Dicotyledons.* John Wiley and Sons, New York.

Schilt, A. R. 1984 *Subsistence and Conflict in Kona, Hawai'i Department of Anthropology Report 84–1*. Bernice P. Bishop Museum, Honolulu.

Sherrod, D. R., J. M. Sinton, S. E. Watkins and K. M. Brunt 2007 Geologic map of the state of Hawai'i. U.S. Geological Survey Open-File Report 2007–1089. U.S. Geological Survey, Washington, DC.

Soil Survey Staff 2014 Kellogg soil survey laboratory methods manual. In *Soil Survey Investigations Report No. 42, version 5.0*, edited by R. Burt and Soil Survey Staff. U.S. Department of Agriculture, Natural Resources Conservation Service, Washington, DC.

Stearns, H. T. and G. A. MacDonald 1942 *Geology and Ground-Water Resources of the Island of Maui, Hawai'i* (Vol. 7). State of Hawai'i, Honolulu.

Sterling, Elspeth P. 1998 *Sites of Maui*. Bishop Museum Press, Honolulu.

Vitousek, Peter M. 2004 *Nutrient Cycling and Limitation: Hawai'i as a Model System*. Princeton University Press, Princeton.

Vitousek, Peter M. and Oliver A. Chadwick 2013 Pedogenic thresholds and soil process domains in basalt-derived soils. *Ecosystems*16(8):1379–1395.

Vitousek, Peter M., Oliver A. Chadwick, Anthony S. Hartshorn and Sarah C. Hotchkiss 2010 Intensive agriculture in Hawai'i: The model system approach. In *Roots of Conflict: Soils, Agriculture and Sociopolitical Complexity in Ancient Hawai'i*, edited by P. V. Kirch, pp. 31–44. School for Advanced Research, Santa Fe.

Vitousek, Peter M, Oliver A. Chadwick, Sara C. Hotchkiss, Thegn N. Ladefoged and Christopher M. Stevenson 2014 Farming the rock: A biogeochemical perspective on intensive agriculture in Polynesia. *Journal of Pacific Archaeology* 5(2):51–61.

Vitousek, P. M., T. N. Ladefoged, P. V. Kirch, A. S. Hartshorn, M. W. Graves, S. C. Hotchkiss, S. Tuljapurkar and O. A. Chadwick 2004 Soils, agriculture, and society in precontact Hawaii. *Science* 304(5677):1665–1669.

Walker, T.W. and J. K. Syers 1976 The fate of phosphorus during pedogenesis. *Geoderma* 15:1–19.

Walker, Winslow M. 1930 *Archaeology of Maui*. [Unpublished manuscript]. Bernice P. Bishop Museum Library, Honolulu.

Yen, Douglas E. 1974 The sweet potato and Oceania: An essay in ethno-botany. *Bernice P. Bishop Mus. Bull* 236:1–7.

Discussion

11 Epilogue

Frances Hayashida

The collected chapters in this volume explore the challenges and possibilities of explaining human-environmental dynamics on a very broad range of topics, including climate and domestication in Southwest Asia (Contreras and Makarewicz; Jones et al.); pastoralist settlement and site organization in Mongolia (Wright); climate and settlement in California (Codding and Jones); land use in Mesoamerica (Borejsza and Joyce); the formation of anthropogenic soils in Amazonia (Browne Ribeiro); irrigation in coastal Peru (Caramanica and Koons) and the American Southwest (Purdue); and soil productivity and political power in Hawaii (Baer). The authors employ a wide array of methodologies grounded in intensive field and laboratory work, spatial analyses, and modeling. Though diverse, they share a concern with establishing stronger correlations as a first step towards identifying the causes of socionatural phenomena and transformations. In my closing comments, I briefly highlight two themes (temporal and spatial scales and resolution) that run through the chapters and Contreras's thoughtful introduction; contemplate culture, politics, and agency; and consider the collective and collaborative work required to build robust arguments that explain complex interactions across the scales of time, space, and human experience.

Temporal and Spatial Scale and Resolution

A number of the authors emphasize the importance of studying and understanding socionatural processes at the scale of human perception and experience, which is more local and of a finer temporal resolution than some of the records used to reconstruct paleoenvironment or define archaeological time periods. Climate records from distant places may not be applicable; regional generalizations cannot capture local variability (Jones et al., Contreras and Makarewicz, Caramanica and Koons, Baer, this volume). Relations of possible causes and effects are obscured when only generalized observations across broad periods of time (e.g., from coarse artifact seriations) can be made (Borejsza and Joyce, this volume). And while archaeological studies of humans and the environment often highlight our ability to step back to look at long-term, large-scale processes (Codding and Jones, Borejsza and Joyce, this volume), several of the chapters demonstrate how we can also zoom in, seeing in microdeposits the patterns of everyday life in settlements (Browne Ribeiro, this volume), or the pulse of sediment-laden water through canals (Purdue, this volume). Ethnoarchaeology also brings us closer to understanding the decisions that people make about where to live with regard to nearby natural features, the microenvironments that they (and their herds) inhabit (Wright, this volume).

Culture, Politics, and Agency

Socionatural processes encompass more than just the relations between people and the environment; human relations also matter. At the same time, the ways that we engage with the natural world are mediated by culture; thus explanations of causation require the same close attention to social and political processes and scales. For example, while the herders discussed by Wright made decisions about campsites with regard to features such as wind, topography, water, and pasture, they may also have considered proximity to kin, conflicts with neighbors, ideas about privacy, and concepts of identity, territory, and history as tied to places (natural and built) on the landscape. Similarly, in Codding and Jones's predictive forager settlement model, deviations from expected outcomes based on climate, demography, and resource characteristics may be explained, as the authors note, by the emergence of cultural institutions and practices that regulated resource access. In other contexts, a decrease in resource predictability may be met with increased ties to near and distant neighbors, what Lepofsky et al. (2005:267) describe as the "social buffering of resource uncertainty" (see also Borck et al. 2015). Thus as we reveal and examine local variations in climate and environment, we should also consider how disparities based on location may have been evened out through information sharing, cooperation, and exchange, or exacerbated through competition, exclusion, and control.

There are other ways that social scales cross-cut spatial scales. Decisions that affect local environments may result from social or political dynamics that extend across much larger areas, or that originate in distant centers of power. Thus farming may be expanded or intensified to meet elite tributary demands resulting in widespread landscape modification, and changes in hydrology, vegetation, and soils, and processes of deposition and erosion (Baer, Boresjza and Joyce, this volume). With increasing centralization, land use decisions made by distant elites for the accumulation of wealth and power were likely very different than those made by farmers primarily concerned with feeding their families. The links between political economy, people, and the environment are a central focus of political ecology, a theoretical framework most often applied to modern contexts (e.g., to untangle the local effects of globalization (Blaikie and Brookfield 1987; Zimmerer and Bassett 2003) but equally relevant to the past.

A perspective centered on human agency affects not only research questions but also research design. As several of the chapters remind us, when evaluating local socionatural dynamics, and particularly when trying to assess human impacts, we ideally compare modified (cultural) to control (natural) contexts, or on-site to off-site areas. But it can be difficult to make these distinctions. The classic example is determining whether an increase in charcoal results from more human-set fires or drier climatic conditions resulting in more natural fires. Examination of multiple paleoenvironmental proxies and comparison of different lines of evidence can help to identify the more plausible explanation (Borejsza and Joyce, this volume).

Defining "off-site" is also problematic given that even small-scale societies profoundly affect the landscapes they inhabit (Thompson and Waggoner 2013). Many examples come from current studies of niche construction, which demonstrate how people modified their surroundings to increase the abundance and predictability of resources (Smith 2011), such as foragers who burn, prune, thin, transplant, or irrigate to encourage the growth of plants that are harvested or that attract game (Bliege Bird et al. 2013; Smith 2014). A similar perspective is held by historical ecologists, who examine the intentional modification of landscapes over time and their long-term effects, with a particular focus on areas that had previously been perceived as pristine (Crumley 1994; Balée 2006; Balée and Erickson 2006). Human actions can also lead, in the short or long term, to unintentional effects such as decreased abundance and diversity, or degraded environments through overhunting, overharvesting, or land modifications that lead to habitat loss, soil nutrient depletion, or widespread erosion (Borejsza and Joyce, this volume). The effects of these dynamics need to be considered in any effort to compare modified and unmodified contexts; in some cases, it may be difficult to identify local unmodified contexts because of the extent, intensity, and long-term effects of past land use. At the same time, for specific times and places, some past processes (e.g., erosion and deposition) may have been relatively unaffected by human actions, while others (e.g., vegetation succession) were altered.

Ways Forward

Several of the chapters discuss ways to scale up high-resolution, micro, or local observations to detect correlations and define possible causes at larger scales. For example, Jones et al. demonstrate the value of "bottom-up" research, using long-term, interdisciplinary, high-resolution observations in a range of localities to see larger-scale processes. A similar perspective is provided by Borejsza and Joyce, who use archaeological and geological data gathered during long-term regional research projects to identify the range of different possible causal explanations for observed phenomena. Their efforts (and particularly their detailed charts of possible causes and effects) make clear the complexity of fully considering competing explanations and multiple causes. This complexity is met by other researchers through modeling

(Kohler et al. 2012; Marean et al. 2015); but again, success depends on access to big datasets incorporating different lines of high-resolution archaeological and paleoenvironmental evidence at different scales.

Achieving this research goal, as other commentators have noted (van der Leeuw and Redman 2002; Crumley 2013), requires time (for intensive and extensive fieldwork and analyses), funding, information sharing and management, and (as Borejsza and Joyce remind us) long-term curation of recovered material for reanalysis. It also requires expanded interdisciplinary exchange, training, and collaboration, which should be possible given that researchers in a variety of fields are increasingly looking to the past to comprehend and address our current, human-induced, global environmental crisis. Yet despite decades of research, archaeological and anthropological contributions have remained largely off the radar of biophysical scientists, including specialists in other fields with common interests in the history of humans and the environment (see, for example, Szabó's (2015) review of the development of historical ecology).

This was made clear in the widely accepted definition of the Anthropocene as a period that began with the Industrial Revolution (Steffen et al. 2011), a designation that archaeologists have questioned and challenged (Erlandson and Braje 2013; Smith and Zeder 2013; Braje 2015; Crumley et al. 2015; Morrison 2015). The debate, centered on the timing and extent of human environmental transformations and how transformations are perceived and studied, echoes earlier exchanges over conservation in areas with long human histories that had long been defined as pristine or wild (Denevan 1992; 2011; Hayashida 2005). Recognition of archaeological and anthropological contributions outside of the discipline has been slow and is incomplete, but is growing, as can be seen in recent publications on paleobiology, paleoclimate, and landscape ecology (Sandweiss and Kelley 2012; Scharf 2014; Kidwell 2015). This is promising, for it will take our collective efforts and expertise to understand, explain, and effectively communicate the complex history of human life on Earth.

References Cited

Balée, William 2006 The research program of historical ecology. *Annual Review of Anthropology* 35:75–98.

Balée, William and Clark L. Erickson 2006 *Time and Complexity in Historical Ecology: Studies in the Neotropical Lowlands*. Columbia University Press, New York.

Blaikie, Piers and Harold Brookfield 1987 *Land Degradation and Society*. Methuen, London.

Bliege Bird, Rebecca, Nyalangka Tayor, Brian F. Codding and Douglas W. Bird 2013 Niche construction and Dreaming logic: Aboriginal patch mosaic burning and varanid lizards (Varanus gouldii) in Australia. *Proceedings of the Royal Society B* 280(1772):20132297.

Borck, Lewis, Barbara J. Mills, Matthew A. Peeples and Jeffery J. Clark 2015 Are social networks survival networks? An example from the late pre-Hispanic US Southwest. *Journal of Archaeological Method and Theory* 22(1):33–57.

Braje, Todd J. 2015 Earth systems, human agency, and the Anthropocene: Planet Earth in the human age. *Journal of Archaeological Research* 23(4):369–396.

Crumley, Carole L. 1994 *Historical Ecology: Cultural Knowledge and Changing Landscapes*. School of American Research, Santa Fe, New Mexico.

——— 2013 The archaeology of global environmental change. In *Humans and the Environment: New Perspectives for the Twenty-First Century*, edited by Matthew I. J. Davies, pp. 269–276. Oxford University Press, Oxford.

Crumley, Carole L., Sofia Laparidou, Monica Ramsey and Arlene M Rosen 2015 A view from the past to the future: Concluding remarks on the 'The Anthropocene in the Longue Durée'. *The Holocene* 25(10):1721–1723.

Denevan, William M. 1992 The pristine myth: The landscape of the Americas in 1492. *Annals of the Association of American Geographers* 82(3):369–385.

——— 2011 The "pristine myth" revisited. *Geographical Review* 101(4):576–591.

Erlandson, Jon M. and Todd J. Braje 2013 Archaeology and the Anthropocene. *Anthropocene* 4:1–7.

Hayashida, Frances M. 2005 Archaeology, ecological history, and conservation. *Annual Review of Anthropology* 34:43–65.

Kidwell, Susan M. 2015 Biology in the Anthropocene: Challenges and insights from young fossil records. *Proceedings of the National Academy of Sciences of the United States of America* 112(16):4922–4929.

Kohler, Timothy A., R. Kyle Bocinsky, Denton Cockburn, Stefani A. Crabtree, Mark D. Varien, Kenneth E. Kolm, Schaun Smith, Scott G. Ortman and Ziad Kobti 2012 Modelling prehispanic Pueblo societies in their ecosystems. *Ecological Modelling* 241:30–41.

Leeuw, Sander van der and Charles L. Redman 2002 Placing archaeology at the center of socio-natural studies. *American Antiquity* 67(4):597–605.

Lepofsky, D., K. Lertzman, D. Hallett and R. Mathewes 2005 Climate change and culture change on the Southern Coast of British Columbia 2400–1200 CAL. BP: An hypothesis. *American Antiquity* 70(2):267–293.

Marean, Curtis W., Robert J. Anderson, Miryam Bar-Matthews, Kerstin Braun, Hayley C. Cawthra, Richard M. Cowling, Francois Engelbrecht, Karen J. Esler, Erich Fisher, Janet Franklin, Kim Hill, Marco Janssen, Alastair J. Potts and Rainer Zahn 2015 A new research strategy for integrating studies of paleoclimate, paleoenvironment, and paleoanthropology. *Evolutionary Anthropology* 24(2):62–72.

Morrison, Kathleen 2015 Provincializing the Anthropocene. *Seminar* 673, New Delhi. http://www.india-seminar.com, accessed 10/26/2015.

Sandweiss, Daniel H. and Alice R. Kelley 2012 Archaeological contributions to climate change research: The archaeological record as a paleoclimatic and paleoenvironmental archive. *Annual Review of Anthropology* 41:371–391.

Scharf, Elizabeth A. 2014 Deep time: The emerging role of archaeology in landscape ecology. *Landscape Ecology* 29(4):563–569.

Smith, Bruce D. 2011 General patterns of niche construction and the management of 'wild' plant and animal resources by small-scale pre-industrial societies. *Philosophical Transactions of the Royal Society B-Biological Sciences* 366(1566):836–848.

——— 2014 Documenting human niche construction in the archaeological record. In *Method and Theory in Paleoethnobotany*, edited by John M. Marston, Jade d'Alpoim Guedes and Christina Warinner, pp. 355–370. University of Colorado Press, Boulder.

Smith, Bruce D. and Melinda A. Zeder 2013 The onset of the Anthropocene. *Anthropocene* 4:8–13.

Steffen, Will, Jacques Grinevald, Paul Crutzen and John McNeill 2011 The Anthropocene: Conceptual and historical perspectives. *Philosophical Transactions of the Royal Society A-Mathematical Physical and Engineering Sciences* 369(1938):842–867.

Szabó, Péter 2015 Historical ecology: Past, present and future. *Biological Reviews* 90(4):997–1014.

Thompson, Victor D. and James C. Waggoner (editors) 2013 *The Archaeology and Historical Ecology of Small Scale Economies*. University Press of Florida, Gainesville.

Zimmerer, Karl. S. and Thomas J. Bassett 2003 Future directions in political ecology: Nature-society fusions and scales of interaction. In *Political Ecology: An Integrative Approach to Geography and Environment-Development Studies*, edited by Karl. S. Zimmerer and T. J. Bassett, pp. 274–295. Guilford Press, New York.

Index

For Product Safety Concerns and Information please contact our EU
representative GPSR@taylorandfrancis.com
Taylor & Francis Verlag GmbH, Kaufingerstraße 24, 80331 München, Germany

www.ingramcontent.com/pod-product-compliance
Ingram Content Group UK Ltd.
Pitfield, Milton Keynes, MK11 3LW, UK
UKHW020013200926
475702UK00014B/1868
9780367872687